Animal Husbandry:

Research, Education and Development

About the Authors

Prof (Dr) P.N. Bhat was awarded Ph.D. in population genetics from Institute of Population Genetics Purdue University, West Laffayette, Indiana, USA in the year 1965. He started his career as teaching assistant in Animal Breeding followed by several post doctoral assignments and visiting professorship in genetics, biotechnology, livestock production systems. He returned to India and joined Punjab Agricultural University at Hisar–Ludhiana followed by several professional assignments in India. He joined as Project co-ordinator (Animal Breeding) at ICAR and subsequently as Head, Division of Animal Genetics at Indian Veterinary Research Institute (IVRI) in 1971. He was responsible for establishing coordinated projects on cattle, buffalo, sheep, goat, pigs and poultry during 1970–74. He was founder Director of Central Institute for Research on Goats. In 1984 he became Vice-Chancellor and Director of IVRI. He joined as Deputy Director-General (Animal Science) in May, 1992 and Animal Husbandry Commissioner, GoI in December, 1992. He is fellow of several National and International Science Academies.

Prof. (Dr.) M.P. Yadav, Ex-Director and Vice Chancellor, Indian Veterinary Research Institute, Izatnagar; Ex-Vice Chancellor, Sardar Vallabhbhai Patel University of Agriculture and Technology, Meerut (UP), India, is an eminent microbiologist having major research contributions in animal health by development of indigenous vaccines against equine influenza A Equi-2 (H3N8), infectious laryngotracheitis (ILT) and colisepticaemia of poultry, and goat pox; isolation and characterization of animal and poultry viruses, namely equine influenza virus, equine infectious anemia virus, equine herpes virus-1 (EHV-1), EHV-3, EHV-4, equine rotavirus, avian adenovirus, buffalo pox virus; followed by development of sensitive and cost effective diagnostic tests/kits; epidemiology of Q Fever, brucellosis, rabies, bovine herpes virus-1, and disease management.

Prof. Yadav worked as Professor & Head Division of Virology at IVRI, Mukteswar, Director of National Research Centre on Equines, Hisar. He is a Fellow of the National Academy of Agricultural Sciences (NAAS) New Delhi and National Academy of Veterinary Sciences, India. Prof. Yadav has guided the destiny of several professional bodies as President, Indian Virological Society (1996-2003); President, Indian Association of Veterinary Microbiologists & Specialists in Infectious Diseases (1999 to date); President, Indian Agricultural University Association (2006-07); President National Academy of Veterinary Sciences (2011-13), Secretary, National Academy of Agricultural Sciences (2014-16) and Patron, Society of Plant Research (2008 to date).

Prof. Yadav is recipient of numerous national/international awards. To mention a few-OIE, International Meritorius Award 2000; World Intellectual Property Organization (WIPO), Geneva Award 2005; ICAR Special Award 1998; Distinguished Veterinarian Award 2002, Dr. P. Richard Masillamony Oration Award 2003; M.S. Swaminathan Agricultural Scientist Award 2009; Dr. P. G. Pandey Oration Award 2010 (IAVMI); Dr. A. Lingard Memorial Award, 2012 by Indian Academy of Environmental Sciences; Dr. M.R. Dhanda Memorial Oration Award, 2012 (IAPH); Excellence in Virology Award - 2013 of Indian Virology Society (IVS) and Prof. K. S. Bhargaba Oration Award- 2016 (IVS).

Animal Husbandry:

Research, Education and Development

Prof.(Dr.) P.N. Bhat
600A, Hamilton Court
DLF Phase –IV,
Gurgaon – 122 002 Haryana

Prof.(Dr.) M.P. Yadav
House No. 365, Sector-45,
Gurgaon - 122003 Haryana
Email: yadav_mp@hotmail.com

SCIENTIFIC
PUBLISHERS

Published by:

SCIENTIFIC PUBLISHERS (INDIA)
5A, New Pali Road, P.O. Box 91
Jodhpur 342 001, India
E-mail: info@scientificpub.com
Website: www.scientificpub.com

Print: 2020

ISBN: 978-93-86652-21-8
eISBN: 978-93-87991-97-2

Printed in India

PREFACE

India became independent on August 15, 1947 and new Government started. The contribution of agriculture sector to the total GDP in 1950–51 was 50.5%, of which the share of animal husbandry was 15.5%. During 1980 to 2012–13, the contribution of livestock to agriculture and allied sectors rose from 18.6% to approximately 26% at current prices in contrast to a reduction of crops agricultural output from 41% to approximately 15%. During 2014–15, the value of output from livestock sector as per the estimates of the Central Statistics Office (CSO) at current prices was about INR 7,33,054 crores which is approximately 28.7% of the value of output from agricultural and allied sector.

If we look at overall world meat production, it is predicted to stagnate at about 321 million tonnes in 2016. Poultry is forecasted to register some growth, followed by bovine and ovine meat, while pig meat output could decline. However, global meat trade is expected to recover, growing by 2.8 percent to 30.6 million tonnes. International dairy product prices remained depressed during the first five calendar months of 2016, due to subdued import demand and plentiful export availability. Milk production continues to increase steadily in many countries, although lower prices are expected to dampen growth in world output in 2016 (FAO- Food Outlook, June 2016).

In India meat production from all sources was 7.0 MT in 2015-16 with an average annual growth rate of 7.51% during the last five years (2011-12 to 2015-16). Buffalo meat has grown at approximately 8% annually. Cattle, buffalo, sheep, goat, pigs and poultry contributed 5.0%, 23%, 7%, 13%, 6% and 46.0%, respectively, to total meat production during 2015-16. The poultry population in India is 729.21 million (2012) which is approximately 3% of the world's population; with the growth in annual egg production being 5.6% between the years 2011-12 and 2015-16, and that of chicken meat approximately 3% of the world's production. The contribution of bullock power for agriculture operations was 71.6% in 1961 when the foundation of the green revolution was laid, though over years it was reduced to about 23% (in 1991) due to farm mechanisation. It has remained at approximately 23% until the 19th Livestock Census 2012.

Presently, the contribution of animal draft power to total motive power in agriculture is about 5.8%.

As per 19th Livestock census (2012), India had 109 million buffaloes, which is more than half (57.3%) of the world's buffalo population. Buffaloes constituted 21.23% of the total livestock population. Buffalo is the 'dairy animal of 21st century' in India. With 57 million adult females, population of breedable buffaloes is one-third. Presently, they contribute about 51% of the total milk produced in the country. The buffalo population has increased approximately 151% between 1951 and 2012 (43.4 million in 1951 to 109 million in 2012). As per the 19th Livestock census (2012), the Indian buffalo population consist of 17.05% pure-bred, 39.58% graded and 43.37% non-descript animals. The river buffalo constitute around 65% of the total world buffalo population and accounts for 92% of the total milk produced. India alone contributes 42% of total milk produced by buffaloes. The largest concentration of swamp buffaloes, found in rice growing countries of Asia, will remain about 5%. The distribution of the buffalo is conspicuous by its being confined principally to the areas where animal husbandry is poorly developed and badly organized. By and large, buffaloes are owned only in small numbers generally by resources-poor farmers. In fairness to buffalo stock owners in these regions, it may be stated that they maintain buffaloes not from ignorance of the potentialities of other large ruminants but because they find that in prevailing agricultural situation no other domestic animal will thrive like the buffalo and at the same time be so useful and economical.

Water buffaloes have been responsible for more than 10% of world milk production for several years, but the potential of these animals has seldom been appreciated or recognized. One of the main reasons for this is that those who have a stake in buffalo rearing are generally poor and underprivileged, and not able to project the impact this animal has on their livelihood and well-being. The water buffalo is an important beast of burden in Asian farming. It is widely used to plough, level land, plant crops, puddle rice fields, cultivate field crops, pump water; haul carts, sleds and shallow draft boats. It is also used to carry people, thrash grain, press sugarcane, haul logs and more. Buffaloes have an advantage over other draught animals in wet or muddy areas, with their large hooves. Their legs can withstand wet conditions better than cattle, although they are not as fast as cattle, horses or mules. Buffaloes have been used as draught animals for centuries. This has led to exceptional muscular development; some animals can weigh more than 1000 kg. Though buffalo is a major source of meat, they have not been used solely for meat production until recently. Most buffalo meat is derived from old animals so not surprisingly the meat is considered to be of poor quality. However, this is not true of meat from younger animals. Buffalo meat from animals properly reared and fed, is tender and palatable. Buffaloes are lean animals. In general, a buffalo carcass has a higher proportion of muscle and a lower ratio of bone and fat than a cattle carcass. Buffaloes are managed under very different conditions over the world. These depend both on the geographic situations and for what purpose the buffaloes are used; any system from multipurpose animals kept in backyards to high yielding milk animals in advanced farms may exist. Over time buffalo rearing has shifted from the backyard to commercial farms and large business enterprises. The immense popularity of buffalo milk and meat products has ensured that buffalo production follow the path of the dairy cattle industry. However, for this species to perform optimally under the pressure of intensive production system, the breeds have to be improved, with clear focus on the desired output. This has not yet happened. Buffalo, although potentially excellent for both milk and meat production still languishes in obscure conditions of poor nutrition and breeding. Management can bring about notable improvements in the productive and reproductive performance of buffalo.

The achievement in agriculture sector are attributed in large part to technology led improvement in productivity and investments in the agriculture R&D with allied sectors. This has been divided into three phases. At the time of independence (1947) the population grew 3.6 times from 361.1 million in 1951 to 1302 million in 2016, while food grain production grew from 51 million tonnes (mt) in 1951 to 273 mt in 2016. India's horticultural production has increased from 25 mt to over 287 mt in the same time, and milk production to 155.5 mt. This has been possible to achieve self-reliance in food security but hunger among the population remains still a cause of worry. While India is still to achieve MDG goal of reducing hunger, the 2016 Global Hunger Index (GHI) report indicate that globally, about 795 million people are chronically undernourished and India has 40% of all such children.

Last century agriculture has shown three distinct phases of growth (i) till 60's it was more dictated where farmer was dictated by food needs of the family and a little for special social requirement, as he did not spend much on produce of such as natural resources or technology, (ii) the second phase was driven by technological production, input driven agriculture, which give high production in food grain crops like rice and wheat under better resources areas and was called Green Revolution, (iii) in third phase emphasis was given to diversification of crop with inclusion of all necessary inputs. This phase was the phase which ensured rural employment, brought in the concept of holistic agriculture and lead to basic agriculture research. In addition, focus on 510 million livestock and 8000 km of coastline will demand agricultural production to lead towards rural agriculture of the country more. India is projected to be the most populous country in the world with about 1.5 billion people by 2030 and

80% of growth will depend on the food production which will be produced, based on research, education, production and human resources.

We started from 1950 and by 2020 we will be holistic and integrated agriculture reckoner. We are now becoming one of the leader in skilled instruction to be no. 1 in education and research and by twenties, our agriculture will be known by broadway, and we are going in the middle of this decade to participate to develop the skill of entrepreneurship and will be more on the path of agriculture and rest of the fast moving India will have good educators of 21st century in the world by 2030. Let God help us.

Prof (Dr) P.N. Bhat
Prof. M.P. Yadav

New Delhi
Date: 26 March, 2018

ACKNOWLEDGEMENT

Thanks are due to Mrs. Aruna T Kumar, Editor (English). Directorate of Knowledge Management in Agriculture for editing the manuscript of the book "Animal Husbandry: Research, Education and Development" both for language and technical content. Thanks are due to Resource persons and contributors for their ideas and discussion.

Thanks are due to Deputy Director General (Animal Science), Indian Council of Agricultural Research for kindly agreeing to draw on the materials/ contributions submitted for the NAAS policy papers.

The hard work put up by Mr. G.P. Gangadharan Pillai, Centre for Integrated Animal Husbandry & Dairy Development, Noida for proof reading of the manuscript and other works related to the editing of this book which is gratefully acknowledged.

ABBREVIATIONS

ADG	Average daily growth
AEZ	Agro-eco-zones
AFS	Australian Friesian Sahiwal
AGID	Agar gel immune diffusion
AHS	African horse sickness
AI	Artificial Insemination
AICRP	All India Coordinated Research Project
AMZ	Australian Milking Zebu
APAU	Andhra Pradesh Agricultural University
APEDA	Agricultural & Processed Food Products Export Development Authority
API	Ayurvedic Pharmacopoeia of India
ARS	Agricultural Research Services
ASMM	Area-specific mineral mixture
ATIC	Agriculture Technology Information Centre
BAH&FS	Basic Animal Husbandry & Fisheries Statistics
BAU	Birsa Agricultural University
BC	Before Christ
BO	Butter Oil
BP	Biological Products
BPD	Business Planning & Development
BQ	Black quarter
BSI	Botanical Survey of India
BUN	Blood urea nitrogen
BVC	Birsa Veterinary College
BVD	Bovine Viral Diarrhoea
BZ	Benzimidazole
CAE	Caprine Arthritis Encephalitis
CAFT	Centre for Advance Faculty Training
CAL	Carcinoscorpius amoebocyte lysate
CARI	Central Avian Research Institute
CB	Cross-bred
CBPP	Contagious bovine pleuro pneumonia
CC	Combination with chloride
CCBF	Central Cattle Breeding Farm
CCPP	Contagious caprine pleuro-pneumonia
CCS	Copper chaperone for superoxide dismutase
CDRI	Central Drug Research Institute
CDV	Canine distemper virus
CE	Capillary electrophoresis
CELO	Chick embryo lethal orphan (virus)
CFB	Complete Feed Blocks
CFTRI	Central Food Technological Research Institute

CIDR	Controlled intra-vaginal drug releasing device
CIE	Counter immuno-electrophoresis
CIRB	Central Institute for Research on Buffaloes
CIRC	Central Institute for Research on Cattle
CIRG	Central Institute for Research on Goat
CLFMA	Compound Livestock Feed Manufacturers Association
CLRI	Central Leather Research Institute
CMVL	Central Military Veterinary Laboratory
Co	Cobalt
CO_2	Carbon dioxide
CP	Control Programme
CPBP	Central Poultry Breeding Programme
CRD	Chronic Respiratory Disease
CSF	Classical Swine Fever
CSO	Central Statistics Office
CSWRI	Central Sheep and Wool Research Institute
Cu	Copper
DAD&F	Department of Animal Husbandry, Dairying and Fisheries
DAH	Department of Animal Husbandry
DARE	Department of Agricultural Research and Education
DCFB	Densified complete feed block
DCS	Dairy cooperative societies
DFRL	Defence Food Research Laboratory
DG	Distillers grains
DIVA	Differentiating infected from vaccinated animals
DNA	Deoxyribonucleic acid
DTMR	Densified total mixed ration
EC	Emulsifying capacity
ECE	Equine coital exanthema
ECP	Estradiol cypionate
EEC	European Economic Community
EHV	Equine herpesvirus
EI	Equine influenza
EIA	Enzyme immunoassay / Equine infectious anaemia
ELISA	Enzyme-linked immunosorbent assay
EMA	European Medicines Agency
ERV	Equine rotavirus
ETT	Embryo transfer technology / Embryo transfer techniques
EU	European Union
EVA	Equine viral arteritis
FAD	Food and Drug Administration
FAO	Food and Agriculture Organization
FAT	Fluorescence antibody technique
FCR	Feed conversion ratio
Fe	Iron
FEC	Faecal egg count
FFS	Food-feed system
FMD	Foot and mouth disease
FROGIN	Forecasting for Rajasthan on ovine gastro intestinal nematodiasis
FSH	Follicle-stimulating hormone
GA	Angora (rabbit) crosses of German
GC	Gas chromatography
GC/MS	Gas chromatography–mass spectrometry

GDP	Gross Domestic Product
GFY	Greasy fleece yield
GHG	Greenhouse gas
GI	Gastrointestinal
GLF	Government Livestock Farm
GMM	Garole × Malpura ×Malpura
GMMP	Garole × Malpura × Malpura × Patanwadi
GMOs	Genetically modified organisms
GMP	Good manufacturing practice
GNP	Gross National Product
GOI	Government of India
GP	Grandparent
GSPA-PHI	Global strategy and plan of action on public health, innovation and intellectual property
H F	Holstein Friesian
HAU	Haryana Agricultural University
HI	Haemagglutination inhibition (assay)
HPL	High Production Livestock
HPLC	High performance liquid chromatography
HPTLC	High-performance thin-layer chromatography
HS	Haemorrhagic septicaemia
HSADL	High Security Animal Disease Laboratory
IARI	Indian Agricultural Research Institute
IB	Infectious bronchitis
IBL	Imperial Baeteriological Laboratory
IBR	Infectious Bovine Rhinotracheitis
ICAR	Indian Council of Agricultural Research
ICDP	Integrated Cattle Development Project
ICDP	Intensive Cattle Development Project
ICMR	Indian Council of Medical Research
ICPC	Indian Cattle Plague Commission
ICRISAT	International Crop Research Institute for the Semi-Arid Tropics
IDA	Indian Dairy Association
IHA	Indirect haemagglutination
ILRI	International Livestock Research Institute
Imp.	Improved
INR	Indian Rupee
IPDP	Intensive Poultry Development Project
IPT	Immuno peroxidase technique
IR	Infrared
ISGP	Indo-Swiss goat development and fodder production project
IVF	In vitro fertilization
IVM	In vitro maturation
IVRI	Indian Veterinary Research Institute
JH	Jamaica Hope
kg	Kilogram
KLDB	Kerala Livestock Development Board
KMF	Karnataka Milk Federation
KV	Key Village
KVS	Key Village Scheme
LAB	Lactic acid bacteria
LAMP	Loop-mediated isothermal amplification
LCMS	Liquid Chromatography Mass Spectrometry
L-DOPA	L-3,4-dihydroxyphenyl-alanine

LDPE	Low-density polyethylene
LLPD	Lakh litres per day
LPD	Litres per day
LPT	Livestock Products Technology
MAB	Monoclonal antibodies
MAS	Marker Assisted Selection
MFAL	Marginal Farmers and Agricultural Labour Development Agency
MFPI	Ministry of Food Processing Industries
mLF	Mithun Lactoferrin
MMT	Million metric tonnes
Mn	Manganese
MOET	Multi-ovulation embryo transfer technology
MPC	Meat protein concentrate
MS	Mass Spectrometry
MSL	Mean Sea Level
MT	Million tonnes
MV	Measles virus
MVC	Madras Veterinary College
MW	Megawatt
MWMP	Modified Worm Management Programme
NABARD	National Bank for Agriculture and Rural Development
NABCONS	NABARD Consultancy Services Pvt. Ltd
NARI	Nimbkar Agricultural Research Institute
NARS	National Agricultural Research System
NBAGR	National Bureau of Animal Genetic Resources
NCA	National Commission on Agriculture
NDDB	National Dairy Development Board
NDP	National Dairy Plan
NDRI	National Dairy Research Institute
NEH	North Eastern Hills
NER	North Eastern Region
NET	National Eligibility Test
NGO	Non-Government Organizations
NIANP	National Institute of Animal Nutrition and Physiology
NMG	National milk grid
NPSI	Network Project on Sheep Improvement
NRCC	National Research Centre on Camel
NRCE	National Research Centre on Equines
NRCM	National Research Centre on Mithun
NRCP	National Research Centre on Pig
NRCY	National Research Centre on Yak
NSP	Non-structural proteins
OECD	Organisation for Economic Co-operation and Development
OF	Operation Flood
OFP	Operation Flood Programme
ONBS	Open Nucleus Breeding System
OPU	Ovum pick up
PAN	Precision animal nutrition
PAU	Punjab Agricultural University
PCR	Polymerase chain reaction
PDP	Project Directorate on Poultry
PFR	Pineapple fruit residue
PIC	Polymorphic information content
PKC	Palm kernel cake

PPR	Peste-des-petits ruminants
PPRV	Peste-des-petits-ruminants virus
PT	Progeny tested
R&D	Research & Development
RAPD	Random Amplified Polymorphic DNA
RCA	Royal Commission on Agriculture
RFI	Residual feed intake
RFLP	Restriction fragment length polymorphism
RP	Rinderpest
Rs	Red Sindhi
RST	Random Sample Test
RV	Rinderpest virus
SAU	State Agricultural University
SFDA	Small Farmers Development Agency
Sh	Sahiwal
SMP	Skim milk powder
SNF	Solids-not-fat
SPF	Specific Pathogen Free
STPP	Sodium tripolyphosphate
SVU	State Veterinary University
TB	Tuberculosis
TCM	Technical Co-operation Mission
TCRP	Tissue-culture rinderpest vaccine
TEM	Transmission Electron Microscope
Thp	Tharparkar
TLC	Thin layer chromatography
TLR	Toll-like Receptor
TMR	Total Mixed Ration
TNAU	Tamil Nadu Agricultural University
TST	Targeted selective treatment
U.P.	Uttar Pradesh
UGC	University Grants Commission
UHT	Ultra-heat treatment
UK	United Kingdom
UMMB	Urea-molasses-mineral block
UNDP	United Nations Development Programme
USA	United States of America
USAID	United States Agency for International Development
USD	US Dollar
USSR	Union of Soviet Socialist Republics
UV	Ultraviolet
VN	Virus neutralisation
VTCC	Veterinary Type Culture Collection
WFP	World Food Programme
WHA	World Health Assembly
WHC	Water holding capacity
WHO	World Health Organization
WRL	World Reference Lab
Zn	Zinc
ZTM	Zonal Technology Management
ZTMC	Zonal Technology Management Centre

Contents

Chapter - 1

Introduction

1.1 ANIMAL HUSBANDRY SCENARIO (GENERAL) 1925

1.1.1. Establishment of the Royal Commission on Agriculture (RCA) 1928

Prior to independence, emphasis in livestock production was on the improvement of cattle for milk in a few dairy breeds and for draught in other breeds of cattle, buffalo and goats for milk, sheep for mutton, and poultry as backyard enterprise. The policy was to achieve these objectives through a large selection process, grading up descript animals by establishing herds of pedigreed cattle, sheep and goats and launch programmes to control losses due to deadly diseases, especially the rinderpest in cattle and buffalo and Ranikhet in poultry.

In 1928, the Royal Commission on Agriculture (RCA) drew special attention to the fact that rearing of bulls would not be remunerative for farmers and that the government should undertake this expensive but essential work of building up herds of pedigreed cattle for production of superior bulls and bullocks in large numbers. Thus, establishment of cattle breeding farms formed an important part of the earlier activities of animal husbandry departments. During 1938–39, there were 53 cattle-breeding farms, mostly concentrated in the Northern provinces. Taking note of some success in the formation and functioning of co-operative cattle-breeding societies in Punjab and similar prospects in Mumbai, the RCA recommended that where cultivators evinced keen interest in their livestock, organisation of co-operative breeding societies should be encouraged by loaning or gifting of a bull and/or by providing grazing grounds on favourable terms. Due to this recommendation, some State Animal Husbandry Departments made efforts to organise such societies.

Another important and major contribution during the late 30s and early 40s was the laying down of description and definition of characteristics of different breeds of cattle and buffaloes by the then Imperial Council of Agricultural Research (later named ICAR: Indian Council of Agricultural Research).

1.1.2 Cattle

Historically, cattle in India have been bred mainly to meet the draught power required for agricultural purposes. This resulted in India having some of the best draught breeds in the world. By and large, up to 1950, milk was generally treated as a by-product. There were dual-purpose breeds such as Sahiwal, Sindhi, Hariana, Kankrej, Tharparkar, Gir and Ongole. Their number did not exceed 10% of the cattle population in entire breed population. The numbers did not exceed 10%–15% of the breed type, on the basis of which the monograph for the RCA has been prepared. Attention was traditionally given more to raising of male than female calves, which were generally neglected.

Introduction of temperate cattle breeds

British tea planters in the hilly areas of the north, north-east and south introduced temperate breeds to produce milk required by them. They conducted their own experiments on the feasibility of breeding exotic breeds in pure form or by crossing their bulls with indigenous cows. European breeds tried were Ayrshire, Shorthorn and Jersey. These exotic cattle rapidly perished as they could not adapt to local climate; however, their cross-breeds survived with enhanced milk production. The military dairy farms established to meet the requirement of milk for army personnel started crossing the local Indian breeds such as Sindhi and Sahiwal with Friesian and Ayrshire bulls to increase milk production. The cross-bred progenies gave encouraging results. Approximately the same time, scientists at the Imperial Dairy Research Institute, Bengaluru; Livestock Research Station, Hosur and Agriculture College, Allahabad, attempted crossing of Indian breeds with Ayrshire, Friesian and Brown Swiss bulls under scientific conditions to determine the level of inheritance that would produce the best results under local conditions. By and large, in all these experiments, the general findings were that the cross-bred progeny could not withstand the local stress factors and diseases beyond 50% of exotic inheritance. Enthused by the results achieved at all these centres the plantation labour and farmers around these places quickly resorted to cross-breeding using locally available bulls. The male cross-bred progeny was extensively used for draught.

Around this period, some enterprising people used the Shorthorn breed for cattle improvement around Patna resulting in the 'Taylor' breed, traces of which are still found in the district; it is otherwise extinct. In Shillong, an enterprising Director of Agriculture used Friesian bulls on local cattle with excellent results, leading to the spread of this technique in and around this place where Friesian crossing continues to this day with outstanding results. Other than in military dairy farms, it is only in Shillong that one can find cross-breeds with the highest level of Friesian inheritance remaining in excellent health and performing at very high levels of production.

Colonel Olver, an animal husbandry expert at the ICAR observed that it was unsound to introduce European breeds of cattle in India and that it would be a sounder policy to effect systematic improvement in the indigenous stock by selective breeding, better feeding, improved management and disease control. He concluded that profit from the sale of bullocks was small compared to the income derived from the sale of milk and other dairy products. Colonel Olver suggested that to derive full benefit from agriculture, it would be necessary to give greater attention to increase milk production in local cows under a system of mixed farming.

Later, Dr. N.C. Wright, while emphasising the importance of improving milk yield in Indian cattle, pointed out that in deciding the breeding policy it was necessary to consider the diverse needs of the dairy industry's various branches. According to him, only high-yielding milch cows could prove economical for city milk supply, whereas draught breeds were required to produce good working bullocks in rural areas. He opined that even from draught breeds it was desirable to obtain as much milk as possible without causing any deterioration in draught quality. He strongly opposed the adoption of large-scale cross-breeding to improve the milk yield of local cattle under the prevailing conditions of animal husbandry and recommended that immediate steps must be taken to select and improve milk strains of indigenous cattle. However, he added that such efforts should not be limited only to milk-type breeds because improvement in milking capacity could also be looked for in most of the other breeds. He recognised that even draught breeds such as Amritmahal and Hariana possessed strains of special milking potentialities. Another important point referred to by Wright was that the breeding policy must consider the environment under which animals had to live and produce. Referring to the attempts that were generally made to improve the size and productive quality of small country-bred stock by supplying large-sized and high-production potential sires, he warned that such attempts might do more harm than good unless these were accompanied by measures to improve the environment of the stock including their feeding. The Animal Husbandry Wing of the Board of Agriculture and Animal Husbandry of India at its Fourth (1940) and Sixth (1945) Meetings discussed the question of mass improvement of inferior indigenous cattle and recommended that grading up of cattle should be taken up (i) in a systematic way, (ii) in selected areas (iii) compulsory castration of stray bulls in these areas and (iv) appropriate feeding of the graded progeny and continuous supply of pure-bred bulls should be ensured.

In 1944, the Government of India invited Dr. R.A. Pepperall for making a rapid survey of the problems facing dairy industry in the country. Pepperall, while recognising that the Indian buffalo as a producer of high-quality milk had good economic value and would always retain an important place as a dairy animal. In his view, it should be considered only complementary and not as a substitute for the cow upon which rests the dual responsibility of supplying a reasonable quantity of milk and draught power. He believed that it was not possible for the farmers to maintain separate animals for milk and draught under the existing conditions of farming. According to Pepperall, there was an urgent necessity for increasing milk production but warned against embarking on a policy of cross-breeding with European bulls. He emphasised that milk yield of indigenous breeds of cattle should be improved rather than attempting large-scale cross-breeding with imported bulls.

Dr. N.C. Wright, an excellent dairy scientist and a few scientists from India such as Dr. S.C. Ray and Dr. P. Bhattacharya, argued the case that out of a few options, in view of a minute amount of data available on crosses, the decision on introduction of exotic breeds could be unproductive with the crossing, producing high-producing cross-bred animals in large numbers in one generation; moreover, the succeeding set-ups would be grading up with the Brown Swiss (American Strain), Jersey (American Strain) and Friesian bull with the best available strain to production of milk and draught power by crossing with these three strains to end up the crosses, with both objectives from the cross-breeds. However, by 1957–59, a larger amount of data was produced by C.B. to draw a major conclusion from the details which could become available to take appropriate and informal discussions.

1.1.3. Buffalo

The situation was different in case of buffaloes where the female calves got better attention. In Northern provinces, buffaloes were the primary milk animals. This was generally true for the entire Indo-Pak subcontinent. The only efforts that were made by the Government in the 50s and prior to it were to improve cattle and buffaloes through pure-breed bulls of superior breeds of cattle and buffaloes. Farmers maintaining such bulls were also provided an annual grant to encourage this procedure. Simultaneously, castration of non-descript bulls were practised in these areas to eliminate indiscriminate breeding.

The buffalo cow was recognised as a dairy milk machine in the northern parts of India including Punjab but not as a replacement for cows. Unfortunately, the emphasis on buffalo cow as that of new products and meat was not focused on quality and quantity of the milk and meat of buffaloes; therefore, the buffalo's advantage as a milk animal was not utilised. Subsequently, the interest in buffalo genotypic identification was initiated, and it was established that there were several buffalo species in South-East associating from Afghanistan, Iran, parts of old south states of USSR such as Azerbaijan. Also India, Pakistan, Bhutan, Bangladesh, and Nepal predominantly had the buffalo *(Bos bubalis)* with chromosome No. 50, H8 (Indian coastal region) and swamps from Iran, Bangladesh, etc. In countries such as Nepal and the Netherlands, in Assam and parts of North Eastern Region (NER) we need to look deeply for other chromosomes under a project involving DNA finger printing and any other parameters for further elucidation of the differences between species.

1.1.4 Sheep

In sheep, during the first decade of the 20th century, cross-breeding of local and Bikaneri breeds with Merino and Romney Marsh rams was undertaken at several places in the former United Provinces, both in plains and hills. Although grading of local and Bikaneri sheep with exotic rams continued up to the second generation, no consistent results were achieved and the experimental work was discontinued for want of financial provision and lack of organisation. Simultaneously, breeding of Bikaneri sheep with Merino was undertaken at the Government Livestock Farm, Hisar and continued over several years. This resulted in the evolution of a new breed named Hissardale. Although there is no authentic record regarding the level of Merino inheritance in the cross-breds interbred to create Hissardale, it is believed that it was approximately 75%. A small flock of this breed is being maintained at this farm. They appear to be well adapted to the extreme heat and severe dry cold climate of north Indian plains.

1.1.5 Constitution of Imperial Council of Agricultural Research

With the constitution of the Imperial (now Indian) Council of Agricultural Research (ICAR) several pilot experiments for breeding superior sheep were undertaken from 1938 to 1949 in different parts of the country. With the experience gained, it was decided to prioritise improvement of indigenous breeds in the plains through selection and restrict cross-breeding to the temperate Himalayan region and sheep-rearing areas of the western part of Deccan plateau. In 1952, the ICAR developed a comprehensive plan for sheep development on a regional basis to evolve superior breeds and study different aspects of sheep husbandry. It also sponsored a full-fledged centre to conduct research on sheep and wool in each of the three regions, *viz.* the temperate Himalayan region, the dry Northern plains and the Southern region. Sheep breeding farms were established at Banihal and Riasi in Kashmir, Gwaldom in Uttar Pradesh, Sarhan in Himachal Pradesh, Darjeeling in West Bengal, Jorbeer (Bikaner) in Rajasthan, Paltan in Gujarat, Pune in Maharashtra, Nilgiris (Ootacamund) in Tamil Nadu and Gaya in Bihar. The breeding programmes included selective breeding within the indigenous breeds or crossing them with exotic fine wool breeds.

1.1.6 Indian Cattle Plague Commission (ICPC)

Importance of protection of livestock against diseases and pests was realised by the Government of India as early as in 1868, when the Indian Cattle Plague Commission (ICPC) was appointed to report on the prevalent cattle diseases and suggest measures on their prevention and control. ICPC and another commission in 1883 recommended the establishment of Imperial Bacteriological Laboratory (IBL), a forerunner of the Indian Veterinary Research Institute (IVRI). The IBL was established in 1889 and a Civil Veterinary Department came into being in 1891. At the time, the IBL was started rinderpest (cattle plague) was widespread among cattle and buffaloes. Thus, early researches were concentrated on rinderpest and an efficacious rinderpest virus vaccine for immunisation of cattle and buffaloes was consequently produced by 1926. This resulted in launching of countrywide programme to control losses due to this deadly disease. The IVRI undertook studies on other viral, bacterial and parasitic diseases of livestock, and developed effective methods and biological products in the form of antisera and vaccines for combating these. These researches made easy the task of combating several animal diseases and helped in taking up programmes of livestock development. The Royal Commission on Agriculture (RCA) in 1928 while appreciating the work done remarked that 'in no sphere has scientific research conferred greater benefits on agriculture than by the provision of means of controlling livestock diseases'.

1.1.7. Livestock and Poultry Diseases and Vaccine Production

During the early period, the main concern of research was to develop suitable therapeutic and control measures to check the colossal losses resulting from high mortality and morbidity inflicted by the livestock diseases, *viz.* surra, anthrax, rinderpest, tetanus, black quarter, haemorrhagic septicaemia, glanders, Ranikhet (New Castle) disease and fowl pox.

The outstanding research accomplishments were the development of a live-attenuated goat adapted vaccine for rinderpest, R_2B vaccine for Ranikhet disease, chicken-egg adapted vaccine for fowl pox and mouse-adapted vaccine for African horse sickness. Large-scale production of the first two vaccines helped in planning and execution of country-wide immunisation programmes against rinderpest and Ranikhet diseases. This resulted in a considerable reduction of economic losses and made animal production possible with small holders and landless labourers. It is a matter of great satisfaction that due to this policy now rinderpest has been eradicated in the year of Golden Jubilee of India's independence. The vaccine developed against African horse sickness played a vital role in its eradication from India.

1.1.8 Livestock Census

The first livestock census was completed in 1951. The total livestock population was 292.8 million, comprising 155.3 million cattle, 43.4 million buffaloes, 39.1 million sheep, 47.2 million goats, 2.86 million equines, 0.6 million camels, 4.4 million pigs and 73.5 million poultry. The human population was 361.1 million, out of which 140.0 million were agricultural workers. Among these, 69.9 million were cultivators, 27.3 million were agriculture labourers and 42.8 million were other workers.

1.1.9 Agriculture Power

Stable motive power for agriculture was achieved through the reduction and control of rinderpest disease, for which an All India Rinderpest Eradication Programme was launched in 1954. Thus, uninterrupted bullock power was provided to agriculture which was 73% in 1960 when the foundation of the green revolution was laid, though over years it was reduced to 23% (in 1995) due to farm mechanisation. It has remained at approximately 23% until the 19th Livestock Census 2012.

1.1.10 Bovine Skins

Exact estimates for 1950–51 with respect to production of bovine and ovine skins and hides contributing towards the leather industry are not available. Production figures for 1983 indicated production of 810 thousand tonnes of bovine and 110 thousand tonnes of ovine skins/hides. In 1980–81, the share of finished leather accounted for 57% of the total foreign exchanges earnings followed by high value-added product group with 31% and the balance accounted by low value added items covering raw material and unfinished leathers. In 1992–93 total export earnings from the footwear industry amounted to more than INR 40 million.

1.1.11 Milk Production

By 2015–16, milk production had increased to 155.5 million tonnes (Fig. 1) from 300 million bovines (2013-14), while 729.2 million poultry provided annual egg production to approximately 74.752 billion in 2013–14 with per capita availability of 61 eggs/year/person against the requirement of 180 eggs and 2.39 million tonnes of poultry meat. The availability of poultry meat is 2.8 kg/head/ year against the requirement of 11 kilograms.

The contribution of animal husbandry to the total GDP was 15.5% during 1950–51. Milk production was 17.40 million tonnes from 67.38 million breedable populations of cattle and buffaloes with per capita availability of 132 g/day. Egg production was 5 million with per capita availability of five eggs/ year/person. Poultry meat production was 55 thousand tonnes with per capita availability of 150 g/year. Wool production was 27.5 million kg from approximately 40 million sheep. In 2013–14, it had grown to 47.9 million kg. Per capita availability of milk was 307 gm in 2013–14 against the world average of 294 gm. According to the CSO estimates, the gross value added from livestock sector at current prices was approximately INR 4,06,035 crores during 2013–14, which is approximately 21.58% of the contribution from agriculture as a whole to the national GDP which was approximately 14.5% in 2014. The livestock sector contributed approximately 4% to the national GDP, expanded by 5.5% during 2013–14 against the 3.7% of growth of total agriculture sector including forestry and fishery.

With the start of Integrated Cattle Development Project (ICDP), some of the key village (KV) blocks were merged with ICDPs and sometime later with Operation Flood (OF) Project was taken up by State Dairy Corporation/Co-operative Dairy Federations.

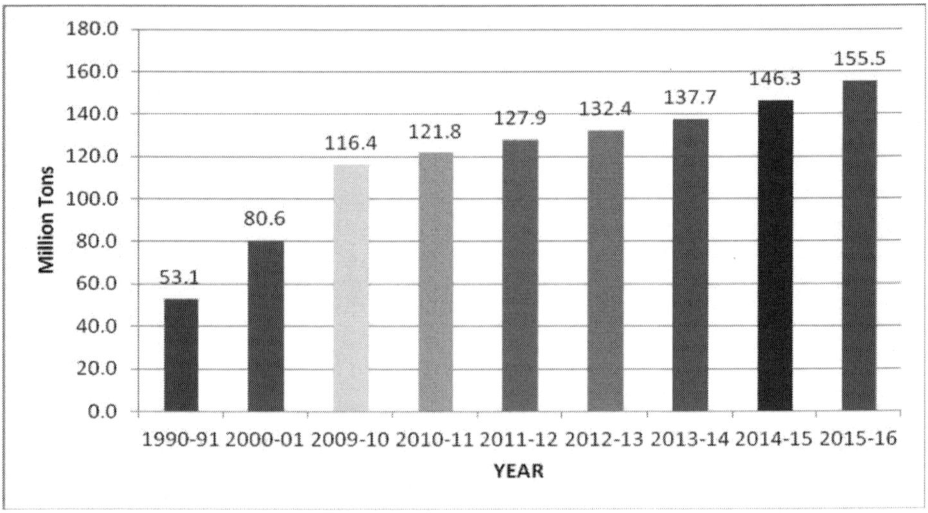

Fig. 1: Production of Milk (Million Tonnes)

Source: Basic Animal Husbandry & Fisheries Statistics 2016, Directorate of Economics and Statistics, Ministry of Agriculture, GOI (BAHFS 2016,http://dahd.nic.in)

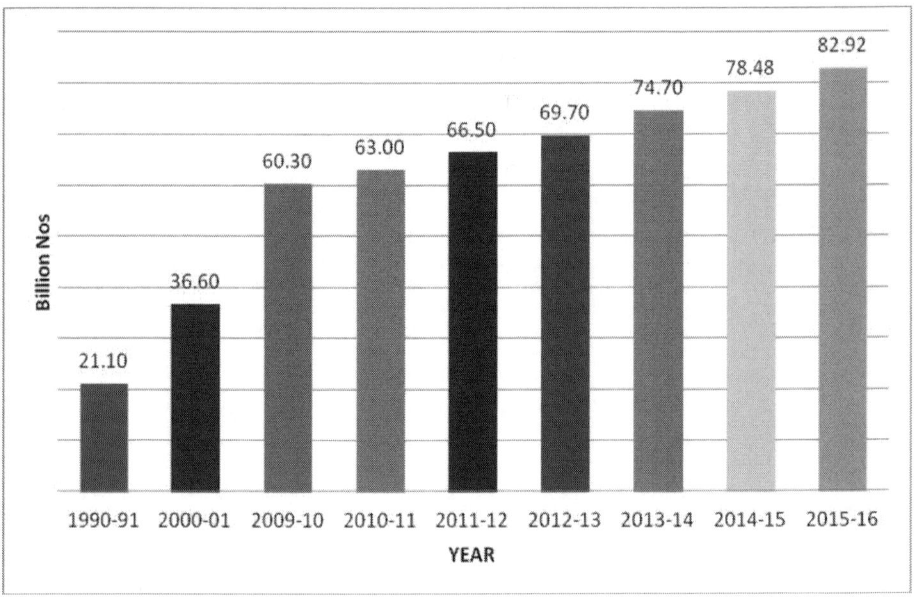

Fig. 2: Production of Egg (Billion)

Source: Basic Animal Husbandry & Fisheries Statistics 2016, Directorate of Economics and Statistics, Ministry of Agriculture, GOI (BAHFS 2016, http://dahd.nic.in)

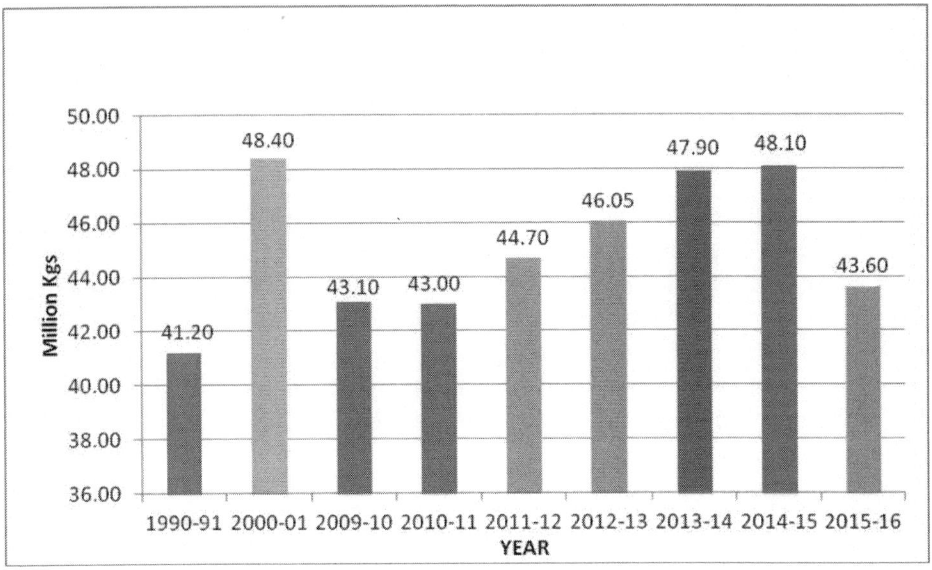

Fig. 3: Production of Wool (Million Kg)
Source: Basic Animal Husbandry & Fisheries Statistics 2016,
Directorate of Economics and Statistics, Ministry of Agriculture, GOI
(BAHFS 2016, http://dahd.nic.in)

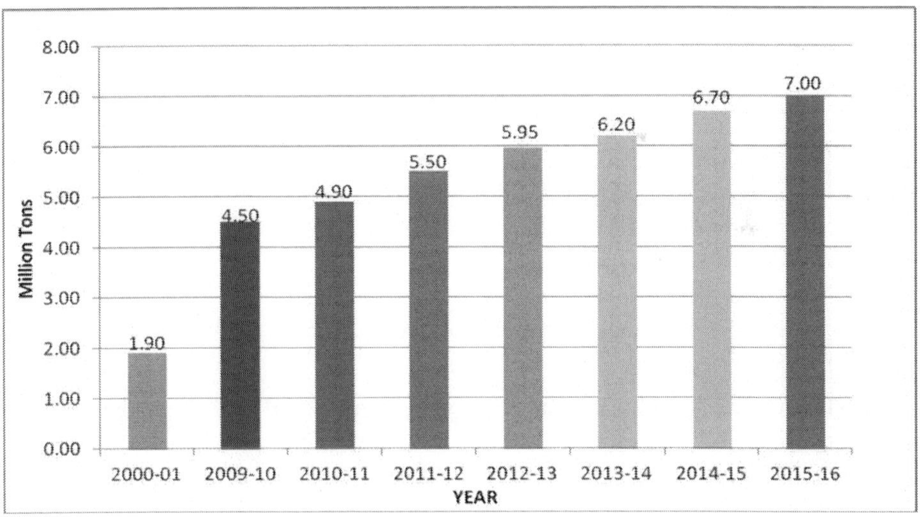

Fig. 4: Production of Meat (Million Tonnes)
Source: Basic Animal Husbandry & Fisheries Statistics 2016
Directorate of Economics and Statistics, Ministry of Agriculture, GOI
(BAHFS 2016, http://dahd.nic.in)

In the Fifth Plan, 100 new Key Village Schemes (KVS) were added. The breeding policy in the ICDP blocks was (i) improvement through selection in descript indigenous breeds; (ii) grading up of non-descript population with superior indigenous breeds in areas with no salubrious environment, inadequate feed resources and with not well developed facilities for marketing of milk; (iii) cross-breeding of non-descript population with exotic dairy breeds in other areas where physical environment and feeding resources were adequate. However, over the years, the breeding policy was modified.

Each ICDP unit was to cover 100,000 breedable cow and buffalo population to make a significant impact in a large area. The project units were in potential areas to ensure satisfactory development and milk production. ICDPs were to provide adequate inputs and services to cover at least 70% population of cattle and increase the milk production by 30% over a period of five years. Each project was to have 100 sub-centres manned by stockman and each sub-centre was to cover 1,000 breedable cows. At the end of the Fourth Plan, there were 62 ICDPs, and it was envisaged that existing projects would have all requisite inputs and 51 new projects would be started in the Fifth plan. Functioning of ICDPs was initially reviewed by the National Commission on Agriculture (NCA) and recommendations were made in 1976. The Government of India (GOI) extended 100% central assistance for the project in the beginning. After two years of establishment, these were transferred to states.

The Programme Evaluation Organisation of the Planning Commission undertook an evaluation study of the ICDPs in 1970–71. The study revealed that there was a considerable lag in providing inputs upto the level as detailed in the model plan.

The programme received a setback even in its earlier stages of impleme-ntation when these projects were transferred to state sector with consequential financial cuts in the central assistance. This resulted in dilution of inputs in later years. The study revealed that (i) 30%–40% of the semen produced was not utilised in most of the projects; (ii) fodder development programme was not making any progress and was integrated with general agricultural development programme; and (iii) the progress made in the organisation of rural milk co-operative societies was poor. The only successful ICDP was found in the Kheda district of Gujarat that had been handed over for implementation to the Kaira District Co-operative Milk Producers' Union.

Based on various studies and analyses of the ICDP and other programmes aimed at increasing productivity of cattle and buffaloes, it had been considered that the ICDP was the major project for cattle improvement being implemented by the state animal husbandry departments. Under the Operation Flood Programme (OFP), the dairy co-operative sector carried out similar activities through the village level co-operatives. In most cases the ICDPs as well as the OFP provided insemination facilities with semen of exotic breeds and gradually adopted use of frozen rather than liquid semen. The ICDPs only provided breeding and health cover but left the activities of milk procurement, processing and marketing to the dairy co-operatives. In fact, in some states, the Dairy Co-operative Federation shelved the responsibility for providing breeding and health cover entirely to the State Department of Animal Husbandry. Significant improvement was observed wherever both facilities were provided to the farmers, and the breeding programmes failed to make any impact when there were no facilities for milk procurement and marketing.

1.1.12 Hill Region Cattle Development Scheme

The ICAR initiated a large cross-breeding project in 1955 in the hilly and heavy rainfall areas for determining the optimum level of exotic inheritance in cross-breds which could sustain productivity after inter-breeding of these cross-breeds under village conditions. The exotic breed used was Jersey using the male germplasm. The project was run at nine locations in the country. Several grades with exotic inheritance of 1/2, 5/8, 3/4 and 7/8 Jersey inheritance were produced. Interse breeding was practical in 1/4, 1/2, 3/4 grades. All the animals in the scheme were identified and milk production was recorded. The results clearly demonstrated that milk in cross-bred cows increased to 200%–300% in different grades as compared to local cattle. Grades with more than 50% exotic inheritance were difficult to be maintained, at the then available health management and disease control programmes, due to high disease incidence and mortality. There was no significant difference in milk production of ½ and ¾ grades. In addition to improvement in milk production in cross-breds, there was an improvement in reproductive performance as reflected by reduction in age at first calving and inter-calving period. These results suggested that, under village conditions, cross-breeds with milk yield upto 2,000 kilograms could be sustained under feeding and management practices followed by a farmer.

State Cattle Breeding Farms: With the implementation of KVS across India, there was a large requirement of breeding bulls of various local and exotic breeds. To meet with this demand, the State Governments set up new farms and strengthened the old. While on one hand the farms did not get good-quality females to breed bulls, on the other hand, the number of bulls produced was so small that the farms could not meet the requirement of the breeding cattle and within the breeding tracts identified under the KVS.

The Government of India also supported the state governments in strengthening already functioning nine cattle and buffalo breeding farms. The central and state governments established breeding farms by importing pure-breed Holstein x Friesian, Jersey and Red Dane cattle. As the number of animals imported, was limited, herd size of the farms was small. Over time, performance of the imported stocks and their progeny deteriorated because there was no selection. Since these were temperate breeds, they did not adapt well to the tropical environment, poor feeding and management. Only a few bulls were made available by these farms for breeding.

A study on state cattle breeding farms revealed that very few of these were maintaining any sizeable population of pure indigenous breeds; mostly kept small herds of exotic, cross-breds and even buffaloes in the same farms. Thus, these farms were acting essentially as multiplication farms and neither any genetic gain was being made through selection nor requisite number of pure-bred males were being produced for use in the genetic improvement programmes of these breeds. Presently, only 25 farms are maintaining pure indigenous breeds located in different states. These farms maintain 15 out of 37 recognised breeds of cattle. Similarly, only 13 farms were maintaining only buffaloes and 20 farms only exotic dairy cattle breeds. The remaining farms were maintaining small herds of cross-breeds and indigenous breeds.

Central Cattle Breeding Farms: During the formulation of the Fourth Five-Year Plan, it was realised that the existing Cattle and Buffalo Breeding Farms were extremely low to meet the requirement of bulls. The Government of India, therefore, decided to set up six large cattle breeding farms on its own to implement the genetic improvement of breeds through selection of bulls on the basis of their progeny performance. Each farm was to specialise in breeding of bulls of selected breeds, *viz.* Red Sindhi, Tharparkar, Jersey, Hostein Friesian breeds of cattle and Murrah and Surti breeds of buffaloes. It was envisaged that each of these farms would develop a herd of at least 300 breedable cows so that 10 bulls at a time could be progeny tested with 10 or more performance recorded daughters. Each farm was expected to have a capacity to maintain a total of 750 animals.

Ministry of Agriculture, Government of India in 1983 and during subsequent years constituted several expert groups and review committees to evaluate the working of these farms. Despite comprehensive work plans developed and recommended from time to time the objective to produce indigenous and cross-bred pedigreed and proven bulls and conservation of indigenous germplasm could not be achieved.

Herd Registration of Cattle and Buffaloes: Herd registration was started in India as a Central Scheme under Ministry of Agriculture, Government of India. The Central Herd Books were started in 1941 for Red Sindhi and Sahiwal breeds, and opened for eight breeds of cattle and two breeds of buffaloes by 1949. The exotic breeds of cattle–namely, Jersey and Holstein Friesian – were subsequently included. The production standards for registration of cattle and buffalo breeds included in the scheme were confirmed from time to time.

Progress of Herd Registration was studied by the National Commission on Agriculture (NCA) and report was submitted in 1976. The report observed that although registration of animals was started in 1941 for Red Sindhi and Sahiwal and in 1949 for other breeds, the progress of the scheme has not been satisfactory. There is little apparent contribution to the genetic improvement of the breeds included, which has not benefitted breeders included in the programme. Several states have not registered even a single animal in the herd book.

A detailed study of the scheme by a Task Force headed by Dr. N.R. Bhasin was conducted, which made a series of recommendations; however, on ground, no advance was made in having an effective animal Herd Book either Nationally or in the states which could lead to an effective breeding programme for the indigenous breeds.

Progeny Testing of Cattle and Buffalo Bulls: Ministry of Food and Agriculture, Department of Agriculture, constituted a Working Group of Experts to review the Breeding Policy for cattle in the country in 1962, which submitted its report in 1963. Thereafter, a Scientists Panel was constituted for drawing up a breeding plan which recommended in October 1965, initiation of 'Scheme for systematic production of progeny tested bulls in cattle breeding farms for improvement of milk production in village cattle'. In the Third Five-Year Plan, a scheme was initiated for progeny testing, and selection of bulls on the basis of performance records of their daughters was started in Hariana cattle and Murrah buffaloes at Hisar (Haryana). During latter half of the Third plan, it was observed that the cattle development programmes started earlier could not make much impact on improvement of the stock for various reasons. In the Fourth Five-Year Plan, it was decided to set up six large breeding farms for some selected milch breeds of cattle and buffaloes of national importance, for undertaking selection of bulls on the basis of their progeny performance.

These large Central Cattle Breeding Farms were set up at Suratgarh in Rajasthan for Tharparkar breed; at Chiplima in Odisha for Red Sindhi; at Dhamrod in Gujarat for Surti Buffaloes; at Koraput in Odisha for Jersey breed; at Alamadhi in Tamil Nadu for Murrah buffaloes. A co-ordinated progeny testing programme was also started in nine State Government Farms. These farms were Chakganjaria in Uttar Pradesh for Sahiwal cattle; at Durg in Madhya Pradesh for Murrah buffaloes and at Hesserghatta in Karnataka for Red Dane cattle. Progeny testing of cross-bred bulls was initiated in the Sixth Five-Year Plan in Andhra Pradesh, Assam, Kerala, Odisha and Maharashtra and later in Tamil Nadu.

During the midterm review of the Sixth Plan, it was decided that attention should be paid to the development of indigenous breeds of cattle and buffaloes. In pursuance of the recommendations of Planning Commission, a centrally sponsored scheme for development of indigenous breeds of cattle and buffalo through genetically superior bulls using associated herd progeny testing programme was started. The programme was initiated in 1992–93 for bringing about rapid genetic improvement in cattle and buffalo production by generating genetically superior bulls through Associated Herd Progeny Testing programmes and utilisation of technology of Embryo Transfer in selected Indian breeds, *viz.* Gir, Hariana, Kankarej and Sahiwal. No useful results have been reported by these farms and almost all of them have discontinued the progeny testing programme for want of funds except at Banavasi and Junagarh. The latter farm was transferred to Gujarat Agricultural University, Anand; subsequently some of these have not been associated in the progeny testing of bulls taken up by the ICAR.

Progeny testing programme in Kerala was undertaken by the Kerala Livestock Development Board (KLDB) during 1977. The programme was taken up on cross-breeds of indigenous cattle with Brown Swiss bulls. It was based on field recording of milk production, associated traits related to growth and reproduction in female progenies of bulls under test. ICDP regions of Mavalikara, Kanjirapally, Vaikom and Kattapana were selected as the breeding areas. Programme ensured that 1,500 test inseminations were performed for each bull to obtain completed first lactation records of at least 50 progenies. One batch of cross-bred bulls was tested every year. The female progenies born out of bulls were identified, registered and followed up at intervals of six months. Initially, the number of bulls under test was low but later increased to >40.

The results indicated improvement in average milk yield in daughters over the years. This would reflect genetic improvement through selection in the population involved and improvement in feeding and management. The programme adopted the selective breeding policy. Top 3% cows out of the milk-recorded herd belonging to the farms were inseminated with the selected cross-bred bulls. The male progenies of such mating, if sound in health and conformation, were purchased and reared in bull stations for production of frozen semen and progeny testing. Approximately 300 elite cows having first standard lactation milk yield not less than 2,400 kilograms under field conditions were selected and used as bull mothers. The breed evolved through *interse* mating has been named 'Sunandini'.

The Central Cattle Breeding Farm (CCBF) in Alamadhi started the progeny testing of Murrah buffalo bulls in 1977–78. Mating of three sets of bulls, breeding values of top two bulls was 1903 and 1813 but were, respectively, only 11% and 6% higher than the herd average of 1,744.4 kilograms. The result was based on the production performance of 19 and 10 daughters, respectively. The programme was discontinued in 1989 due to reduction in the herd strength. The farm was associated in 1993 with the Central Institute for Research on Buffaloes (CIRB) for progeny testing of bulls. Progeny testing unit for Murrah buffaloes at Banawasi in Uttar Pradesh was started in 1972 with a batch of five bulls and 160 females, the former were selected on the basis of their dam's yield.

Animal science research under the ICAR and NARS focuses on development of new technologies to support production enhancement, profitability, competitiveness and sustainability of livestock and poultry sector for food and nutritional security. The mission is to facilitate need on the basis of priority research in livestock and poultry sector in ongoing and new emerging areas to support productivity increase, thereby reducing the gap between potential and actual yield to meet the challenges ahead in the current era of globalisation.

Impact of Improved Germplasm on Productivity: To overcome the low per-animal productivity of Indian livestock, a massive programme was taken up for cross-breeding of non-descript cattle, mainly utilising semen of two exotic breeds: namely, Holstein Friesian (for irrigated areas and for farmers with adequate fodder resources) and Jersey (for dry/hilly areas and farmers having low fodder resources).

The Frieswal Project conceptualised and designed by Dr. R.M. Acharya, Dr. P.N. Bhat and Brig. Dilbagh Singh in 1984 in association with Directorate of Military Farms as a permanent activity of Project Directorate on Cattle, presently renamed as Central Institute for Research on Cattle (CIRC), envisages evolving a national milch breed 'Frieswal', a Holstein x Sahiwal cross, yielding 4,000 kilograms of milk with 4% butter fat in a mature lactation of 300 days. In addition, two strains of cattle — namely, Karan Swiss (average milk production per lactation of approximately 3,600 kg/lactation of 305 days with highest recorded yield of 7200 kilograms) and Karan Fries (average milk production of approximately 4,000 kilograms per lactation of 305 days with highest recorded yield of 8,338 kg) developed by cross-breeding of Sahiwal cows with Brown Swiss (America) and Holstein Friesian have greatly contributed to milk production enhancement in the country.

Emphasis has also been on indigenous milch breeds of cattle such as Sahiwal, Gir, Tharparkar, Kankrej, Ongole, and Deoni for their improvement, conservation and propagation. In buffaloes, the programme for genetic improvement of well-defined breeds and upgrading of non-descript buffaloes using semen of superior milch breeds such as Murrah, Nili Ravi, Jaffrabadi and other important breeds and gradual spread of artificial insemination. This has considerably improved the profitability of livestock owners. Elite herds of Murrah, Niliravi, Jaffrabadi, Pandarpuri, Surti, Bhadawari and swamp buffaloes are maintained at CIRB, NDRI and some other centres located at SAUs and ICAR institutions under Network Project on Buffalo Improvement for undertaking associated progeny testing. Herds of elite Murrah buffaloes have been developed through research efforts, producing on an average of 2,445 kg milk/lactation.

Number of technologies on nutrition intervention resulted in economical ration formulations involving conventional and fibrous feeds for milk and meat production and on area-specific mineral mixture for obtaining improved reproductive efficiency, convinced the farmers for rearing superior germplasm and commercial livestock farming.

Front line research on embryo biotechnologies including embryo transfer, cloning and *in vitro* fertilisation and cutting-edge technologies on stem cell research, genomics and metagenomics, proteomics and bioinformatics are being pursued in several institutions. Buffalo cloning, which has application for faster multiplication of elite individuals, resulted in the birth of first cloned calf 'Garima' produced from skin fibroblast using the hand-guided cloning technique. Later, 'Garima 2' and 'Shreshta' (the first cloned male buffalo) were also produced through cloning using fetal fibroblast and embryonic stem cells of buffaloes at the NDRI. Further effort for producing clones from elite male buffalo semen yielded the birth of male calf 'Swarn'. This technology will go a long way in the multiplication of superior elite animals in India.

National Project for Cattle and Buffalo Breeding: Appreciating that bringing about genetic improvement is a long-term activity; the Government of India initiated National Project for Cattle and Buffalo Breeding in October 2000 for a period of 10 years. The project envisaged genetic upgradation and conservation of important indigenous breeds. The project was to begin with delivery of improved artificial insemination service at the farmers' doorstep as well as to bring all breedable females among cattle and buffalo under organised breeding either through artificial insemination or natural service using high-quality bulls. Managing the production and supply of genetic inputs as well as liquid nitrogen was entrusted to specialised autonomous and professional state implementing agencies.

The bull production programme proposed that bulls would be selected through field performance recording programmes by the State implementing agencies, dairy co-operatives, breeders' associations and other agencies. The activities under this component were identification and tagging of superior germplasm and rearing of male calves suitable for breeding up to breeding age in case of Murrah buffaloes and one year in case of cattle bulls. The other features planned for the participating farmers included production of quality exotic and cross-bred bulls from imported embryos and semen under the component. This task was to be accomplished through reorientation and intensification of Progeny Testing Programmes established during Phase I, establishment of Progeny Testing Programmes in new areas/milk pockets, setting up of Open Nucleus Breeding System (ONBS) and strengthening of bull mother farms. The scheme included increased and improved coverage of breedable females; approximately 80% of total adult females among cattle and buffaloes to be brought under organised breeding activity (AI or natural service); bring about large scale replacement (20 million) of low-producing non-descript cattle and buffaloes by improved animals; and development of several indigenous cattle and buffalo breeds. The results from some states were encouraging but did not reach the expectations for want of total commitment on behalf of the State Animal Husbandry staff as well as the services and goods support.

National Dairy Plan: National Dairy Plan (NDP) Phase-I is a central sector scheme for the period of 2011–12 to 2016–17 with total investment of INR 2,242 crores. It is a scientifically planned multi-state initiative with objectives to:

(i) Help increase productivity of milch animals and thereby increase milk production to meet the rapidly growing demand of milk and

(ii) Help provide rural milk producers with greater access to the organised milk-processing sector.

The objectives would be pursued through adoption of focused scientific and systematic processes, provision of technical inputs supported by appropriate policy and regulatory measures. Under NDP, it is planned to meet the 100% bull replacement needs of 'A' and 'B' graded semen stations by the end of NDP I and thereafter largely through producing high genetic merit bulls by progeny testing and pedigree selection programmes and a small number through import of exotic pure-breed bulls or equivalent embryos.

As per the estimate, to breed approximately 35% breedable animals by the end of NDP I, the country would need approximately 100 million high-quality disease free semen doses and make available approximately 900 high genetic merit bulls for replacement of bulls maintained at all the 'A' and 'B' graded semen stations. Based on the current and future demand of high-genetic merit bulls of different breeds for semen production, the breeds identified for establishing progeny testing programme include Holstein Friesian, HF cross-bred (CB), Jersey CB, Sunandini cattle and Murrah and Mehsana buffaloes and the breeds identified for pedigree selection programmes to begin with include Rathi, Kankrej, Tharparkar, Gir, Sahiwal and Hariana cattle and Nili-Ravi, Jaffarabadi, Banni and Pandharpuri buffaloes.

Adequate bio-security measures would also need to be ensured at the village level as well as at pre-quarantine, quarantine and rearing station levels. Therefore, institutions capable of meeting the afore-mentioned eligibility criteria would be entrusted to implement progeny testing or pedigree selection programmes to produce the required number of bulls. Both progeny-testing and pedigree-selection programmes are to be implemented through independent project teams and governed by a Management Committee.

Conclusion

Since the beginning of the planned development in India, undertaking genetic improvement of Indian cattle and buffalo breeds have been at the centre of the objective of all the programmes. The Key Village Scheme and Intensive Cattle Development Programme have been major programmes covering/intending to cover a very large population of cattle and buffaloes. Their failure to achieve the targeted genetic improvement has many causes and deficiencies. None of these programmes have been found faulting as far as planning is concerned; the deficiency is in the execution of the programmes at the field level that hindered the achievement of aims and objectives. The main deficiencies are(i) the lack of field organisational structures to implement to obtain output should be given as input equal to expected output; (ii) the lack of trained manpower to reach the farmers and organise them into either co-operatives or self-help groups; (iii) lack of infrastructure for generating technical inputs required to be provided to the farmers such as the good quality of bulls, semen, facilities for artificial insemination, (iv) lack of facilities to produce feed and fodder, (v) lack of sterilisation of stray and unused bulls during the programme and (vi) lack of facilities to procure, process and market milk. Above all, the major deficiency has been the lack of facilities to record the various production parameters that would ultimately determine whether genetic improvement is taking place.

In support of field-based programmes such as KVS and ICDP, governments and the ICAR setup State and Central cattle and buffalo breeding farms and cross-breeding of indigenous cattle, undertook progeny testing of bulls of identified cattle breeds, buffaloes and cross-breds of Indian and exotic breeds. Most progeny-testing projects have unable to establish availability of progeny tested bulls. The only project that has been able to provide a documented result in the progeny testing programme is the one undertaken by the Kerala Livestock Development Board. So far, approximately 10 batches of bulls have been recorded as progeny tested. This improved lactation yield of Sunandini breed from 1,408±484 kg to 1,829±564 kg over the lifetime of the project. This increase is extremely inadequate and could be accounted for better nutrition and health control management. Results based on progeny testing, have also identified some outstanding buffalo bulls (AICRP on Buffalo) and cross-bred bulls (CIRC Meerut, NDRI Karnal).

Establishing a new breed through cross-breeding has been always a matter of scrutiny and controversy. It has been considered that inter-breeding cross-breds lead to a decline in their performance because inter-breeding among this population splits the heterotic gene combinations, leading to a decline in their performance. This decline was largely due to lack of selection in F_1 that can also result in some decline in milk yield in F_2. It is well known that neither the Indian cattle population nor the exotic bulls come from highly inbred populations. Both the populations are random bred. Therefore, the cross-breeding between two breeds leads to gene segregation when the F_2 cross-breds are interbred. The inter-breeding, on the one hand, results in breakdown of heterotic gene combinations and, on the other hand, results in large variability in F_2. However, this remains a controversial issue and no serious attempt has been made to resolve it. The international experience on cross-breeding suggests a continued increase among the cross-breds if the F_1 generation is backcrossed to one of the high-potential parents to gear greater productivity and reduce the effect of gene dilution. There is a need to undertake elaborate breeding experiment, which should produce pure-breed exotic and their native reciprocal F_1–F_2 cross-breds. The analysis of their differences would help in determining the various components of genetic variation.

1.2 CATTLE AND BUFFALOES

The historical perspective of the British Policy framework reflected that cattle in India have been bred mainly for draught purposes to meet the requirements of draught power in agriculture, transport and defence and buffaloes for milk. This resulted in India having some of the best draught breeds of cattle in the world. The Government in early part of the 20th Century had made available to farmers, the services of pure-breed bulls of superior breeds of cattle and buffaloes to improve their stocks by stationing these bulls at the veterinary hospitals for providing service on payment; encouraged farmers to maintain superior bulls by providing an annual grant for their upkeep and simultaneously castrated non-descript bulls in these areas to eliminate indiscriminate breeding with bulls of inferior quality.

1.2.1 Breeding Policy for Cattle and Buffaloes

The number of pedigreed buffalo bulls produced at the Government farms was minute compared to the requirements. To bridge the gap, the State Governments adopted the system of distribution of approved pure-breed bulls from breeding tracts. This method was the cheapest available and several bulls could be obtained, particularly for grading up of non-descript cattle. Due to this policy, the number of pedigreed and approved bulls increased in rural areas. However, little impact could be made in changing the type and production of stock except in limited pockets as there was no (i) planned castration, (ii) controlled breeding, (iii) timely replacement of bulls, and (iv) while several scrub bulls were present. Farmers were reluctant to get their young stock castrated as it was feared that such a practice would adversely affect the growth of male calves into good bullocks. Castration of scrub bulls was performed by the State Veterinary Departments but regular periodical castration campaigns were not appropriately organised. Particularly when large numbers of uncastrated bulls were present, all these factors largely nullified cattle and buffalo development programmes.

The Animal Husbandry Wing of the Board of Agriculture and Animal Husbandry of India at its Fourth (1940) and Sixth (1945) Meetings discussed the question of mass improvement of inferior indigenous cattle and recommended that grading up of cattle should be systematically taken up in selected areas, compulsory castration, appropriate feeding of the graded progeny and continuous supply of pure-breed bulls should be ensured.

In 1949, the Goseva Sangh at Wardha advised the Government of India that the ultimate aim in the development of cattle should be the production of dual-purpose animals and that the buffalo might be treated as a dual-purpose animal in areas where male buffaloes were used for cultivation. The ICAR considered the above suggestions and recommended adoption of the following course for the improvement of cattle:

(i) A large percentage of our cattle population is non-descript. Therefore, it is essential to produce a general utility animal with draught and milk qualities.

    In areas where **specific types** (as distinct from **well-defined breeds**) exist, the policy should be to effect improvement both for milk and work qualities by selective breeding. The objective should be to put in as much milk in them as possible without materially impairing work quality.

(ii) In case of well-defined milch breeds, where numbers have considerably decreased after the country's partition, the Committee is of the considered view that it will be in the larger interest of the country as a whole to develop their milking capacity to the maximum by selective breeding and principally utilise them for the development of cattle in underdeveloped areas through upgrading.

The 12th and 13th meetings of the Animal Husbandry Wing of the Board of Agriculture and Animal Husbandry recommended that breeding programmes for the cows must place maximum emphasis on milk production and that the breeding of milch animals should be encouraged in rural areas adjoining towns. Regarding buffaloes, it recommended that every effort should be made to improve their milking qualities by judicious breeding and popularising improved practices of feeding and management.

1.2.2. Key Village Scheme

The Key Village Scheme (KVS) was launched as the main attempt for cattle and buffalo development during the First Five-Year Plan. The Gosadan Scheme was another initiative launched during this Plan period for removal of useless cattle from areas of natural grazing to tracts where fodder was in abundance and remained unutilised. The Gaushala Development Scheme was initiated for developing gaushalas into centres of economic milk production. The breeding policy during the **First** and **Second** Plans was to develop **dual-purpose breeds** which could provide good bullocks and increased quantities of milk. As an experimental breeding programme, it was also decided to undertake cross-breeding in high altitude and/or heavy rainfall areas using bulls of exotic breeds. A farm with a herd of Jersey cattle was set up in Kataula, Himachal Pradesh to meet the requirement of Jersey bulls for the cross-breeding projects. The need for registration of cattle conforming to certain prescribed standards in the main breeding tracts was recognised as an important means for development of cattle breeds. Accordingly, a scheme to organise breeders' societies for registration of cattle, milk recording and supply of breeding stock of **Hariana, Gir** and **Ongole** breeds of cattle and **Murrah** breed of buffalo was initiated.

1.2.3. Intensive Cattle Development Projects

In the Third Plan, a scheme for progeny testing was started with Hariana cattle and Murrah buffaloes at Hisar. During the latter half of the Third Plan's period, it was realised that cattle development programmes started earlier could not make much impact on improvement of stock owing to lack of inputs and absence of tie up of these programmes with an appropriate marketing system. An area development approach for cattle development, popularly known as Intensive Cattle Development Project (ICDP), was formulated as a part of the Special Development Programme. A new cattle breeding policy was evolved according to which cross-breeding was introduced in areas covered by ICDP and Key Village blocks that lay in the milk-sheds of dairy projects.

An Expert Committee was set up by the Government of India in 1959 to review and evaluate the work done under the All India KVS. The Committee recommended that while draught quality in cattle was important, every effort should be made to increase milk production as quickly as possible. For achieving this objective, the Committee suggested that improvement of buffaloes should be undertaken wherever they were found suitable and in non-descript cattle even cross-breeding with exotic breeds might be resorted to where climatic and other conditions were favourable. In well-defined breeds or specific types, it was considered essential that their milking quality should be improved through selective breeding. The Committee was of the view that in areas where specialised breeds already existed, no cross-breeding should be permitted.

The future approach to the problem of cattle development was discussed by the Gosamvardhana Council in 1960, and particular attention was drawn to the fact that the cow in India had lost its due place in the integrated agricultural pattern owing to years of neglect. An urgent need was felt for giving immediate attention to the improvement of the cow under developmental programmes.

There was a gradual shift in the approach to cattle development from a specific policy of developing draught cattle. The approach shifted to develop dual-purpose breeds of cattle for milk and draught, and effectively interweave dairy development programmes with the economy of the surrounding villages. In large city milk supply schemes, effort was made to link dairy plants with the countryside to help enriching the rural economy through development of cattle with the twin objectives of increasing milk production and supplying good animals for draught purposes. For the hill areas, a policy of cross-breeding with Jersey breed was advocated.

The Animal Husbandry Wing at its 14th meeting (1961) considered the need for effecting a rapid increase in milk production in cattle and suggested certain changes in the existing cattle breeding policy to permit cross-breeding of non-descript cattle with exotic breeds. It was recommended that the existing breeds might be reclassified and animals of breeds considered unimportant in the context of the present-day needs of draught might be graded up with other improved indigenous breeds. To supplement the work of the government cattle-breeding farms to produce superior bulls, it was proposed that private cattle breeders should be encouraged to undertake breeding work on scientific lines and be given financial assistance for this purpose. It was further recommended that cross-breeding should be taken up in areas having non-descript cattle, if these areas were otherwise suitable for maintenance of cross-bred animals and were connected to large milk-consuming centres. It was suggested that a limited number of exotic breeds should be selected on the basis of experience gained from the research work already conducted in India and in other tropical countries, and that not more than one exotic breed should be introduced in any one area.

The cattle-breeding policy in general and the policy to be adopted concerning cross-breeding with exotic breeds were examined in 1961 by a Committee set up by the Central Council of Gosamvardhana. The Committee recommended that cross-breeding programmes should be taken up in an intensive manner using two or more exotic breeds in three selected areas, one each in the plateaus, the plains and the hills. It also suggested that while taking up cross-breeding on a large scale in new areas, the places should be carefully selected so that cross-breeding might not conflict with the needs and preferences of the local farmers. It was further proposed that in the plains cross-breeding should be started in areas where the holdings were small, agriculture was intensive, cattle were stall-fed and farmers were interested in breeding cows for milk production. Dr. P. Bhattacharya and his students were the key persons responsible for bringing these changes in policies.

The ICAR started a major project on cattle at Haringhata, West Bengal for evolving a new breed of dairy cattle by crossing grade Hariana cows with Jersey bulls. This scheme was merged in 1968 with a more comprehensive project supported by the United Nations Development Programme (PL 480), the Central Government and the Food and Agriculture Organization of the United Nations. This programme included three exotic breeds, *viz.*, Holstein × Friesian, Brown Swiss and Jersey with 1,500 grade Hariana type cows to produce a new dairy breed suited to humid tropics.

1.2.4 All India Coordinated Research Projects on Cattle/Buffalo Breeding

In quick succession, a Coordinated Research Project on Cattle Breeding was also started with the objective of evolving a breed of dairy cattle by crossing indigenous cows of Hariana, Ongole and Gir breeds with bulls of exotic superior dairy breeds that would have reasonable adaptability to local environments and high breeding efficiency. A Central Frozen Semen Bank was set up at Hessarghatta (Karnataka) for collection, processing, storage and transport of frozen semen. With a view to intensify cross-breeding work in selected ICDPs, four more frozen semen stations were set up one each at Amristar, Gurgaon, Indore and Bengaluru with the assistance from the Danish Government. Considering the contribution of buffaloes to milk production, an All India Coordinated Research Project on Buffaloes was also started by the ICAR.

To supplement the research agenda on the eve of the Fourth Plan, it was decided to set up six large cattle breeding farms for some selected milk breeds of cattle and buffaloes for undertaking selective breeding and progeny testing of bulls. Three large Central Cattle Breeding Farms one each at Suratgarh (Rajasthan) for Tharparkar breed, Chiplima (Odisha) for Sindhi breed, and Ankaleswar (Gujarat) for Surti buffaloes were set up. Work on two more Central Farms one at Koraput (Odisha) for Jersey breed and another at Alamadhi (Tamil Nadu) for Murrah buffaloes was initiated. During the Fourth Plan period, a cattle-breeding programme based on selection and progeny testing of bulls was also started in nine government farms in the States.

The Scientists' Panel on Animal Husbandry set up by the Union Ministry of Agriculture and Irrigation reviewed the cattle breeding policy in 1965 and suggested that selection among the indigenous pure-breeds, grading up of non-descript cattle with established indigenous breeds and cross-breeding with exotic dairy breeds should be undertaken in an intensive and co-ordinated manner, simultaneously ensuring provision of suitable environmental conditions. The Panel emphasised that bulk of the exotic inheritance should be obtained through

Jersey breed and that cross-breeding with Brown Swiss and Holstein x Friesian might also be tried to a limited extent. It was suggested that for some time to come cross-breeding programme should be limited to obtaining half-breds and to follow *interse* breeding among these half-breeds. However, there could be scope for producing cross-breds in areas where feed and fodder availability was satisfactory and farmers were progressive.

1.2.5 City kept Cattle/Buffalo

The Fourth Five-Year Plan laid great emphasis on cross-breeding of cattle for milk production. It was rightly stressed that the success of cross-breeding programme would depend on the quality of cross-bred bulls used for *interse* breeding and that cross-breeding should be implemented in a planned and controlled manner. It was during this period the concept of economic viability of cattle and dairy enterprise was recognised to be a necessary part of the integrated link in the chain of development. The concept of dairy complexes as centres of milk production around city centres with attendant problems of movement of *Khattals* to these centres in Bombay (Mumbai) and Madras (Chennai) was experimented and its futility was realised.

1.2.6 Operation Flood Programme

The approach to cattle and dairy development projects in the Fifth Plan saw a shift from organised milk colonies in urban centres of consumption to production by small and marginal farmers through co-operation. Thus, the *Anand pattern* was born and Operation Flood (OF) was launched. The need for extending facilities to the farmers for improved breeding of their stock and to provide an adequate and fair price for their produce through an organised marketing system was emphasised. Dairy cattle rearing were recognised as an instrument of social change to improve the economy of small and marginal farmers and agricultural labourers. An aggressive cross-breeding programme aiming at a target of 5.7 million inseminations/year was laid down. For achieving this objective, frozen semen, breeding bulls and foundation female stock of exotic breeds were imported. It was decided to increase the number of exotic cattle breeding farms from 25 to 35, and establish 50 more ICDPs in the milk sheds of existing and new dairy projects. Moreover, large-sized integrated cattle-cum-dairy development projects were proposed to be taken up around milk supply schemes in cities such as Kanpur, Bengaluru, Bhopal, Nagpur, Hyderabad and Jaipur, on the lines of milk production enhancement and marketing programmes taken up under the Operation Flood (OF) Project (Kurien 1987).

OFI was launched in 1970 following an agreement with the World Food Programme (WFP), which undertook to provide as aid 126,000 tonnes of skim milk powder (SMP) and 42,000 tonnes of butter oil (BO) for financing the programme.

The programme involved organising dairy co-operatives at the village level, providing physical and institutional infrastructure for milk procurement, processing, marketing and production enhancement services at the Union level and establishment of city dairies. The main thrust was to set up dairy co-operatives in the milk sheds, to link them to the four metro cities of Bombay (Mumbai), Calcutta (Kolkata), Delhi and Madras (Chennai), in which a commanding share of the milk markets were to be captured. The overall objective of operation flood I was to lay the foundation of a modern dairy industry in India, which would adequately meet the country's need for milk and milk products.

Funds for operation flood I were generated by the sale of skim milk powder and butter oil. A total of INR 116.4 crores was invested in the implementation of the programme. By its end, approximately 13,300 dairy co-operative societies (DCS) in 27 milk sheds were organised, enrolling 18 lakh farmer members. It achieved a peak milk procurement of 34 lakh litres/day (LPD) and marketing of 28 LPD.

The background of the institutional framework of OFII essentially comprised the successful replication of the Anand Pattern, a three-tier co-operative structure of societies, unions and federations. OFII was designed to build on the foundation already laid in OFI and the IDA-assisted dairy development projects in Karnataka, Rajasthan and Madhya Pradesh. The programme was approved by the Government of India, for implementation during the Sixth Plan period, with an outlay of INR 273 crore. Approximately USD150 million was provided by the World Bank and the balance in the form of commodity assistance from the European Economic Community (EEC).

OFII helped to market milk in approximately 148 cities and towns, with a total population of 15 million, through a national milk grid, linking these towns and cities to 136 rural milk sheds. The project expanded the number of village co-operative societies to 34,500, covering 36 lakh farmer members. The peak milk procurement increased to a level of 79 lakh litres/day (LLPD) and milk marketing of 50 lakh litres/day (LLPD).

Dairy co-operatives have been marketing milk in 200 cities including metros and approximately 550 small towns. The third phase has now led to National Dairy Plan 2020. One of the major achievements of Operation Flood has been the installation and efficient functioning of the national milk grid (NMG). Through a network of insulated rail and road tankers, milk was transported by the dairy federations from milk surplus to milk deficit areas. Milk travelled as far as 2,000 kilometres from Anand to Kolkata. This helped overcome regional imbalances in milk production, procurement and marketing. Similarly, the network of feeder-balancing dairies helped in converting liquid milk to milk powder during the milk surplus season and converting the milk powder into recombined liquid milk for marketing during milk deficit periods of the year. Operation Flood helped transform India into a single, homogeneous, seasonally and regionally balanced market.

It has been rightly realised that unless cross-breeding of indigenous cattle with exotic dairy breeds is undertaken in a big way, it will be well-nigh impossible to bridge or even narrow the gap between the availability and demand for milk in the country. The emphasis was on (i) increasing milk production for improving the dietary requirement of the people in general and to meet the ever-increasing demand for milk and milk products of the urban population and the growing industrial complexes and (ii) improving the working efficiency of bullocks in rural areas. Breeding for milk production should be concentrated in milk shed areas that could be conveniently linked up with the existing dairy development projects. Production of milk should be on a commercial basis and it should be sufficiently remunerative to attract the farmers to milk production enterprise. This should be attempted through a system of planned cross-breeding, selective breeding and grading up of indigenous cattle and selective breeding and grading up of buffaloes depending upon the suitability of each system under the prevailing local conditions. This, however can only be achieved if selection of males (genetic merit) is based on performance of daughters and surplus cattle and buffaloes (200 million) females are removed from the system.

While the policy for breeding dual-purpose cattle remains, the emphasis has shifted from draught to milk and integrated development of milk sheds on the basis of co-operative milk production scheme of operation flood and other models of dairy development in the industrial sector.

1.3 SHEEP

Sheep play an important role in the Indian economy by providing food, fibre, manure and rural employment in hilly, tribal, arid and semi-arid regions of India. They are important source of livelihood for small and marginal farmers, landless labourer, migratory farmers and farmers who follow transhumance for raising sheep. There are 65 million sheep in India which produced 48.10 million kg of wool during 2014-15 (Fig. 3).

There is large genetic variability in sheep as is reflected by 40 diverse breeds found in varying agro-climatic areas which have been classified into four major zones, *viz.* (i) Northern temperate; (ii) North-Western; (iii) Southern Peninsular and (iv) Eastern. Sheep in the Northern temperate region produce wool of good apparel quality. In climatically similar areas of southern hills, Nilgiri sheep also produce wool of similar quality. In North-Western region sheep produce wool of superior carpet quality. The sheep of Southern peninsular region either produce no wool or little wool of 36S quality. They are primarily used for mutton production. In the eastern region, owing to extremely high humidity, sheep are found in only small valleys with low rainfall such as Sikkim.

The sheep population as per the first Census in 1951 was 39 million. It has now increased to 65 million, registering an increase of 66.6%. However, sheep meat (mutton) has registered an increase of 42.7%.

With the re-constitution of the Imperial (now Indian) Council of Agricultural Research (ICAR) in 1929, pilot experiments for breeding superior sheep were undertaken from 1938 to 1949 in different parts of the country, primarily to focus attention on improvement in wool production. Based on the available knowledge and experience, it was decided that the improvement of indigenous breeds in plains through selection should receive priority, and that cross-breeding with European breeds should be restricted to temperate Himalayan region and to sheep-rearing areas in the western part of Deccan plateau. Grading up with superior indigenous breeds (Bikaneri) was undertaken in Andhra Pradesh, Karnataka, Tamil Nadu and Uttar Pradesh. The results were not very encouraging in the Southern states. There was, however, an improvement in wool production and quality, as reflected by increase in average fleece weight and decrease in average fibre diameter in crosses.

1.3.1 Cross-breeding of Sheep

The East India Company initiated programmes for crossing indigenous sheep with exotic fine wool breeds to improve the quality and quantity of wool to meet the requirements of the woollen mills in the United Kingdom. These were probably the earliest attempts in this direction. No definite information regarding the cross-breeding programmes, the exotic breeds used and the number of sheep imported is available. A few old records indicated that some success was achieved in cross-breeding work, particularly with Cape Merino around Poona (Pune) where 4,000–5,000 cross-bred sheep were raised. The cross-bred sheep produced wool superior in quality and more than four times in quantity than the indigenous sheep. However, the good work done faded away within 10 years as the programme could not be sustained. The Royal Commission on Agriculture (1928) attributed this to lack of organisational setup. Some trials were undertaken at the Amritmahal Farm in the former Mysore State, but they also failed. It appears that some of these sheep were taken to Nilgiri hills and are still available there in small numbers. However, their numbers have increased in recent years due to the efforts of co-ordinated project on wool of the ICAR and Tamil Nadu Agricultural University (TNAU). Several sporadic cross-breeding trials were also undertaken in the first half of the 19th century in Punjab, Bengal and Madras. Southdown and Cape Merino were the exotic breeds used in these trials. The Cape Merino crosses thrived better than those from the British breeds.

The importation of exotic sheep started as early as the beginning of the 19th century, but a definite record of breeds or number of sheep imported is not available. However, most of these imports were extremely low in number and imported animals were primarily used for cross-breeding with indigenous breeds for improving their wool yield and quality. There are records of importation of Southdown and Cape Merinos during this period. In the earlier part of the 20th century, Merino and Romney Marsh were imported for United Provinces. Merinos were also imported for Punjab and maintained at the Government Livestock Farm, Hissar. Sporadic importation of Merino sheep was also effected in Mysore (now Karnataka) State. Five rams and 10 ewes of Merino from South Africa were imported in 1938 for the Sheep Breeding Research Station, Poona (Pune). These did not survive for more than five years for apparent lack of adaptation to the new environment. In 1940, rams of Cheviot, Border Leicester and Weinsleydale from Britain and Merinos from Australia were imported. The British breeds did not perform well. In 1951, one Delaine Merino ram and 10 Rambouillet sheep were imported. In 1952 a small flock of Merino comprising five rams and I0 ewes was imported by the Uttar Pradesh Government for their Sheep Breeding Research Station, Pipelkoti. No serious problems in the maintenance of the exotic breeds were faced in Jammu and Kashmir. Since 1953–54, there has been continuous importation of exotic breeds such as Corriedale, Border Leicester, Polworth and Rambouillet into Uttar Pradesh. Corriedale and Border Leicester were imported only in 1953 as they did not perform well. Polworth and Rambouillet were imported over several years. Their performance was considerably satisfactory.

Cross-breeding trials on indigenous sheep with different exotic breeds showed some merit. Regional centres were set up in Uttar Pradesh hills, Rajasthan and Deccan Plateau. During the Second Plan, three sheep-breeding farms were established for production of breeding rams, and sheep and wool extension centres were started in some States for carrying out improvement in rural areas through the supply of superior stud rams. A wool testing laboratory was set up at each farm to help in the assessment of wool improvement through the sheep and wool extension activity. A large-scale sheep shearing, wool-grading and marketing programme was initiated in Rajasthan with the assistance of Food and Agriculture Organization (FAO) of the United Nations. The Central Sheep and Wool Research Institute (CSWRI) was established in 1962 at Malpura. At the beginning of the Fourth Plan, there were 84 sheep breeding farms, 602 sheep and wool extension centres and 52 supplementary sheep breeding centres, in operation in the entire country.

In 1964, a flock of 300 ewes and 100 rams of Rambouillet breed was imported for the Central Sheep and Wool Research Institute (CSWRI) and in 1968 a flock of 26 rams, 90 ewes and 71 lambs of Corriedale was imported for the CSWRI Substation at Mannavanur.

1.3.2 Sheep Breeding Policy Committee

The *Ad hoc* Sheep Breeding Policy Committee, constituted by the Government of India in 1970, suggested import of 20,000 exotic sheep of fine wool breeds for producing breeding rams. It also suggested production of an equal number of good quality stud rams of indigenous breeds. For achieving these objectives, a Central (Indo-Australian) Sheep Breeding Farm was set up at Hissar during the Fourth Plan with a foundation stock consisting of 3,000 ewes and 300 rams of Corriedale breed. The Farm has now a strength of approximately 8,000 Corriedale sheep. Two large sheep-breeding farms were also set up in the private sector —one in Maharashtra and the other in Gujarat — for producing pure- and cross-bred rams. The Government of India initiated action to set up large sheep-breeding farms, one each in Jammu and Kashmir, Rajasthan, Andhra Pradesh, Karnataka, Madhya Pradesh and Bihar during the Fourth Plan.

The *Ad hoc* Sheep Breeding Policy Committee recommended that in the northern temperate zone, cross-breeding with Rambouillet and/or Merino should be undertaken for improvement of fine wool. The Committee suggested that the level of exotic inheritance should be maintained at 50% in the field, except in case of sheep-breeding farms and selected areas. The experience in this area has indicated that cross-breds with 75% exotic inheritance do perform better and pose no serious disease or management problems.

A research project designed to evolve new wool breeds suitable for arid and semi-arid regions was started at the CSWRI in 1964. It involved the crossing of Rambouillet with three diverse indigenous types (i) medium-fine carpet/medium apparel type: Chokla; (ii) medium carpet wool: Jaisalmeri; and (iii) coarse hairy carpet wool: Malpura. Sheep with three levels of Rambouillet inheritance, *viz.* 50%, 62.5% and 75%, were interbred. The results indicated that there was little improvement in body weight and greasy wool production beyond 50% exotic fine wool inheritance. There was no serious decline in performance of the progeny produced from interbreeding of cross-breds. The half-breeds posed no serious management and disease problems. Higher crosses, however, showed lower adaptation to high temperature and greater susceptibility to diseases.

Rambouillet x Chokla half-breeds produce in a year approximately 2.5 kilograms of greasy wool of 58–64 counts suitable for manufacturing medium quality apparel. Rambouillet x Malpura half-breeds produce in a year approximately 2 kg of greasy wool which can be used for superior quality carpet or woollen apparel. Economic gains from improvement in wool alone were approximately 25% in Rambouillet x Chokla half-breeds over Chokla. Rambouillet x Jaisalmeri half-breeds, though superior to Jaisalmeri in wool production and wool quality, pose serious problem due to black patches on the fleece which reduce considerably the return from the sale of wool. At 75% of exotic inheritance, the indigenous types involved in crossing did not show any significant difference in their performance.

1.3.3 All India Coordinated Research Project on Sheep

During the Fourth Plan (1971) the ICAR sponsored two All-India Coordinated Research Project (AICRP) on Sheep for Fine Wool and another AICRP for Mutton. It was realised that unless the feed environment changed in central and north central area, cross-breeding for wool or mutton would not succeed. While in temperate areas, new breeds were developed, e.g. Kashmir Merino. The objective of these projects was to develop superior fine-wool breeds suited to different agro-climatic regions of the country. The main emphasis in the projects was to study breed combinations, level of exotic inheritance, heterosis in production traits and problems involved in interbreeding of cross-breds for new breed development. Ancillary studies on nutrition, physiological response to environment, reproduction and disease susceptibility were coordinated with success.

Conclusion

(i) On the basis of these findings, it was concluded that cross-breeding with exotic breeds could produce improvement in fleece quantity and quality at 50% exotic inheritance with improvements in diseases management and nutrition.

However, the improvement was significant in temperate eco-system which leads to development of Kashmir Merino.

(ii) There was no significant improvement in meat characteristics when compared with Indian breeds.

However significant improvements could be made in indigenous sheep through selection in the long term if followed in a closed-circuit system (ONBS).

(iii) None of these innovations could, however, be implemented as neither the farmers nor the government was keen to invest in the improvement.

1.4. POULTRY

1.4.1 Poultry Research Division at IVRI

A Poultry Research Division was established in 1938 at the Imperial (now Indian) Veterinary Research Institute (IVRI) to bring in development in poultry. The poultry farm of the Division could maintain 4,000 layers, and had facilities for hatching, brooding and rearing of birds. The farm was stocked with exotic breeds such as White Leghorn, Rhode Island Red, Australorp, Black Minorca, New Hampshire and Light Sussex. The breeding programme in the early stages was limited to multiplication of pure stock and some cross-breeding for improved egg production of local birds. The IVRI also supplied to the neighbouring villagers hatching eggs for obtaining better quality chicken and cockerels of improved breeds for improving indigenous stock. The most important single contribution of the IVRI for poultry development in India was the preparation in 1942 by its Pathology and Bacteriology Division of an effective vaccine against Ranikhet disease, the greatest obstacle to successful poultry farming in the country. With this singular achievement, several model poultry farms were established during the 40s by different State Governments for demonstration, training and multiplication of improved poultry stock. Most of the states appointed Poultry Development Officers to promote poultry production. With financial assistance extended by the ICAR, a cadre of Poultry Disease Investigation Officers was created in several states. The ICAR sponsored research projects on poultry at the IVRI and in some states. In keeping with these developments, the IVRI initiated postgraduate training programmes in poultry science for State Animal Husbandry Officers.

During the Second World War, the demand of the army for eggs and table birds increased manifold. To meet this demand, the military authorities set up 10 poultry farms with improved exotic stock, each of 10,000 birds capacity. From time to time, these military poultry farms distributed some poultry stock and eggs for hatching to the neighbouring farmers for raising poultry stock of improved variety. These steps helped popularise the rearing of better-quality poultry stock among small sections of rural people.

The cumulative effect of these different activities was a greater appreciation by the Central and State Governments of the possible advantages of poultry production and development. Demonstrations and extension work by different poultry farms in states created interest among farmers in localised rural areas in producing poultry of superior quality as a subsidiary occupation for economic benefit. Organisation of poultry and cattle also shows popularised poultry rearing and created enthusiasm among poultry breeders.

As part of the post-war development programmes, a few officers from the IVRI, State Departments and other organisations were sent abroad for advance studies in poultry science. Education and training of these officers proved beneficial for poultry development work initiated after Independence.

The foundation on which poultry farming grew in India was truly laid during the 50s with the advent of planned development. The ICAR and some of the State Governments had initiated a few poultry breeding research programmes as part of the 'Grow More Food Campaign'. Under the First Plan, an All-India Poultry Development Programme aiming at establishing of Poultry Extension-cum-Demonstration Centres was launched. In the beginning, 33 centres having facilities of modern poultry houses and staff were developed under the State Animal Husbandry Departments. Some State Government Poultry Farms were also strengthened to take up a programme of large-scale multiplication and acclimatisation of exotic stock and subsequent propagation. The scope of the programme was enlarged during the Second Plan by establishing five Regional Central Government Poultry Farms. Substantial aid was received from the US Technical Co-operation Mission (TCM) which supplied 30,000 day-old chicks of White Leghorn, Rhode Island Red breeds and poultry equipment. The TCM chicks were distributed among the Central and State Government poultry farms to augment substantially of breeding programmes that were already in progress. Consequently, the number of Poultry Extension-cum-Demonstration Centres increased 10 fold. These centres, each with 100 layers of improved breeds, were set up to (a) demonstrate the modern methods of poultry keeping and the economic benefits obtainable from poultry farming with improved varieties of birds; (b) impart training to the farmers in raising of poultry starting from day-old chicks to the layer stage; (c) supply hatching eggs, day-old chicks, adult birds, poultry equipment and construction material at subsidised rates or free of cost in some cases, to start small poultry units and (d) provide custom hatching facilities. Trained extension officers were attached to these centres to make technical knowhow available to the farmers on a continuous basis. All these steps taken during the Second Plan period made a substantial contribution to the development of poultry production. Awareness was created among farmers about the role scientific poultry farming could play in improving their economy.

1.4.2 Intensive Poultry Development Project

The 60s proved to be the turning point in the history of poultry production in the country. It is during this period that poultry farming was accepted as a source of livelihood. The concept of backyard poultry keeping yielded place to poultry farming as a commercial enterprise. Initiation of an Intensive Egg and Poultry Production-cum-Marketing Programme, popularly known as the Intensive Poultry Development Project (IPDP), with an area development and package approach, was the most important factor for bringing about this favourable development.

Introduction of deep litter system of management and further importation of high-quality stock and mass preventive vaccination against common avian diseases, income tax exemption for the poultry sector in agriculture, progressive industrialisation with resultant higher purchasing power of people, were among other major contributing factors.

Intensive Egg and Poultry Production-cum-Marketing Centres were established in selected urban areas in different states. Encouraged by the progress achieved under the project, the state governments decided on a more ambitious programme and extended the area of operation to 81 other centres. Under this programme all the required inputs and services such as good-quality birds and equipment, training of farmers, effective disease control measures and improved management knowhow were made available to 300 farmers per centre, helping them to have poultry units of 100 layers each. Besides, facilities were provided to market their outputs with an assured and a reasonable margin of profit.

Due to the concerted drive, poultry production started increasing and gaining momentum. To ensure the momentum private entrepreneurs were encouraged to establish franchise hatcheries of foreign-based poultry breeding organisations for production and supply of high egg producing hybrid chicks. Anticipating a possibility of seasonal surplus of eggs the establishment of a dehydration plant at Tiruvella, Kerala for egg powder was sanctioned. During this period the Central Government launched a special project (SFDA/MFAL) to help small and marginal farmers and agricultural labourers. Poultry development formed an important component of this project. By the 80s, the ICAR has developed two research institutes solely devoted to poultry.

Seven poultry-breeding companies were registered and India was self sufficient in poultry germ plasm of both layer and broiler types.

A separate unit for Random Sample Test (RST) of egg laying stock was established in 1969 by the Central Government at the Hessarghatta farm. With the development of this facility, it was considered desirable to determine the egg laying potential of the stock evolved under the CPBP. Random Sample Tests were, therefore, conducted with strains of White Leghorn developed at Hessarghatta and other farms under the Project. In addition to these, strain and breed-cross birds evolved at Hessarghatta and birds from other Government and private farms and commercial hatcheries were subjected to such tests. Four such RSTs have so far been conducted.

This thrust by the Government came when they considered Poultry Production as integral part of food production programme. The industry created a new model which involved an estate approach, which started with day-old chicks upto rearing of the broiler birds and a layer bird all to be paid for by the company and finally processing and marketing to have a chain which produced all the input. This chain could outsource the raw material at a back stage. Thus, the present industry scored and exercise is absolutely organised.

Conclusion

The present situation in this sector is considerably comfortable as the government has allowed the industry benefits which they needed. The industry has developed the infrastructure in latest technology transfer in breeding stock developed globally. The management technology and marketing are such that it has developed in an integrated model in which all the sectors including the farmers can prosper and grow.

■■■■

Chapter - 2

Science, Technology and Interventions

INTRODUCTION

The intention of the authors is to discuss in greater detail the research and development experiments done on the issues which need the attention and find solutions in the context of researches conducted in the country in the last 25 years; to usher in the developments to increase wealth, make the livestock rearing more productive and comfortable for resource poor farmers who use these livestock for their own good as well as for the society. It also considers the technologies developed to reduce human drudgery. We have considered the needs of the most productive elements needed to produce food from cattle (milk and its products). Cattle were used first for providing agricultural power through the bullocks until the discovery of fuel engine early in the 20th century when the priority shifted to production of milk—a complete food.

2.1. CATTLE

Cattle occupy a pivotal position in our national life and are indispensable to the rural economy. According to the 2012 Census, the estimated population of cattle was 190.9 million, whose numbers had decreased at the rate of >1%. They provide together with working buffaloes motive power for various agricultural operations (ploughing, irrigation and transport), approximately 5.8% of the power required, produce dung for manure (60% fuel and 40% as fertiliser) and supply milk and milk products; 155.5 million tonnes milk (2015–16), the only source of animal proteins in the diet of pre-dominantly vegetarian population of this country.

The total livestock population consisting of cattle, buffalo, sheep, goat, pig, horses, ponies and mules, donkeys, camels, mithun and yak in the country was 512.05 million in 2012, besides 729.2 million poultry. The total livestock population has decreased by approximately 3.3% over the previous census (2007).

Livestock population has substantially increased in Gujarat (15.36%), Uttar Pradesh (14.01%), Assam (10.77%), Punjab (9.57%), Bihar (8.56%), Sikkim (7.96%), Meghalaya (7.41%) and Chhattisgarh (4.34%).

The number of milch animals, cows and buffaloes (in-milk and dry) has increased from 111.09 million to 118.59 million, an increase of 6.75% to produce 137.7 million tonnes/year (2012).

The number of animals, cows and buffaloes in milk and dry has increased from 77.04 million to 80.52 million showing a growth of 4.51%.

The female cattle (cows) population increased from 77.04 million to 80.52 million showing a growth of 4.51%. The female cattle (cows) population has increased by 6.52% over the previous census (2007), and the total number of female cattle in 2012 was 122.9 million.

The exotic/cross-bred milch cattle increased from 14.4 million to 19.42 million, an increase of 34.86%.

Indigenous milch cattle increased from 48.04 million to 48.12 million, an increase of 0.17%.

Indian breeds have been developed essentially for draught power, while this role was assigned to horses in Europe and America. The Indian cattle breeds are known better for their suitability to hot, hot-humid tropics, better heat tolerance and presumably their resistance to biotic and abiotic stresses and survival in their environment as compared to their European counter parts.

There are 26 well-defined indigenous breeds of cattle in India. Broadly, they have been classified into (a) milch breeds such as Sahiwal, Sindhi, Gir and Tharparkar; (b) dual-purpose such as Hariana, Ongole, Gaolao, Krishna Valley and Kankrej, (c) draught breeds such as Nagori, Kenkatha, Malvi, Kherigarh, Hallikar, Amritmahal, Kangayam and Khillari and (d) dwarf breeds such as Poonganoor, Vechur and Siri.

There was a well-defined breeding structure, when all the elements of environment were in harmony, especially their numbers and those of humans. With overall degradation in agricultural productivity, which started in ca. 1500 A.D. and had become uncontrolled by 1800 A.D., the system of controlling numbers and breeding by rotating bulls between families had broken down aggravating the already bad situation besides there was shortage of feed and fodder. Cattle numbers had increased beyond the carrying capacity of available land resources.

To stop this steep down trend, several British experts were requested to examine this problem. The Royal Commission on Agriculture (RCA) was set up which gave far reaching recommendations. On the basis of the work of Dr. N.C. Wright and R.A. Pepperall, the Government realised that the bulls need to be improved in quality if genetic improvement is to be obtained and sustained on long-term basis. The government took the policy decision(i) to establish 53 cattle-breeding farms to produce superior bulls for genetic improvement of local cattle through selection; (ii) as the number of pedigreed bulls being produced and supplied by the government farms was small as compared to the requirement, breeding bulls were purchased from the breeding tract from the progressive breeders for grading up the non-descript cattle; (iii) efforts were made to describe different breeds defining their characteristics by the then Imperial Council of Agricultural Research. As this also would be insufficient to meet the requirement, a scheme was launched which was popularly called the Premium Bull Scheme.

During pre-independence period, the bulls were supplied either as a gift or on loan to village panchayats. This was the cheapest method to put good bulls for improvement. The number of pedigreed and approved bulls increased in the rural areas. This practice unfortunately had a few limitations, *viz.* (i) controlled breeding could not be practiced, as timely replacement of bulls was not ensured and (ii) use of large numbers of scrub bulls by the farmers due to non-adoption of castration of young bulls was simultaneously continued.

Review of Cattle and Buffalo Breed Improvement Programmes-Historical

Historically setting up of the Royal Commission on Agriculture (RCA) in 1928 and the National Commission on Agriculture (NCA) in 1976 were landmark decisions which gave the basic mandate to planning and execution of various cattle and buffalo breeding programmes for improving the genetic potential of specified breeds, improving non-descript cattle and buffaloes to increase productivity of cattle and buffaloes as well as to increase milk production in India.

The breeding policy during the first and the second Five-Year Plans was to develop dual-purpose breeds of cattle to provide good bullocks and have higher milk production. In principle, this policy framework was for both breeds/groups/populations to bring in genetic improvement through selection and cross-breed, in selected areas of non-descript cattle by exotic breed—preferably Jersey, which was taken for cross-breeding due to an association that it was an efficient converter of feed had good reproductive traits and higher fat in milk.

Some important activities during the pre-planning phase of development undertaken by the Animal Husbandry Departments in establishment of cattle breeding farms, Gosadan Scheme for removal of useless cattle, and *Gaushala* Development Schemes for developing *Gaushalas* into centres for economic milk production, have somehow continued to date without contributing to the cause they were supposed to answer: high production of milk and removal of surplus cattle.

Major emphasis on cattle and buffalo development was started during the First and Second Five-Year Plans, with the establishment of Key Village Scheme. It was a comprehensive programme of cattle improvement in selected compact areas of the breeding tracts of the recognised breeds.

Key Village Scheme (KVS): The basic objective of the KVS was multiplication of superior germplasm from the established farms and its use in selected breeding tracts of important breeds of cattle and buffaloes. The scheme comprised a block with one main centre and six sub-centres and covered 10,000 breedable cattle/buffaloes. Breeding through artificial insemination (AI) was an integral part of the technical programme. Other aspects of the scheme were to propagate improved feeding and scientific management practices, organised marketing facilities in milk shed areas and provision of effective health cover.

Functioning of the KVS was studied in detail by the NCA in 1976. Experience with the KVS showed that, by and large, this had failed to produce desired impact on milk production as there was an increase of only 3 million tonnes of milk in 10 years of the implementation of KVS. This could not adequately support the dairy plants for processing and marketing of milk.

The fact that this programme was conceptualised without considering either the market for product or the infrastructure such as feed, milk and fodder availability or measures for disease control, it could not succeed in obtaining its targets.

The KVS programme was replaced by Intensive Cattle Development Project (ICDP) which started in the latter half of Third Five-Year Plan. It was emphasised that all ICDPs should be linked with fluid milk purchase, processing and marketing activities. It was considered that cattle improvement for milk production and marketing of milk should function as complimentary to each other.

This scheme was a typical scheme of the Government lacking in allocation of appropriate funds for infrastructure development; moreover, since bulls of some indifferent quality available in state government farms gave semen, which was passed onto farmers who had a few cows, which were not productive and no support as credit or incentives was made available.

Intensive Cattle Development Projects (ICDP): With the start of ICDP, some of the KV blocks were merged with ICDPs and some later with OFP taken up by State Dairy Corporations/Co-operative Dairy Federations. In the Fifth Plan, 100 new KVS were added. The breeding policy in the ICDP blocks was (i) improvement through selection in descript indigenous breeds, (ii) grading up of non-descript population with superior indigenous breeds in areas with no salubrious environment, inadequate feed resources and without well-developed facilities for marketing of milk, (iii) cross-breeding of non-descript population with exotic dairy breeds in other areas where physical environment and feeding resources were adequate. However, over the years, there were modifications in the breeding policy.

Each ICDP unit was to cover 100,000 breedable cows and buffalos with a view to make a significant impact in a large area. The project units were in potential areas to ensure satisfactory development and milk production. ICDPs were to provide adequate inputs and services to cover at least 70% population of cattle/buffaloes and increase the milk production by 30% over a period of five years. Each project was to have 100 sub-centres manned by stockmen and each sub-centre was to cover 1,000 breedable cow and buffalo females. At the end of Fourth Plan, there were 62 ICDPs and under the Fifth Plan, it was envisaged that existing projects would have all requisite inputs and 51 new projects would be started. Functioning of ICDPs was initially reviewed by NCA and recommendations were made in 1976. The GOI initially extended 100% central assistance for the project. After two years of establishment, these were transferred to the states on permanent basis.

The Programme Evaluation Organisation of the Planning Commission undertook an evaluation study of the ICDPs in 1970–1971. The study revealed that there was a considerable time lag in providing inputs up to the level as detailed in the model plan. The programme received a setback even in its earlier stages of implementation when these projects were transferred to the state sector with consequential financial cuts in the central assistance. This resulted in dilution of inputs in later years. The study revealed that (i) 30%–40% of the semen produced was not utilised in most of the projects; (ii) fodder development programme was making no progress and should be integrated with general agricultural development programme and (iii) the progress made in the organisation of rural milk co-operative societies was poor. The only ICDP that had been implemented with success was in the Kheda district of Gujarat that had been handed over for implementation to the Kaira District Co-operative Milk Producers Union.

On various studies and analyses of the ICDP and other programmes aimed at increasing productivity of cattle and buffaloes it was found that the ICDP was the major project for cattle and buffalo improvement being implemented by the state animal husbandry departments. Under Operation Flood the dairy co-operative sector conducted similar activities through the village level co-operatives. In most cases, the ICDPs as well as the OF provided insemination facilities with semen of exotic breeds and gradually adopted the use of frozen semen. The ICDPs only provided breeding and health cover but left the activities of milk procurement processing and marketing to the dairy co-operatives. In fact, in some states, the Dairy Co-operative Federation shelved the responsibility for providing breeding and health cover entirely to the State Department of Animal Husbandry. There was significant improvement wherever both facilities were provided to the farmers; however, in the absence of facilities for milk procurement and marketing the breeding programmes failed to make any impact.

Hill Region Cattle Development Scheme: The ICAR initiated a large cross-breeding project in 1955 in the hilly and heavy rainfall areas for determining the optimum level of exotic inheritance in cross-breds, which could sustain productivity after inter-breeding of these cross-breds under village conditions. The exotic breed used was Jersey. The project was run over nine locations in the country. Several grades with exotic inheritance of 1/2, 5/8, 3/4 and 7/8 Jersey inheritance were produced. *Interse* breeding was practiced in 1/4, 1/2, 3/4 grades.

All the animals in the scheme were identified and milk recorded. The results clearly demonstrated that milk in cross bred cows increased to 200%–300% in different grades as compared to local cattle. Grades with more than 50% exotic inheritance were difficult to maintain, at the then available health management and disease control programmes, due to high disease incidence and mortality. There was no significant difference in milk production of 1/2 and 3/4 grades. In addition to improvement in milk production in cross-breds, there was an improvement in reproductive performance as reflected by reduction in age at first calving and inter-calving period. These results suggested that cross-breds with milk yield up to 2,000 kg could be sustained with feeding and management practices followed by the farmers under village conditions.

State Cattle Breeding Farms: With the implementation of KVS across India, there was a large requirement of breeding bulls of various local and exotic breeds. To meet this demand, the State Governments set up new farms and strengthened the old. While on one hand, the farms did not get good quality of females to breed bulls, the number of bulls produced was so small, on the other hand, that the farms could not meet the requirement of breeding cattle and buffaloes within the breeding tracts identified under the KVS.

The Government of India also supported the state governments in strengthening already functioning nine cattle and buffalo breeding farms. The central and state governments established breeding farms by importing pure-breed Holstein Friesian, Jersey and Red Dane cattle. As the number of animals imported, were limited, herd size of the farms was small. Over time, the performance of imported stocks and their progeny deteriorated due to absence of selection. Since these were temperate breeds, they did not adapt well to the tropical environment and the poor standards of feeding and management. Only a few bulls were made available by these farms for breeding.

A study on the state cattle breeding farms revealed that very few farms were maintaining any sizeable population of pure indigenous breeds, as most of them were keeping small herds of exotic, cross-breds and even buffaloes in the same farms. Thus, these farms were acting essentially as multiplication farms as neither requisite number of pure-bred males were being produced for use in the genetic improvement programmes of these breeds nor were they used in the farms. There are only 25 farms maintaining pure indigenous breeds located in different states. These maintain 15 out of 37 recognised breeds of cattle. Similarly, there are only 13 farms, which are maintaining only buffaloes and 20 farms only exotic dairy cattle breeds. Rest of the farms were maintaining small herds of cross-breds and indigenous breeds.

Central Cattle Breeding Farms: During the formulation of the Fourth Five-Year Plan, it was realised that the existing cattle and buffalo breeding farms were negligible to meet the requirement of bulls. The Government of India, therefore, decided to set up six large cattle breeding farms on its own to implement the genetic improvement of breeds through selection of bulls based on their progeny performance. Each farm was to specialise in breeding of bulls of selected breeds, *viz.* Red Sindhi, Tharparkar, Jersey, Holstein Friesian breeds of cattle and Murrah and Surti breeds of buffaloes. It was envisaged that each of these farms would develop a herd of at least 300 breedable cows so that 10 bulls at a time could be progeny tested with ten or more performance recorded daughters. Each farm was expected to have a capacity to maintain 750 animals.

Ministry of Agriculture, Government of India in 1983 and during subsequent years constituted several expert groups and review committees to evaluate the working of these farms. Despite comprehensive work plans developed and recommended from time to time, the objective to produce indigenous and cross-bred pedigreed and proven bulls for conservation of indigenous germ-plasm could not be achieved.

Herd Registration of Cattle and Buffaloes: Herd registration was started in India as a Central Scheme under Ministry of Agriculture, Government of India. The Central Herd Books were started in 1941 for Red Sindhi and Sahiwal breeds. By 1949 herd books had been opened for eight breeds of cattle, two breeds of buffaloes and two exotic breeds of cattle, Jersey and Holstein Friesian were subsequently included. The report on progress of the Herd Registration studied by the National Commission on Agriculture (1976) revealed that although the registration of animals was started in 1941 for Red Sindhi and in 1949 for other breeds, yet its progress has not been satisfactory. There is little apparent contribution to the genetic improvement of the breeds included and it has not benefited the breeders included in the programme. Several states have not registered even a single animal in the herd book.

A detailed study of the scheme by a Task Force headed by Dr. N.R. Bhasin made a series of recommendations; however, no advance on ground was made in having an effective animal Herd Book either nationally or in the states, which could lead to an effective breeding programme among the indigenous breeds.

Progeny Testing of Cattle and Buffalo Bulls: Ministry of Food and Agriculture, Department of Agriculture constituted a Working Group of Experts to review the Breeding Policy for Cattle in the country in 1962, which submitted the report in 1963. Thereafter, a Scientists' Panel was constituted for drawing up a breeding plan, which recommended in October 1965, initiation of 'Scheme for systematic production of progeny tested bulls in cattle breeding farms for improvement of milk production in village cattle'. In the Third Five-Year Plan, a scheme was initiated for progeny testing and selection of bulls on the basis of performance records of their daughters in Hariana cattle and Murrah buffaloes at Hissar (Haryana). During the latter half of the Third Plan, it was observed that the cattle development programmes started earlier could not make much impact on improvement of stock. In the Fourth Five-Year Plan, it was decided to set up six large breeding farms for some selected milch breeds of cattle and buffaloes of national importance for undertaking selection of bulls on the basis of their progeny performance.

These large CCBF were set up for Tharparkar breed at Suratgarh, Rajasthan; at Chiplima, Odisha for Red Sindhi; Dhamrod, Gujarat; at Koraput, Odisha; Murrah buffaloes, Jersey breedand at Alamadhi, Tamil Nadu. A coordinated progeny testing programme was also started in nine State Government Farms. These farms were at Chakganjaria (Uttar Pradesh) for Sahiwal cattle; Durg (Madhya Pradesh) for Murrah buffaloes; Bharatpur (Rajasthan) for Hariana cattle; Gauriakarma (Bihar) for Jersey cattle; and at Hesserghata (Karnataka) for Red Dane cattle. Progeny testing of cross-bred bulls was initiated in the Sixth Five-Year Plan in Andhra Pradesh, Assam, Kerala, Odisha, Maharashtra and later in Tamil Nadu.

During the mid-term review of the Sixth Plan, it was decided that attentions should be paid to the development of indigenous breeds of cattle and buffaloes. In pursuit of the recommendations of Planning Commission, a centrally sponsored scheme for development of indigenous breeds of cattle/buffaloes was initiated in 1984–85 which continued in the Seventh Plan. National Bull Production Programme was planned in 1991 and was aimed at bringing improvement in cattle and buffalo through genetically superior bulls using associated herd progeny testing programme. The programme was, in fact, initiated in 1992–93 for bringing about rapid genetic improvement in cattle and buffalo production by generating genetically superior-bulls through Associated Herd Progeny Testing Programmes and utilisation of technology of Embryo Transfer in selected Indian breeds, *viz.* Gir, Hariana, Kankarej and Sahiwal. No useful results have been reported by these farms and almost all farms have discontinued the progeny testing programme for want of funds except at Banavasi and Junagarh, the latter farm was transferred to Gujarat Agricultural University subsequently. Some of them have now been associated in the progeny testing program of bulls taken up by the ICAR.

Progeny testing programme in Kerala was undertaken by the Kerala Livestock Development Board (KLDB) during 1977. The programme was taken up on cross-breds of indigenous cattle with Brown Swiss breed of bulls. It was based on a field recording of milk production and associated traits related to growth and reproduction in female progenies of bulls under test. The ICDP regions of Mavelikara, Kanjirapally, Vaikom and Kattapana were selected as the breeding areas. Programme ensured that 1,500 test inseminations were conducted for each bull to obtain completed first lactation records of at least 50 progenies. One batch of cross-bred bulls was tested every year. The female progenies born out of bulls are identified, registered and followed up at intervals of six months. The number of bulls under test was initially low but later increased to>40.

The results indicated improvement in the average milk yield in the daughters over the years. This would reflect genetic improvement through selection in the population involved and improvement in feeding and management. The programme adopted selective breeding policy. Top 3% cows out of the milk recorded herd belonging to the farmers were inseminated with the selected cross-bred bulls. The male progenies of such mating, if sound in health and conformation, were purchased and reared in bull stations for production of frozen semen and for being subjected to progeny testing. Approximately 300 elite cows having first standard lactation milk yield not less than 2,400 kilograms under field conditions were selected and used as bull mothers. The breed evolved through *interse* mating has been named 'Sunandini'.

The CCBF in Alamadhi started the progeny testing of Murrah buffalo bulls in 1977–78. Mating of three sets of bulls gave the breeding value of top two bulls of 1903 and 1813 but these were only 11% and 6%, respectively, higher than the herd average of 1,744.4 kilograms. The result was based on the production performance of 19 and 10 daughters per bull of 1903 and 1813, respectively. The programme was discontinued in 1989 with reduction in the herd strength. The farm has now been associated since 1993 with the CIRB for progeny testing of bulls. Progeny testing unit for Murrah buffaloes at Banawasi in Uttar Pradesh was started in 1972 with a batch of five bulls and 160 females, the former were selected on the basis of their dam's yield.

National Project for Cattle and Buffalo Breeding: Appreciating that bringing about genetic improvement is a long-term activity; the Government of India initiated 'National Project for Cattle and Buffalo Breeding' in October 2000 for a period of 10 years. The Project envisaged genetic upgrading and conservation of important indigenous breeds. The project was to begin with delivery of improved artificial insemination service at the farmers' doorstep as well as to bring all breedable females among cattle and buffalo under organised breeding either through artificial insemination or natural service using high quality bulls. The job of managing production and supply of genetic inputs as well as liquid nitrogen was entrusted to specialised autonomous and professional state implementing agencies.

The bull production programme proposed that bulls would be selected through field performance recording programmes by the State implementing agencies, dairy co-operatives, breeders' associations and other agencies. Under the component, identification and tagging of superior germplasm and rearing of male calves suitable for breeding up to breeding age in Murrah buffaloes and one year in case of cattle bulls will be done. Participating farmers were to be given incentive. Production of quality exotic and cross-bred bulls from imported embryos and semen will also be taken up under the component. The project proposed to accomplish this task through reorientation and intensification of Progeny Testing Programmes established during Phase I, establishment of Progeny Testing Programmes in new areas/milk pockets, setting up of the Open Nucleus Breeding System (ONBS) and strengthening of bull mother farms. The scheme included, among other things, increased and improved coverage of breedable females; approximately 80% of the total adult females among cattle and buffaloes to be brought under organised breeding activity (AI or natural service); bring about large scale replacement (20 million) of low producing non-descript cattle and buffaloes by improved animals; and development of several indigenous cattle and buffalo breeds. The results from some states were encouraging but did not reach the expectations for want of total commitment on behalf of the state Animal husbandry staff as well as the services and goods support.

National Dairy Plan: National Dairy Plan Phase-I is a central sector scheme for a period of 2011–12 to 2016–17 with a total investment of INR 2,242 crore. It is a scientifically planned multi-state initiative:

(i) Help increase productivity of milch animals and, thereby, increase milk production to meet the rapidly growing demand of milk.

(ii) Help provide rural milk producers with greater access to the organised milk-processing sector.

The objectives would be pursued through adoption of focused scientific and systematic process in provision of technical inputs supported by appropriate policy and regulatory measures. Under the NDP, it is planned to meet the 100% bull replacement needs of 'A' and 'B' graded semen stations by the end of NDP I and, thereafter, largely through producing high genetic merit bulls by progeny testing and pedigree selection programmes and a small number through import of exotic pure-breed bulls or equivalent embryos.

It is estimated that to breed approximately 35% of breedable animals by the end of NDP I, the country would need to produce some 100 million high-quality disease-free semen doses and make available approximately 900 high-genetic-merit bulls for replacement of bulls maintained at all 'A' and 'B' graded semen stations. On the basis of current and future demand of high genetic merit bulls of different breeds for semen production, the breeds identified for establishing progeny testing programme are Holstein Friesian, HF cross-bred, Jersey CB, Sunandini cattle and Murrah and Mehsana buffaloes and the breeds identified for pedigree selection programmes to begin with are Rathi, Kankrej, Tharparkar, Gir, Sahiwal and Hariana cattle and Nili x Ravi, Jaffarabadi, Banni and Pandharpuri buffalo.

Adequate bio-security measures would also need to be ensured at the village level as well as at the pre-quarantine, quarantine and rearing station level. Therefore, the institutions that can meet the afore-mentioned eligibility criteria would be entrusted to implement progeny testing or pedigree selection programmes to produce the required number of bulls. Both progeny-testing and pedigree selection programmes are to be implemented through independent project teams and governed by a Management Committee.

Conclusion

Since the beginning of planned development in India undertaking genetic improvement of Indian cattle and buffalo breeds have been at the centre of the objective of all the programmes. The Key Village Scheme, the Intensive Cattle Development Programme have been major programmes covering/intending to cover an extremely large population of cattle and buffaloes. Their failure to achieve the genetic improvement has many causes and deficiencies. None of these programmes have been found faulty as far as planning has been concerned. It is the deficiencies in execution of the programmes at the field level that hindered the achievement of aims and objectives. Among many deficiencies important ones have been: the lack of field organisational structures to implement the programme; the lack of trained manpower to reach the farmers and organise them into either co-operatives or self-help groups; lack of infrastructure for generating technical inputs required to be provided to the farmers such as the good quality of bulls, semen, facilities for artificial insemination; lack of facilities to produce feed and fodder; lack of facilities to procure milk, process and market it. Above all, the major deficiency has been the lack of facilities to record the various production parameters that would ultimately help in determining whether genetic improvement is taking place.

In support of the field-based programmes such as KVS and ICDP, the governments and the ICAR set up State and Central cattle and buffalo breeding farms, cross-breeding of indigenous cattle, undertook progeny testing of bulls of identified cattle breeds, buffaloes and cross-breds of Indian and exotic breeds. Most of the progeny testing projects have yet to establish availability of progeny tested bulls. The only project that has been able to provide a documented result is the Progeny Testing Programme undertaken by the Kerala Livestock Development Board. So far, approximately 10 batches of bulls have been recorded as progeny tested. This has led to the improvement in lactation yield of Sunandini breed from 1408±484 to 1829±564 kilograms over the lifetime of the project. Surely this increase is very inadequate. Results based on progeny testing has also identified some outstanding Buffalo bulls (AICRP on Buffalo) and cross-bred bulls (CIRC, Meerut/NDRI, Karnal).

Establishing a new breed through cross-breeding has been always a matter of scrutiny and controversy. It has been considered that inter-breeding cross-breds leads to a decline in their performance. This is because interbreeding among the cross-bred progeny breaks down the gene combinations and this leads to decline in their performance. The decline in performance has been largely due to lack of selection of F_1 bulls, this can result in decline in the mean yield in F_2. The breakdown of heterotic gene combination has usually been indicated as the cause of such a decline. It is well known that neither the Indian cattle population nor the exotic bulls come from highly inbred populations. Both populations are randomly bred. Therefore, the cross-breeding between two breeds leads to gene segregation when the F_1 cross-breds are interbred. This interbreeding on the one hand results in breakdown of heterotic gene combinations while on the other hand results in large variability in F_2. In fact, the breakdown of gene linkages on the same and different chromosomes and genetic environmental interaction and correlation may be more important in causing such a decline rather than the presence of heterosis. However, this is still a controversial issue and no serious attempts have been made to resolve.

The international experience on cross breeding suggests a continued increase in milk among the cross-breds if the F_1 generation is backcrossed to one of the high potential parents to gear greater productivity and reduce the effect of gene dilution. There is need to undertake elaborate breeding experiment which should produce pure-breed exotic and native, their reciprocal F_1, F_2 and cross-breds and the analysis of their differences would help in determining the various components of genetic variation.

The Frieswal Project was conceived and designed by Dr. R.M. Acharya, Dr. P.N. Bhat and Brig. Dilbag Singh in 1984, envisages evolving a national milch breed 'Frieswal', a Holstein x Sahiwal cross, yielding 4,000 kg of milk with 4% butter fat in a mature lactation of 300 days, in association with Directorate of Military Farms as a permanent activity of PD on Cattle. Moreover, two strains of cattle, namely, Karan Swiss (average milk production per lactation of approximately 3,600 kg/lactation of 305 days with highest recorded yield of 7,200 kg and Karan Fries (average milk production per lactation of approximately 4,000 kg per lactation of 305 days with highest recorded yield of 8,338 kg developed by cross-breeding of Sahiwal cows with Brown Swiss (American) and Holstein Friesian has greatly contributed to milk production enhancement in the country. Emphasis has also been on milk breeds of indigenous cattle such as Sahiwal, Gir, Tharparkar, Kankrej, Ongoleand Deoni for their improvement, conservation and propagation.

Military Farms: The first Military Farm was established in 1889 at Allahabad to meet the needs of Defence Forces with respect to milk and milk products. The successful functioning of this farm led to the establishment of more farms. The breeding policy followed was organised cross-breeding of indigenous cattle with European breeds to improve milk yield in indigenous breeds, reducing late maturity and long calving interval. Adoption of this policy resulted in great improvement in milk yield.

The earliest organised cross-breeding programme involving Ayrshire bulls and Hariana cows was taken up at the Imperial Dairy Institute, Bengaluru in 1910. In 1919 it was extended by involving two more indigenous breeds, *viz.* Sindhi and Sahiwal, at the Hosur Farm (Madras). The cross-breds showed distinct superiority in milk production, early maturity and in having reduced dry period (Little Wood, 1933).

The Agriculture Institute at Allahabad procured Holstein-Friesian, Brown Swiss, Guernsey and Jersey breeds from 1924 to 1934 and undertook a cross-breeding experiment with Sindhi, Sahiwal, Kankrej, Gir and Hariana breeds. After a review, from 1934 only Jersey was used on Red Sindhi cows followed by repeated backcrossing with Red Sindhi bulls. Red Sindhi half-breeds with Jersey were superior to other grades.

During post-Independence, there was spurt in development activities. Prof. H.D. Ray, Director, National Institute of Research in Dairying, Reading, U.K. was invited in 1947 by the Government of India to review the situation regarding dairy education, research and development in India. Prof. Ray reiterated Dr. Wright's recommendations and suggested early establishment of the National Dairy Research Institute including Dairy Science College at/or near Delhi. It was finally decided to set up this Institute in the premises of the Cattle Breeding-cum-Dairy Farm, Karnal. The farm had 800 ha of land and over 1,000 heads of pedigreed Tharparker and Sahiwal cattle. The foundation stone of the Institute was laid by the then Minister of Food and Agriculture, Shri Ajit Prasad Jain on August 7, 1955.

This Institute has developed two new strains of cross-bred cattle, one drawn from Brown Swiss gene line now called Karan Swiss and another from American Friesian lineage called Karan Fries. These two tropically adopted new breeds are of high producing cattle. In recent years, the Institute has expanded considerably. The Institute has been given the status of a Deemed University. The Institute has two regional stations one at Bengaluru in Karnataka and second at Kalyani in West Bengal.

Another step taken immediately after independence was to strengthen cattle farms, established earlier for the generation of pure-breed stock, on scientific lines to produce superior bulls. Five large cattle-breeding farms were set up in the states to implement a coordinated breeding project for genetic improvement of Tharparkar, Sindhi, Surti and Jersey herds. The progress at these farms has been slow and below expectation. None of the farms could develop the requisite herd size to start progeny-testing programme. The Department of Animal Husbandry, Ministry of Agriculture during the Fourth Five-Year Plan sponsored a planned progeny-testing programme at the six farms involving Sahiwal at Indian Agricultural Research Institute (IARI), Delhi, Harianaat the Government Livestock from Hissar (Haryana), Jersey at Sunabeda near Koraput (Odisha) and Cattle-Breeding Farm, Palampur, Dist. Kangra (Himachal Pradesh); Red Sindhi at Cattle-breeding farm, Chiplima (Odisha), Gir at Cattle-Breeding Farm, Rajkot (Gujarat), and Cattle-Breeding Farm, Junagarh (Gujarat); Red Dane at Indo-Danish Cattle-Breeding Farm, Hessarghatta, Bengaluru, (Karnataka) and Cattle-Breeding Farm, Hosur, Dist. Dharampuri (Tamil Nadu). The person behind this ambitious programme was Dr. P. Bhattacharya, the then Animal Husbandry Commissioner to the Government of India. He realised that cross-breeding could only succeed if superior cross-bred bulls, with proven merit based on progeny-testing data were used in *interse* matings.

2.1.1 Cross-breeding during the 50s

Isolated attempts to carry out cross-breeding in the field were made in some other areas during the 1950s. In Himachal Pradesh in 1953, cross-breeding was undertaken with Jersey x Sindhi bulls obtained from the Agriculture Research Institute, Allahabad. Later Jersey bulls were obtained from Australia and cattle breeding farm was established at Kataula in Mandi District. The Himachal Pradesh Government also obtained some Dexter cattle from Jammu and Kashmir in 1956 and used them for cross-breeding. The Mysore Government took up cross-breeding to increase milk production in the urban areas of Bengaluru and Mysore during 1955–56. The semen of the Jersey bulls for the purpose was obtained from the National Dairy Research Institute, Bengaluru.

The Animal Husbandry Wing of the Board of Agriculture and Animal Husbandry of the Government of India, in the meeting, held at Pune in 1953 decided that a pilot project on cross-breeding should be taken up on a field scale in selected hilly and heavy rainfall areas of the country. Accordingly, 12 ICAR sponsored cross-breeding units were set up between 1955 and 1957 at Palampur (Punjab), Darjeeling (Bengal), Choharpur near Dehradun (Uttar Pradesh), Ranchi (Bihar), Shillong (Assam), Imphal (Manipur) Vishakhapatnam and Hyderabad (Andhra Pradesh), Coorg (Karnataka), Ootacamund (Tamil Nadu) and Neyyattinkara and Chalakudy (Kerala). The Jersey cross-bred produced under this programme yielded 1,200–1,800 kilograms of milk per lactation (Bhat 1984). The Animal Husbandry Wing at a meeting held in Pune also recommended that cross-breeding might be taken up for research to answer questions, which needed to be settled before going further.

2.1.2. Cross-breeding during the 60s and 70s

The 14th Meeting of the Animal Husbandry Wing of the Board of Agriculture and Animal Husbandry held at Bengaluru in 1961, reiterated that cross-breeding with non-descript cattle should be done in areas where feeding and management conditions were suitable for the maintenance of cross-bred cows. The meeting further recommended that only a few selected exotic breeds suitable for cross-breeding in India available under similar environmental conditions elsewhere should be imported and not more than one exotic breed should be used in any one area.

During March 1961, the Central Council of Gosamvardhana decided that there was no objection to taking up cross-breeding on an experimental basis. A subsequent meeting of the Central Council of Gosamvardhana in 1961 set up a sub-committee under the chairmanship of Raja Bajrang Bahadur Singh of Bhadri to consider the entire cattle breeding policy, including cross-breeding with exotic breeds. The Committee generally favoured the cross-breeding approach for cattle improvement and recommended that intensive cross-breeding should be taken up in selected areas, one each in the plateau, the plains and the hills. It also supported the proposal for taking up research-oriented cross-breeding project at Haringhata (West Bengal) and at the National Dairy Research Institute, Karnal. The Committee suggested that a thorough analysis be done of the data on cross-bred cattle generated at Military Farms.

In 1965 the Scientific Panel on Animal Husbandry constituted by the Ministry of Agriculture and Irrigation reviewed the cattle breeding policy and suggested the undertaking of an intensive coordinated cross-breeding programme. The Panel recommended that Jersey should be the main breed for cross-breeding in India and Holstein-Friesian, Brown Swiss and Red Dane may be used to a limited extent. The Panel also suggested that exotic germplasm be maintained at 50% level through *interse* mating of half-breds except that cross-breds with higher levels of exotic germplasm may be produced in areas with particularly favourable conditions of feeding and management. Ayrshire animals were imported by Maharaja Karan Singh *Sadre-e-Riyasat* Jammu and Kashmir and several cross-bred cattle were produced around Srinagar.

The cross-breeding project at Haringhata started in 1956. It was considerably expanded in 1963 and 1968 with the financial support from the ICAR, the United Nations Development Programme and PL-480. Initially the programme was crossing of Jersey bulls with Hariana cows but later extended to crossing of Hariana cows with Holstein-Friesian and Brown Swiss bulls in addition. The logic of using Jersey as the improver breed was based on the experience of development of Jamaica Hope breed, its small size and higher fat content. This was questioned at Harringata, in light of the data from other countries and from India, where it was shown that Friesian crosses out bet the other breeds on all counts (Bhat, 1967). The enlarged project was the first organised research scheme on cross-breeding in the country.

Two ICAR-sponsored programmes involving cross-breeding of Hariana with Holstein-Friesian, Brown Swiss and Jersey bulls were undertaken during 1967 at the Punjab Agricultural University (PAU), Hissar, and at the IVRI, Izatnagar. These two centres and the Haringhata Project became part of the All-India Co-ordinated Project on Cattle with effect from 1969. The additional centres of the project were set up at Jawaharlal Nehru Krishi Vishavidyalaya, Jabalpur (Madhya Pradesh), Mahatma Phule Krishi Vishva Vidyalaya, Rahuri (Maharashtra) and Andhra Pradesh Agricultural University, Lam, Guntur (Andhra Pradesh). In addition to Hariana breed, Gir and Ongole breeds of cattle were included in the programme. Dr. P.N. Bhat became the first coordinator of this major effort in which 7,500 females were involved in a crossing of six breeds (Three temperate and three Exotic).

The National Dairy Research Institute, Karnal, took up cross-breeding with the American Brown Swiss semen using Sahiwal and Red Sindhi cows. The project was later expanded to include Holstein-Friesian on the male side and Tharparkar on the female side. Cross-bred strains, *viz.* Karan Swiss and Karan Fries, have been evolved from this cross-breeding programme.

2.1.3. Bilateral International Cross-breeding Projects

Starting in the early 70s, five cross-breeding programmes receiving assistance under the bilateral international agreements were taken up in the field. These were the Indo-German Agricultural Development Projects in Himachal Pradesh (Mandi) and Uttar Pradesh (Almora), the Indo-Swiss projects in Kerala (Mattupetty) and Punjab (Patiala), and Indo-Danish Dairy Project in Karnataka (Hessarghata). The Kerala Indo-Swiss Project which was started in 1963 envisaged the production of Brown Swiss x local cross-breds, with 62.5% Brown

Swiss inheritance. The Patiala project, taken up in 1971 aimed at cross-breeding the population of local cows with Brown Swiss semen followed by repeated matings with Sahiwal x Brown Swiss half-bred bulls to ultimately generate a breed with 50% Brown Swiss and 50% Sahiwal inheritance. The Indo-Swiss Project at Patiala was discontinued in 1979 because the farmers preferred cross-breds of Holstein-Friesian breed. The Indo-Danish Project gave assistance to import and build up a foundation herd of Red Dane breed at Hessarghata. The cross-breeding programme under the Indo-German Project involved crossing of local cattle with bulls of German Highland Spotted breed in Mandi and with Brown Swiss in Kumaon.

The results of the AICRP on Cattle indicated the following:

(i) Irrespective of the local Zebu breed involved and ecology in which they were to produce Holstein crosses performed the best.

(ii) In cattle, when two breeds of diverse origin, one of superior merit and the other relatively inferior, are crossed in a scheme of forward crossing, there is no linear increase in production level beyond 50% with the increase in genes of superior parent. To stabilise a new breed type from such a cross-bred foundation, *interse* mating cross-breed with selection using proven bulls is the only alternative (Bhat et al 1978a).

The Cattle and Buffalo Breeding Policy has remained almost constant, the selection of bulls on the basis of performance of progeny and culling of those which are surplus and not as per the selection norms.

One of the basic problems with genetic improvement of cattle and buffaloes, which has not been discussed, is the surplus number of cattle and buffaloes to our requirement. How can any selection operate if no animals are removed from the system, which have been culled in the process of selection? Hence, no selection is possible if culled animals are not sent to slaughter. The enormity is so large that 200 million cattle and buffaloes together are surplus to the requirement for today. They are eating on an average the feed and food resources of farmers to the extent to INR 4,983,873.5 million annually.

A fundamental change has taken place in the cattle development programmes since the formulation of breeding policy. Cross-breeding was to be taken up in a restricted manner, and in areas of low producing cattle, has now spread indiscriminately all over the country, including the tracts of well-established improved indigenous breeds. Large-scale cross-breeding programmes are being undertaken through programmes such as intensive cattle development projects, Operation Flood and other bilateral projects. There is a belief that animal power should also be developed on scientific lines so that in energy crisis, we may need to depend on animal draught power again. In view of these changes, it is essential that a fresh look at the cattle breeding policy of the country be made.

The major strategy for development of indigenous cattle for milk has been to cross-bred with improved European dairy breeds. Initial cross-breeding attempts were not encouraging because of diseases such as rinderpest and other such killer diseases. With the control of these diseases with prophylactic vaccines, planned cross-breeding experiments with various Euro-American breeds (Holstein, Ayrshire, Jersey, Guernsey, Red Dane, Brown Swiss) were taken up in different parts of the country. Cross-bred grades with different levels for exotic inheritance from one or two exotic breeds have been produced and their performance tested under different agro-climatic conditions. The following conclusions emerge from these experiments:

(i) Exotic inheritance of approximately 50% is the most ideal for growth, reproduction and milk production, and the yield in higher crosses falls short of theoretical expectations. The grading up, therefore, to a total replacement of genes will not lead to higher production in cattle (Taneja and Bhat 1972, Bhat et al 1978a, b, Taneja et al 1979, Rao and Taneja 1982).

(ii) The crosses of temperate with improved indigenous breeds (Sahiwal, Red Sindhi, Gir and Tharparkar) attained the same level of performance under uniform feeding and are superior to crosses from other native cattle.

(iii) Holstein crosses were superior to other temperate breed crosses for growth and production while Jersey crosses have better reproductive efficiency (Bhat 1974, 1983).

(iv) Decline in milk yield from F_1 to F_2 generations owing to *interse* mating among F_1 cross-breds is small (Taneja and Bhat 1978). In some experiments, the large decline is due to poor quality of cross-bred bulls used.

These results indicated that in areas with good feed resources, specially irrigated cultivated fodder, cross-breeding of indigenous low producing cattle with Holstein and stabilisation of exotic inheritance at 50% through inter-breeding and further improvement through selection may be adopted. Such cross-breds would produce approximately 3,000 litres of milk per lactation and have improved reproductive performance.

Several strains such as Taylor, Jersind, Karan-Swiss, Karan-Fries and Sunandini have been evolved using cross-bred populations as the base foundation. In addition, five cross-bred genotypes (two and three breed crosses) with Hariana, Gir and Ongole as the indigenous breeds, were produced under the All India Coordinated Research Project on Cattle.

Vrindavani Cattle in India

Vrindavani is a new synthetic produced at IVRI, Izatnagar by crossing of a mixed population of cross-bred carrying 50% to 75% inheritance of Holstein-Friesian,

Brown Swiss, Jersey and indigenous inheritance of Hariana cattle at 25% to 50% is now under 3rd generation of selection.

The mean birth weight of Vrindavani calves was 22.13±0.12 kg. The mean lactation milk yield, 305-day milk yield and peak yield was 3,219.75±41.09, 3,047.42±33.8 and 16.58±0.16 kg, respectively. The average age at first successful service, age at first calving, service period and dry period was 746.28±8.94, 1,012.14±9.32 and 149.54±4.55 and 99.65±5.75 days, respectively.

Current Action Programmes

India is a vast country with a large variation in climate, agriculture and economic conditions. A large part of the area in the country is with low inputs and would, therefore, need well adapted indigenous breeds both for milk and draught animal power. Cross-breds have shown high performances where plenty of green fodder and other essential inputs such as health care are available. The breeding programmes, therefore, for improvement of both indigenous breeds and cross-breds need to be drawn up. A system needs to be set up which would allow selection within the population to be effective. While developing strategies for genetic improvement, location specific parameters especially inputs must be given first preference.

The indigenous breeds to be improved through selection based on progeny performance need to be identified. The magnitude to which these are to be used in grading up of local cattle and their numbers required should be worked out in detail so that appropriate programmes for their multiplication, improvement and production of quality bulls are drawn up. The existing set up of small independent farms of these indigenous breeds has not given any worthwhile desired results.

To overcome the problem of small herd size and make the within population selection more effective, the programme of associated herd testing has been introduced. This has so far been attempted for Sahiwal which is one of the high yielding dairy breeds of the Indian Sub-continent. The breeding tract of this breed is in Pakistan and only a few herds are available in India. The herd strength on most of these farms varies between 50 and 300. These farms were established for producing quality bulls for use in the developmental programmes. Most of these herds did not register any improvement in milk over their period of existence. This was because of small herd size; use of sires selected on the basis of their dam's record and conformation resulted invariably in a negative genetic trend. It was observed, that only in a few herds was there an improvement in the milk yield over the years, which was due to use of sons of outstanding sires.

Recognising the importance of indigenous breeds for draught power and the need for developing animal power on scientific lines, the Indian Council of Agricultural Research has initiated a much larger programme for Hariana and Ongole breeds. The major objective is to conserve and improve these breeds for milk and draught qualities. Hariana is a dual-purpose breed and is widely used in the Indus-Gangetic plains. Initially, six breeding farms with a total population of more than 1,200 Hariana breedable females were included in the programme. The semen freezing facilities are being developed at the Germplasm unit at Hariana Agricultural University, Hissar. A total of 20 bulls coming from high-yielding dams and selected on the basis of body conformation will be used in each cycle. The bulls will be brought to the Germplasm unit and semen frozen for use in six herds. Each bull will be allotted 60 cows for breeding. The female progeny born will be raised at the respective farms for performance recording after appropriate identification, while five male calves each for breeding and draught (work) per bull will be brought to the Germplasm unit. Draught studies on these selected bull calves will be made. Bulls will be finally ranked on the basis of their daughters' milk production and draught qualities of their sons and the top 2–3 bulls identified for use in production of future bull calves and improvement in the herds. Correlation between milk yield and draught power will also be studied. As soon as the programme commences, approximately 10,000 females registered with the Central Herd Registration Scheme, operational in Rohtak, Ajmer, Mahendragarh and Bhiwani districts, will be involved in the programme. In addition, 15%–20% of total animals available with small and landless farmers could also be associated in the programme by developing the field recording system for progeny testing. The bulls, thus selected after the progeny test, will be used in the Hariana breeding tract and other adjoining areas where the Hariana breed has been used for upgrading the local cows.

A similar programme for Ongole was proposed during the Seventh Plan. Ongole is a dual-purpose breed. Its breeding tract extends over parts of Krishna, Kistna, Guntur, Nellore and Vizagapatnam in Andhra Pradesh. The centre of the Ongole area is embedded with rivers. The banks of these rivers form excellent grazing land, as they are less cultivated due to the fear of floods. Presently, very few farms of this breed are available. The existing Ongole farms are being strengthened by adding more animals to have at least 800 breedable cows at the farms. The Ongole animals on the farms, and those registered with the breeders' association, are proposed to be involved in an improvement programme on the pattern suggested for the Hariana breed.

Genetics of cross-breeding

Artificial selection to improve economic characteristics in domesticated livestock was practiced even before genetics as science came into being. A scientific approach to such efforts however, ushered in due to the advent of principles given by Gregor Mendel in 1865 and the reconciliation between biometry and genetics in the hands of Fisher in 1918 through his infinitesimal model based on numerous genes with small effects and environmental modification. The heritability of a trait as a proportion of the phenotypic variation that is due to genetic causes became a prime indicator to make selection decisions for genetic improvement of economic characteristics. The prediction of response to selection in terms of intensity and accuracy of selection and genetic variability was found to be successful across several crop plants, livestock, poultry and fishery. The relationship between phenotype and genotype was, however, considered like a black box wherein an inferential approach based on statistical principles was the only way to look into it.

In developing high yielding plant varieties, the concept of inbred lines/isogenic lines and their interbreeding and utilisation became a new tool and was named as hybrid breeding. The principles of crossing of hybrid varieties to utilise heteroisis to produce higher yields and repeated back-crossing of isogenic lines/highly inbred lines to higher yielding phenotypes would get converted to the better parent in view of the facts that both the lines are isogenic/highly inbred in nature.

Applying similar principles with cattle and other livestock within the same or similar breed populations did not however respond similarly as in plants. It was, however noted that the Mather's principles applied to isogenic lines of plants where highly inbred plant lines (isogenic lines) did response, as lines were homozygous, where all the genes were fixed when back cross-bred with parents of superior type, genes of original parents became fixed in those of original parents. This does not happen in livestock particularly in cattle.

When these phenomena were compared based on the experiment done on livestock, it become obvious that in the case of livestock, when you cross-bred the two random breeding segregating populations the F_1 so produced had the 50% complement of genes from each population to produce a hybrid population (F_1) which were theoretically similar to an F_1 cross of two isogenic plant lines. Once these F_1 progeny are interbred, there is random segregation of genes on chromosomes and new genotype are formed, which on average at population level may carry half the genes from each line but at genotypic/phenotypic level are totally different from each other varying from genotype of 1st parent on one side to almost genotype of the 2nd parent on the other.

In cattle when two breed populations one inferior with respect to milk yield and other superior where crossed the resultant F_1 population had half the genes from each of the parental populations, when the F_1 generation was interbred the resultant population (F_2) was segregating as per Mendel's Principles of random segregation, i.e. on an average the F_2 population had half the genes of male parent across the population but individually genes from an inferior gene type to the individual genes (F_2) of superior parent quality and all the best would fit in between these two limits of the present population.

If F_1 offspring of this population was back-crossed continuously with superior genotype the position became more complex resulting in ¾, 7/8, 15/16, 31/64 etc. Theoretically, the genes under ¾ grade or 7/8 grade on an average have 7/8 genes limit on population level are due to mendelian segregation but the individual would have any gene combination as dictated by random segregation. The situation that environment has large component of the effect on product (output of milk) the position becomes more complex because F_2 was only genetic component which could be used for genetic improvement of economic traits in cattle given the breed type (new synthetic) is result of phenotype, i.e. (P=G+E+GE).

The most cross breeding experiments in the topics are comprised of crossing indigenous cattle with the better developed dairy breeds of European origin. The F_1's thus generated would have half the number of genes coming from the better parent and the second half coming from indigenous parent. The fact of the case is that this cross involves two segregating populations where males from one population having higher gene frequency for production are being crossed with females having low gene frequency for that trait. A cross-bred population when further mated would then show segregation and linkage between blocks of genes affecting production traits which would then generate a series of genotypes with different proportions of inheritance from either parent population.

The situation becomes more complex when backcrossing with the better parent is continued to generate genotypes with a higher proportion of exotic inheritance namely ¾, 7/8, 15/16, 31/32 and 63/64. In fact, the very validity of calling a certain grade 3/8, ½, 5/8 or ¾ etc. is questioned because individual within that grade may not in reality have that level of exotic inheritance.

In a somewhat planned crossed breeding conducted with cattle (*Bos taurus* x *Bos indicus*) at Military dairy farms in India, it has been generally observed that with an increase in exotic inheritance, the advance made in production in later crosses has fallen short of theoretical expectations (Bhat and Khanna 1970; Taneja 1973). This led to backcrossing with the indigenous parent. Due to forward and backward crossing, a series of different genotypes having varying levels of exotic inheritance were generated. Therefore, the problem is to identify each grade with respect to another assuming that these two are from different populations, if any, in the phenotypic expression of economic traits and their genetic basis.

The data for the study were taken from breeding records of animals maintained at eight Military Dairy Farms: Ambala, Bareilly, Dehra Dun, Jabalpur, Jullunder, Lucknow, Meerut and Mhow. The breeding policy in operation prior to 1952 was that of criss-cross breeding. Since 1953 a policy of backcrossing with Sahiwal bulls was introduced with a view to reduce the exotic inheritance gradually.

A four-way fixed effects model with grades, farms, periods and seasons was used for analysis following the procedure given by Harvey (1966). Using least-square constants, the records were adjusted for significant effects due to farms, periods and seasons. The adjusted records were used to estimate genetic divergence among the grades.

The results of this investigation reveal, that in cattle when two breeds of diverse origin, different in their production levels, one of superior merit and the other relatively inferior, are crossed in a scheme of forward crossing, there is no linear increase in production level with the increase in the level of inheritance of the superior parent. The genetic reasons for this are not obvious. In view of this, it cannot be assumed that by grading up a total replacement of genes can be obtained, at least in cattle.

The result of the above study has given several leads about the behaviour of genes of livestock, in special reference to randomly segregating populations of several cattle breed types and their crossing behaviour. Based on these results and those of similar nature done under similar tropical environment, several breeding plans can be proposed to stabilise various breed types:

India

Cross-breeding of cattle started as early as 1857 when the 'Taylor breed' of cattle was said to have been evolved near Patna using crosses of Taurus bulls (Ayrshire bulls from U.K) by Mr. Taylor, an Englishman, the then commissioner of Patna division. Since then the cross-breeding work on organised pattern with various Euro-American breeds (Holstein, Ayrshire, Jersey, Guernsey, Red-Dane, Brown Swiss) has been going on. The information on the crosses generated has been analysed and results reported by several workers. Mostly two-breed crosses were produced barring a few planned experiments which envisaged production of three-breed crosses also.

Military dairy farms

In 1889, military authorities started a dairy at Allahabad to provide milk to the British troops and their families. European cattle brought by British Regimental personnel from the UK were the fore-runners of military farms on the European model. Shorthorn Ayrshire and Friesian breeds were used in the initial years (1907–1925). Cross-breds with Friesian were found to be superior to other breeds; hence, cross-breeding with Friesian was continued. The level of Friesian inheritance has been maintained between 3/8 and 3/4 through a scheme of alternatively using Friesian and Zebu bulls. Currently, approximately 15,000 cows are available in 16 farms representing four agro-climatic zones.

The data maintained at military dairy farms from 1930 to 1980 were extensively investigated (Naidu and Desai 1965, Amble and Jain 1967, Bhat et al 1978a, b, Taneja et al 1979). The results suggested that the Holstein inheritance between 1/2 and 3/4 was most suitable for growth, reproduction and milk production; however, further increase in milk yield in higher grades fell short of theoretical expectations (Bhat et al 1978a, Rao 1979). Various factors such as undesirable genetic effects and lack of adaptability to tropical environment have been attributed to poor performance of higher grades, though these observations are not supported with any data. The differences among Holstein x Zebu half-breds for important economic traits were non-significant, suggesting that improved Indian dairy breeds have nothing differential to contribute in cross-breeding. These findings have also been substantiated using the divergence analysis based on important economic traits (Taneja and Bhat 1974).

Table 2.1: Least squares means for first lactation traits in Holstein-zebu crosses at military dairy farms

Breed groups	Age at First Calving (Months)	Lactation yield in 300 days (Kg)	Lactation length (days)	Calving Interval (days)
3/8 H −5/8 Sh	35 ± 0.4	2,286 ± 68	316 ± 6	424 ± 11
1/2 H −1/12 Sh	33 ± 0.3	2,787 ± 44	326 ± 4	424 ± 7
5/8 H −3/8 Sh	33± 0.2	2,654 ± 38	318 ± 4	419 ± 6
3/4 H −l/4 Sh	34 ± 0.3	2,632 ± 45	322 ± 4	439 ± 7
7/8 H −l/8 Sh	34 ± 0.4	2,647 ± 71	338 ± 6	446 ± 11
1/2 H −l/2 Thp	33 ± 0.8	2,760 ± 115	315 ± 11	446 ± 18
1/2 H −l/2 Rs	32 ± 0.6	2,528 ± 94	312 ± 9	392 ± 14
1/2 H −112 Gir	32 ± 0.5	2,526 ± 62	313 ± 5	414 ± 11

H: Holstein Friesian; Sh: Sahiwal; Thp: Tharparkar; Rs: Red Sindhi

Agriculture Institute, Naini, Allahabad

At the Agriculture Institute, Naini, small herds of five dairy breeds (Kankrej, Gir, Hariana, Sahiwal and Red Sindhi) were established to study their performance. Dr. Schneider developed a cross-breeding experiment involving Friesian, Brown Swiss, Jersey and Guernsey in 1924. On the basis of the results evaluated in 1934, it was decided that Jersey x Red Sindhi crosses were most suitable owing to small size, better adaptability and high fat percentage. The breeding plan for this and subsequent evaluation were detailed by Schneider (1944). This involved back-crossing of high producing cross-breds with Jersey. Stonaker et al (1953) analysed production and reproduction records of these crosses and observed that cows with Jersey inheritance between 3/8 and 5/8 had higher milk production and better reproduction efficiency. They recommended that these should be interbred, selected and named as Jersind (Singh, 1982) using records of Jersey x Red Sindhi crosses concluded that 1/2 and 5/8 grades were most ideal for reproduction and not 3/8 as suggested by Stonaker et al (1953) if Jersey is to be used (Table 2.2).

Table 2.2: Least squares means for first lactation traits in Jersey x Red Sindhi crosses at Naini (India)

Breed Groups	Age at first calving (days)	Calving Interval (days)	Lactation Length (days)	Milk Yield (Kgs)	Milk Yield Per day of Calving (kg)
Red Sindhi	1250 ±24	497 ± 16	385± 13	1338 ± 60	2.7 ± 0.2
1/8 J–RS	1160 ± 43	532 ± 25	438 ± 21	1314 ± 95	2.5± 0.4
2/8 J–	1083 ± 31	446 ± 19	376 ± 16	1554 ±72	3.6 ± 0.2
3/8 J–	1063 ± 35	421 ± 22	379 ± 18	1557 ± 83	3.8 ± 0.2
4/8 J–	977 ± 26	427 ± 16	355 ± 13	1783 ± 61	4.2 ± 0.2
¾ J–	1935 ± 49	472 ± 33	385 ± 28	1247 ± 128	2.8 ± 0.3
5/8 J–	975 ± 44	412 ± 27	340 ± 22	1861 ± 99	4.6 ± 0.3

National Dairy Research Institute, Karnal

Cross-breeding of Sindhi with Brown Swiss was initiated in 1963 on the advice of Dr. S. Salisbury and Touchberry of Illinois University, USA to evolve a new dairy breed. Semen of bulls with a progeny test index between 6,000 and 7,000 kg of milk was used. The *interse* mating in these crosses was practised for three generations and cross-breds named as 'Karan Swiss' (Bhatnagar et al 1980). The first lactation yield was higher in F_1 crosses than in F_2. This was mainly due to non-selection of F_1 cross-bred sires. In another pilot experiment (Nagarcenkar et al 1981), Tharparkar (T) females were crossed with Friesian (F), Brown Swiss (B) and Jersey (J) to produce half-breeds; $3/4^{th}$ and $3/4^{th}$ with two exotic breeds. The mortality between birth and six months was higher in 3/4 grades (8.4%–13.3%) than half-breeds (2.4%–7.7%). Similarly, the age at first conception was lower in half-breeds (18.4–21.1 mo) than $3/4^{th}$ with two exotic breeds (22.8–25.1 mo). Milk yield (305 days) in first lactation was 3,392 2,755 and 2,714 kg respectively in Friesian half-breeds, Brown Swiss half-breds and Jersey half-breds. The yield in 3/4 Friesian crosses (3,049 kg) was higher than 3/4 grades with two exotic breeds (2,376 kg for FJT and 2,705 kg for FBT). These results suggested that $3/4^{th}$ with two exotic breeds were not superior to 3/4 straight breeds for most of the traits.

All-India Coordinated Research Project (AICRP) on Cattle

AICRP on cattle was initiated in 1969 as a major project to evolve new dairy breeds in tropics. Bhat (1971) discussed in detail the genesis, breeding programme and questions which need to be answered from experimental cross-breeding — several half-breeds, *viz.* ½ F, ½ L(FL), ½ B, ½ L (BL), ½ J, ½L (JL) and $3/4^{th}$ with two exotic breeds, *viz.* ½F, ¼ B, ¼ L (FBL),½B, ¼ F, ¼ L(BFL), ½F, ¼J. ¼ L (FJL) and ½ J. ¼ F. ¼ L (JFL)—have been produced at six different agro-climatic locations. Friesian half-breds were excelled in growth, production and reproduction with all the local (L) breeds (Table 2.3). The differences among the three half-breeds at different locations were almost of the same magnitude except at Rahuri, where the overall performance was good due to favourable climate, higher inputs in feeding and management (The variations from farm to farm were large).

Table 2.3: Means with standard errors for various traits in cross-breds under AICRP in cattle

Breed groups	Birth Weight	Weight at 18 months (kg)	Age at first calving (days)	300 day 1st lactation yield (kg)	Calving interval (days)	Fat (%)
	Hariana (H)					
FH	26 ± 0.2	252 ± 2.4	936 ± 10	2,438 ± 38	428 ± 7	4.8
BH	27± 0.3	225 ± 6.9	1029 ± 17	2126 ± 59	412 ± 5	4.3
JH	22 ± 0.2	225 ± 2.8	892 ± 10	1897 ± 29	401 ± 19	4.4
FBH	29 ± 0.4	239 ± 3.4	1051 ± 17	2061 ± 49	461 ± 15	4.1
FJH	24 ± 0.3	213 ± 6.3	1012 ± 17	2233 ± 57	472 ± 20	4.2
BHF	29 ± 1.6	231 ± 3.6	1108 ± 24	2073 ± 53	487 ± 18	4.1
JFH	24 ± 0.4	219 ± 3.3	975 ± 18	2085 ± 58	455 ± 19	4.1
	Gir (G)					
FG	27 ± 0.2	323 ± 2.8	805 ± 7	3238 ± 57	410 ±3	4.4
JG	21 ± 0.3	279 ± 2.4	757 ± 43	2522 ± 50	399 ± 6	4.5
FJG	25 ± 0.3	317 ± 2.5	814 ± 8	2856 ± 66	420 ± 6	4.0
BFG	30 ± 0.3	322 ± 5.0	861 ± 12	2799 ± 70	434 ± 9	4.0
JFG	23 ± 0.7	300 ± 2.9	842 ± 14	2428 ± 75	424 ± 9	4.1
	Ongole (O)					
FO	27 ± 0.3	213 ± 3.7	1003 ± 13	2117 ± 26	424 ± 9	4.3
BO	27 ± 0.4	200 ± 4.2	1037 ± 13	1853 ± 52	420 ± 11	4.2
JO	22 ± 0.4	203 ± 14.1	915 ± 14	1546 ± 53	401 ± 14	4.2
FBO	28 ± 0.3	231 ± 18.5	1090 ± 60	1324 ± 76	561 ± 31	4.4
FJO	23 ± 0.4	189 ± 8.0	1004 ± 21	1311 ± 68	503 ± 16	3.8
BFO	29 ± 0.3	207 ± 14.9	1124 ± 27	1487 ± 103	496 ± 18	4.2
JFO	23 ± 0.3	183 ± 7.9	1050 ± 27	1265 ± 50	468 ± 18	4.7

The 3/4th with two exotic breeds had lower body weights than half-breeds at all the units, except Rahuri. The birth weight in 3/4th was higher. None of the 3/4th excelled Friesian half-breeds in milk yield. The differences among the 3/4th grades were negligible except that JFL had lower mature weights and better reproductive efficiency. There is, however, a basic defect in interpretation of these results because the half-breeds and 3/4th were not contemporary and the 3/4 grades were exposed to severe feeding and managemental stress due to increase in herd size without proportionate increase in Infrastructure. The half-breeds and 3/4th under challenge feeding showed much higher performance and 3/4 grades barring JFL were almost equal or even better than Friesian half-breeds (Table 2.4). The Jersey inheritance in 3/4th contributed lower mature weights and higher reproductive efficiency. The percent fat in milk was highest in Jersey half-breeds (4.9%) while there were no differences among the other half and 3/4 grades.

Table 2.4: Milk yield (kg) in various cross-bred grades under challenge of feeding (AICRP on cattle)

Genetic Groups	Lactation			
	I	II	III	IV
FH	2586 ±62	2849 ±76	2959 ±78	2728 ±72
BH	2438 ±88	2529 ±100	2615 ±95	2430 ±96
JH	2030 ±87	2258 ±106	2505 ±110	2141 ±112
FBH	2588 ±103	2849 ±157	3084 ±217	3291 ±388
BFH	2690 ±111	2721 ±119	2665 ±192	2135 ±268
FJH	2697 ±179	2994 ±235	3202 ±356	2908 ±451
JFH	2258 ±105	2780 ±159	2588 ±187	2137 ±246

Exotic x non-descript crosses under field conditions

A programme of crossing low yielding non-descript animals in hilly and heavy rainfall areas with Jersey suggested by Dr. P. Bhattacharya was initiated in 1955 at 15 locations in different states. A total of 11,282 records on local, 3,493 on ½ Jersey, 547 on ¾ Jersey and 95 on ¼ Jersey were analysed. The milk yield of half-breeds was 1,512 kg, local cows was 692 kg, and ¾ Jersey 1,935 kg. The Brown Swiss half-breeds from local cows in Kerala gave a first lactation yield of 1,958 kg as against 716 kg in local cows. The 3/4th gave a higher yield of 2,499 kg. A significant reduction in age at first calving in F_1 crosses was observed (Nair, 1973 and Bhat, 1983). Red Dane x Red Sindhi cross-breds under village conditions around Bengaluru gave a lactation yield of 3,180 kg in 305 days as against 2,335 kg by Red Dane x Hallikar crosses. This difference was attributed to production difference in the two indigenous breeds used (Madsen 1975). The Holstein x non-descript cross-breds at the National Dairy Research Institute,

Karnal attained body weights of 274 and 351 kg at 18 and 24 months respectively. These calved for the first time at 32.2 months with 1.9 services per conception (Taneja et al 1981). These results suggested that cross-breeding of non-descript cows with exotic breeds led to significant improvement in all the economic traits. Exotic inheritance between ½ and ¾ was optimum for maximum production depending upon the level of animal husbandry practices availability of inputs and climatic conditions.

Sri Lanka

The Sinhala, an indigenous breed of Sri-Lanka, is a small size animal with poor milk potential (450 kg in 305 days). The F_1 crosses of this breed with Red Sindhi, Friesian, Jersey and Shorthorn in different ecological-zones in the country showed a marked increase in milk production (Buvanendran and Mahadevan 1979). Friesian half breds were better than Jersey half-breeds (Table 2.5). Milk yield declined significantly (30%–10%) in F_2 crosses and was attributed to heterosis.

Table 2.5: Milk yield (kg) in different crosses in Sri Lanka

High Potential Areas		**Semi-Arid Coastal and Dry Areas**	
Breed	**Milk Yield**	**Breed**	**Milk Yield**
Friesian x Sinhala		Jersey (J) x Red Sindhi	
F1	1482 (387)	F1	1929 (295)
F2	981 (333)	F2	1115 (265)
F3	957 (332	5/8 J	884 (265)
		¾ J	1700 (317)
Jersey x Sinhala		Jersey x grade Sindhi	
F1	1076 (377)	F1	1209 (303)
F2	809 (273)		
Sinhala	234 (186)		

Figures in parentheses are those of corresponding lactation lengths

Thailand

In an experiment, cross-breds of native cows with Red Dane inheritance between 37.5% and 67.5% were produced (Madsen and Vinther 1975). The performance of these grades showed that mortality up to 6 months in Red Dane (7.3%), higher grades (7.3%–9.2%), lower grades (9.1%) and Indian milch cows (15.5%) was higher than cross-breds with inheritance approximately 50% (3.3%–5.9%). Age at first calving in Indian milch breeds was 35 months and among cross-breds was

26–30 months. Milk yield in cross-bred grades increased with increase in level of exotic inheritance (Table 2.6). The high mortality in higher grades suggested that best proportion of Red Danish inheritance would be between 60% and 80%. The performance of Red Danish cattle (both imported and born in Thailand) for milk and fat production was better than of the cross-breds which was due to high levels of feeding. The high frequency of abortions, high post-natal mortality and long calving intervals in Red Danish cattle, however, suggested that pure-breeds are less adapted to climate and management conditions than cross-breds.

Table 2.6: Performance of different crosses in Thailand

***Bos taurus* Inheritance**	**Foundation Females**	**Age at First Calving (months)**	**Lactation Yield (kg)**	**Lactation fat Yield (kg)**	**Calving Interval (days)**
0	Indian Milch	34.8 ±0.8	987 ±155	44.7 ±7.1	467 ±9
37.5	Native/Imp. Native	28.8 ±0.6	1,000 ±115	47.3 ±5.3	421 ±14
50	Native	28.2 ±0.5	1128 ±89	61.9 ±4.1	410 ±8
50	Imp. Native	27.3 ±0.5	1256 ±85	66.0 ±3.9	419 ±8
50	Indian Milch	28.0 ±0.8	1960 ±149	87.5 ±6.9	443 ±15
62.5	Native/Imp. Native	29.6 ±0.6	1554 ±99	73.3 ±4.5	444 ±11
75	Native	29.0 ±0.5	1689 ±87	79.6 ±4.0	446 ±9
75	Imp. Native	28.8 ±0.4	1875 ±74	87.0 ±3.4	441 ±8
87.5	Native/Imp. native	29.7 ±0.6	1928 ±106	86.6 ±4.9	464 ±10
100	Red Danish 1	30.2 ±0.8	2305 ±133	95.4 ±6.1	525 ±15
100	Red Danish 1	29.0 ±1.0	3445 ±182	133.6 ±8.4	480 ±14

Imp., improved; 1, born in Denmark

Jamaica

Jamaica Hope (JH), a synthetic cross-bred having 80% Jersey, 15% Sahiwal and 5% Holstein breeding for the Bodles herd averaged 3,216 kg milk in 305 days as compared to 2,755 kg in the farmer herds: however, milk yield per day of calving was similar. Age at first calving was 1.7 month higher in farmers' herds (35.9 month). The 305-day yield in farmer's herd averaged 2,905, 2,000 and 1,623 kg, respectively, for good, mediocre and poorly managed herds (Wellington et al 1970). Schneeberger et al (1962) reported that yearly average milk yield declined from 3,890 kg in 1969 to 2,678 kg in 1975. These figures for registered Jamaica Hope cattle were 3,257 kg in 1969 and 2,676 kg in 1975. The decline in yield over years has been attributed to marked rise in herd size without a corresponding increase in feed supplies and slow genetic progress from reduced culling. Non-JH cows had higher yield than JH cows. Most of the JH sires put in the test had few daughters. Of the total 62 sires tested, only 9 exceeded 100 kg in sire values and the average difference between sire values was small 9 kg. However, the Jamaica Hope had developed a remarkable degree of tolerance to anaplasmosis and piroplasmosis. In recent years, numerous Holstein x Friesian cattle have been imported to Jamaica from North America. Unless Jamaica Hope breeders are prepared to meet the demand for more breeding stock and improve their production levels using rigorous selection of sires, there is every possibility that this breed might die in the years to come.

Kenya

A National Dairy Cattle Breeding Policy on the basis of research findings, requirements of people and potential of different ecological zones has been formulated (Meyn and Wilkins 1975). In high potential areas, rapid upgrading with Ayrshire, Friesian, Guernsey and Jersey is going on. Sahiwal x Jersey crosses from a large well-managed farm had higher yield in first through fourth lactation (1,446–2,051 kg) than the Jersey (1,044–1,780 kg). Similarly, Friesians (2,495 kg) although had slightly higher yield than Sahiwal x Friesian crosses (2,341 kg), the cross-breds had lower mortality (3.5%). higher per cent live calves weaned (89.0) and shorter calving interval (410 days). The performance of crosses on small holdings, however, was lower.

In semi-arid areas, Meyn and Wilkins (1973) reported the first lactation milk yield for Sahiwal, Ayrshires, Ayrshire-Sahiwal and Sahiwal-Ayrshire as 1,519; 2,152; 2,233 and 2,466 kg respectively. In coastal areas, the lactation milk yield in a well-managed herd for Ayrshire, 3/4 Ayrshire—1/4 and 1/2 Sahiwal—1/2 Ayrshire cows was 3,024; 3,151 and 3,004 kg with calving interval being 424, 405 and 404 days respectively, thus favouring the 3/4 Ayrshires. In arid areas, Sahiwal along with other indigenous breeds is being used both as pure-breed and in crosses with various European breeds. The Sahiwal cows gave a first lactation yield of 931 kg in 234 days with butter fat of 5.15%. The overall herd average increased from 1,042 kg milk per cow per lactation in 1965 to 1,630 kg in 1972. With reasonably high variation in milk yield and heritability, estimate of 0.33, rapid progress in this National Sahiwal herd can be achieved (Meyn and Wilkins 1975). Progeny testing programme for Sahiwal is in progress.

Iraq

Cross-breeding of native Genubi cows with Friesian was started at various stations, as early as 1941 by Dr. Winters and crosses with 50, 75, 87.5 and 95% Friesian inheritance have been produced. These showed a large variation in reproduction and production traits. A significant increase in milk yield in cross-bred grades was observed over Genubi cows (Table 2.7). The differences, however, among cross-bred grades were non-significant. Similar trend was noted for milk yield per day of calving interval. These results suggested that exotic inheritance between 50% and 75% was ideal for maximum milk production. The higher grades, however, were reported to be susceptible to various diseases (Singh 1982).

Table 2.7: Least squares means for some economic traits in Friesian (F) × Genubi crosses

GRADES	Age at calving (days)	Milk Yield (kg)	Lactation Length (days)	Calving interval (days)	Milk Yield per day of Calving
Genubi (G)	1163 ±77	609 ±182	277 ±26	454 ±43	1.3 ±0.37
½ F −1/2 G	991 ±67	2277 ±181	383 ±26	545 ±41	4.6 ±0.35
½ F −1/4 G	1147 ±54	2382 ±145	358 ±21	545 ±34	4.8 ±0.37
½ F −1/8 G	1060 ±54	2144 ±145	325 ±21	489 ±35	4.5 ±0.38
½ F −1/16 G	1041 ±58	2060 ±159	335 ±23	449 ±38	4.8 ±0.41

Israel

Israeli Friesian breed has been developed by continuous backcrossing of local, mainly Damascus cows, to Friesian bulls and the proportion of Damascus genes in the present Israel breed is very small. Hirch and Schindler (1957) observed that F_1 cross was intermediate between the two for body measurement and by further back-crossing, the grades approached the Friesian measurement. The first cross produced 29% more milk than the pure Damascus cows, though the fat content was less. The 3/4 Friesian x 1/4 Damascus produced more than the pure Friesian: further backcrossing did not show any further improvement. Contemporary batches of half and 3/4 Friesians were compared in nine progeny groups and in eight of these the 3/4 Friesians were superior. The increase of Friesian level from 3/4 to 7/8 showed rise in milk production in four of nine progeny groups while rise from 7/8 to 15/16 gave an increase in three, a decrease in six and no change in one progeny group. The back crossing of Friesian crosses was continued so that the proportion of Damascus genes in the present Israel Friesian breed is very small. Due to use of progeny tested sires, the milk production in this population has continuously risen.

Egypt

Crosses of Friesian and Shorthorn with Egyptian native cattle have been produced at Government/Institutional farms. European breeds and their crosses were superior to native cattle in all aspects except calving interval which was lowest in native (407 days). Age at first calving in native, ½ Friesian, ¾ Friesian and pure Friesian was 42.4, 32.3, 33.7 and 34.2 month. Friesians (2,700 kg) and the cross-breds (½ F −2,106 kg, ¾ F-2,303 kg) were markedly superior in milk production over the native (1,256 kg). The milk yield in ½ Friesian was close to mid-parent value while ½ Shorthorn crosses produced more than pure Shorthorn and ¾ (El-Itriby and Asker 1958). This suggested the presence of heterosis in Shorthorn crosses. The mortality between birth and two years in cross-breds was much lower than in the pure-breeds, it being 10.3%, 19.3% and 11.5% for Friesian cross, for Shorthorn cross and native, 30.3% and 43.6% for Friesian and Shorthorn calves (Shahin et al 1966).

Southern United States

A cross-breeding experiment involving two zebu breeds (Red Sindhi and Brahman) and three temperate breeds (Jersey, Holstein-Friesian and Brown Swiss) was initiated in Southern United States to examine the possibility of overcoming the low milk yield and growth rate of temperate cattle in this region. The results were discussed by Branton et al (1966). Age at first oestrus and at calving increased with increased proportion of zebu blood. A marked decrease in milk yield, lactation length and persistency was noted with increasing Sindhi inheritance, irrespective of type of temperate breeds used. Similar trends were observed for Jersey x Brahman crosses. The reduction in average milk yield and lactation length was however larger. Many of these changes were attributed to detrimental effects of progressing proportions of Zebu inheritance on dairy temperament and milk let down. An increase in age at first calving and decline in milk yield, fat per cent and lactation length was observed from F_1 to F_2 generations in Jersey. Regarding fat percentage, a marked increase with increase in proportion of Sindhi inheritance was observed in Holstein and Brown Swiss crosses while the effect in Jersey crosses was irregular.

Australia

One third of land in Australia lying north of tropic of Capricorn has problems of poor pastures, high temperature and cattle ticks. To meet the challenge of these problems, two dairy breeds, the Australian Friesian Sahiwal (AFS) and Australian Milking Zebu (AMZ) have been developed. The AMZ developed by CSIRO contains between 20 and 40% *Bos indicus* blood (Sahiwal, Red Sindhi) and remaining is from Jersey. The formation of the breed has been described by Hayman (1974). Little difference for milk production between Jersey (1,944 kg) and AMZ (1,917 kg) heifers under conditions of optimum management was observed. Moderate estimates of heritability for milk (0.23) and total fat (0.27) for AMZ population suggested that selection would be effective. The results of comparison of AMZ with Guernseys and Friesians for milk and total fat showed that AMZ (3,304–146 kg) were comparable with Guernsey (2,913–124 kg) and Friesian (4,165–138 kg). The increase in grain supplement from 2.7 to 5.5 kg per day produced no significant response in production in AMZ, while Friesians consumed all extra ration and responded to it but did not out yield AMZ. There is also some difference in degree of heat tolerance between AMZ and *Bos taurus* cattle as is evident from the fact that exposure in climatic room at a temperature of 36°C for Friesian and 40.5°C for AMZ resulted in depression of milk yield by less than 5% in AMZ and by 30% in Friesian (Hayman 1977).

The AFS was developed by Queensland Department of Primary Industries as a tick resistant heat tolerant breed. It has Sahiwal and Friesian inheritance at 50% level and was produced by initial matings of Sahiwal bulls and Friesian cows. The AFS under Queensland conditions (Tierney 1985) produced approximately 75% of the milk (2,749 kg) produced by Holstein x Friesian (3,670 kg) in the same herd and 82% of fat (Table 2.8). The production of AFS cows improved under wet tropical conditions and these produced 124% of the milk and 141% of fat produced by their *Bos taurus* herd mates. These results suggested that AFS would not be a viable alternative to the Holstein Friesian in most areas of Queensland but would be an alternative to most other *Bos taurus* breeds. However, under hot humid conditions both in Australia and other tropical countries, the AFS provided a real alternative for economic dairying.

Table 2.8: Production of different breeds in Queensland herds (1983–84)

Breed	No. of Cows	Lactation Length (days)	Milk (litres)	Fat (%)	Fat (Kg)
AFS	567	273	2749	4.2	115
AMZ	144	286	2457	4.4	108
Ayrshire	1114	272	2869	3.9	112
Guernsey	2863	278	2765	4.3	119
Holstein	42435	282	3670	3.8	141
Illawarsa	14290	273	3228	3.8	122
Jersey	8194	278	2688	4.7	125

Conclusion

Trends of growth in human population in developing world suggest that more and more land would be used for production of cereals and feed resources are likely to be further reduced. The breeding objective, therefore, is to increase the productivity per animal. This can be achieved through (i) Intense selection within the native breeds, (ii) cross-breeding of native with temperate dairy breeds, and (iii) replacement of native with improved temperate breeds. There is an urgent need to reduce the numbers so that the existing feed resources could be efficiently utilised for higher productivity of livestock and poultry.

The native breeds by and large are late maturing and poor producers of milk but over the generations developed qualities of resistance to disease and pests and adaptation to harsh climates. Systematic scientific breeding programmes to improve the productivity of these native cattle have not been undertaken. Although large genetic variability for important economic traits in these breeds is available and high heritability estimates for milk yield have been reported, the results of selection have not been rewarding (Bhat, 1977). Assuming an annual genetic gain of 1%, it would take approximately 100 years to double the milk production of a herd with an average lactation yield of 1000 kg. Selection therefore, within indigenous breeds would not meet the food requirements of ever growing human population in this region. There is, however, a need to identify special genes for fast growth, efficient breeding and resistance to diseases and draught power while selecting for milk in these native breeds. For this, breed comparisons should be undertaken within geo-climatic regions and these programmes are coordinated at a national level. International agencies such as FAO could assist in these comparisons. The breeds with advantages of disease resistance, fast growth, regularity of breeding need be conserved and suitable programmes for their improvement taken up.

Possibility of replacement of these low producing native breeds with high yielding temperate genotypes does not seem to be practicable in view of problems of disposal/slaughter in some countries, non-availability of temperate cattle in large numbers and problems of their survival and adaptability in tropical environment. In humid tropics, even the most efficient genotypes despite satisfactory inputs are unable to produce efficiently (Bhat, 1983). This is evident from the fact that performance of temperate genotypes in tropical environment has been 30%–10% lower than in countries of their origin. Long term studies on their reproductive rate have shown that rate of replacement is slower than rate of culling (bad reproduction, death, accident and disease) with the result that these herds have disappeared within 10–15 years after establishment.

Cross-breeding with various temperate dairy breeds has been undertaken to combine the hardiness, disease resistance and heat tolerance of native cattle with high milk yield of improved temperate dairy breeds. Cross-breds have performed better and are more adaptable. The cross-breeding experiments were designed to answer questions such as (i) Importance of native breeds in cross-breeding, (ii) Relative superiority among exotic breeds, (iii) Optimum level of exotic inheritance and (iv) Effects of inter-breeding among cross-breds. The crosses of temperate, with improved indigenous breeds (Sahiwal, Red Sindhi, Gir and Tharparkar) attained the same performance levels under uniform feeding and were superior to crosses from other native breeds or nondescript cattle (Madsen and Vinther 1975, Buvanendran and Mahadevan 1979, Rao 1979). This suggested that improved native breeds have some role to play in cross-breeding. The comparison of crosses of various European with indigenous breeds revealed that the Holstein crosses were superior to other temperate breed crosses for growth and production while Jersey crosses had better reproductive efficiency (Bhat, 1974, 1977, 1983, Katpatal 1978). Exotic inheritance of approximately 50% was ideal for growth, production and reproduction and production in higher grades in most of the studies fell short of theoretical expectations. Taneja et al (1979) showed that when two breeds of diverse origin, one of superior merit (temperate dairy breeds) and other relatively inferior are crossed in a scheme of forward crossing, there is no linear increase in milk production levels above 50%

with increase in genes of superior parent. They concluded that grading up to a total replacement of genes will not lead to higher levels of production in cattle.

A decline in milk yield from F_1 to F_2 generation was observed in some experiments. This has been attributed to heterosis. In most of these experiments, progeny tested superior sires were used to produce F_1 progeny, while no selection among F_1 sires was practised. Part of the decline, therefore, due to non-selection of F_1 sires was expected. Only two sets of data are available in support of this argument, one from Sri Lanka and other from NDRI, Karnal. The data from NDRI showed that out of the 11 bulls used for producing F_2, 10 were below herd average. The decline was thus expected not due to heterosis but due to poor quality of bulls used. Similar situation could explain the results of the experiment of Buvanendran and Mahadevan (1979) in Sri Lanka. Bhat (1984) demonstrated that difference in variability in F_2 over F_1 for milk yield, even if it is thought to be governed by 10 loci would be as small as 70 kg, which is almost impossible to detect with numbers generally available in cattle breeding experiments. It is therefore suggested that numerous cross-bred bulls should be put under progeny testing and only tested ones used for inter-breeding F_1 populations.

Several breeds such as Taylor, Jersind, Jamaica Hope, Australian Milking Zebu, Australian Friesian Sahiwal, Karan Swiss and Sunandini have been evolved using cross-bred foundations. Barring AMZ, AFS, JH and Sunandini, most of the breeds have existed only at organised farms and selection in them has not been rewarding mainly because of small herd size. In most countries, the field recording is not in practice and therefore, progeny testing of sires is restricted only to Government/Institutional farms. This is a serious limitation in accurately assessing the sire-values and achieving desired selection intensity because of use of few sires. This fact is born out from decline in performance of cattle over the years where sire selection has been restricted only to organised herds. There is thus a need to produce cross-breds in large numbers under field conditions. The animals should be identified, registered and breed societies formed which should take up the performance recording if they must replace the local cattle.

The option of using cross-bred for developing new breeds in tropical countries may not be practicable due to change climate and economic conditions may not be practicable. By the time, breed is evolved farmers may need types of cattle different from those developed. In situations, where breeding objectives are likely to change quickly, it will be desirable to continue the system of criss-crossing using both European and tropical breeds and the grades produced depending upon the potential of the area. This would also have advantage of genetic progress made through selection in parental breeds used in crossing. The performance results of various temperate breeds and their crosses suggest that Holstein and their crosses have out yielded all other breeds and Jerseys which at one time were thought to have advantage of size, early maturity and high fat are getting replaced by Friesians in view of high performance of Holstein and changing socio-economic conditions. It would be essential that cross-breds have larger proportion of Holstein breeding and other breeds such as Jersey should be introduced in small proportions to widen the diversity of gene pool and to take the advantage of early maturity and improved reproduction. Such a cross-bred base should be used for development of new breeds.

Examination of stabilised and criss-cross-breds for various traits suggests that there does not seem to be anyone breed which is superior enough to be used for upgrading the other breeds. Similarly, formation of new breeds is not rewarding because of presence of heterosis for most of beef traits. Thus, the system of continuous crossing to have advantage of heterosis would be the best for beef production.

2.2 BUFFALO

The buffalo is our main dairy animal. India has more than half (57.3%) of the world's buffalo population (109 million buffaloes with 57 million adult females). The population of breedable buffalo is one-third of the total female breedable bovine population in India. Buffaloes contributed more than half of the total milk produced (68.8 million metric tonnes (MMT) by 2012.

2.2.1 The Indigenous Buffalo Diversity

The Indian buffaloes are mostly riverine type (2n=50). However, some swamp type (2n=48) buffaloes are available in north-eastern region of the country. As per 2012 census, the country had 109 million buffaloes, which is more than 50% of the world's buffalo population. Buffalo is the 'dairy animal of 21st century' in India. Presently they contribute about 51% of the total milk produced in the country.

The buffalo population has increased approximately 151% in last six decades (43.4 million in 1951 to 109 million in 2012). The Indian buffalo population consists of 30.2% pure bred, 13.7% graded and 63.5% non-descript animals. The diversity of Indian buffalo is reflected in terms of 13 breeds registered by National Bureau of Animal Genetic Resources (NBAGR), Karnal. The livestock census 2007, reflected the population size of 21 breeds varying from 20.5 million (Murrah) to 1000 animals of Chilika buffalo in the State of Odisha. Population size of some of the breeds such as Mehsana (3.73 million), Surti (2.98 million) and Jaffarabadi (1.84 million) are good but the population size of some breeds such as Chilika, Manda, Kujang are very small to the extent of few thousands only. FAO has listed a total of 177 buffalo breeds in the world of which 122 belong to the Asia-Pacific region and 23 are in India. Among the different buffalo breeds, Murrah is the best dairy breed, however, the production capacity of Nili Ravi and Jaffarabadi buffalo is also very good. Besides milk production, buffaloes are used for meat production as well as for draft.

Conflicts between Conservation and Improvement

The National Register, as per NBAGR classification and data provides a comprehensive window of the available breeds of buffaloes in their eco-region and the respective population size of each of these breeds. Distinct regions for Bhadawari, Nagpuri, Jaffarabadi, Surti and Mehsana have been outlined. Whether we follow the natural service or Artificial Insemination (AI) programmes for breeding, the breed and the quality of the male have no **certified identification** and even a regularised procedure for ensuring the specific breed application is missing. All organised programmes of buffalo improvement (especially through AI) are involving, invariably, Murrah or Murrah (type) males due to superiority as improver breed and availability. The only progeny tested buffalo semen with identified superiority (P.D. Index) is available for Murrah animals and hence it is this semen which is extensively used. The major agencies using buffalo semen for AI are sourcing the semen from places where only Murrah or Murrah type animals are available. The superiority of Bhadawari and Jaffarabadi buffaloes for fat, is thus distinctly lost. The organised Surti and Mehsana breed improvement programmes where numerous females are available, involve male germplasm other than their own breed which also distinctly contributes to this conflict in improvement programme. The regional breeding programs which are Murrah based need to be delineated and followed strictly.

Buffalo Milk Excellence: The Lost Advantage

Although approximately 51% of the available milk in the country is from buffalo, we do not yet have a focus on separate processing of buffalo milk and utilise its various quantitative differences from cow milk.

Our all techniques, processes and procedures in vogue have not exploited buffalo milk with distinct compositional advantage in terms of its fat, protein and SNF. Our researchers have compared buffalo and cow milk for their composition to enlist the differences. Special characteristics of buffalo milk have not been exploited. The education as well as the R&D focus has never been to innovate upon the strength of our country. The large volume of over 68 million tons of milk, the single largest distinct commodity, has never been addressed to develop a niche product line and/or market franchise. This mindset must change.

It is high time that when we are endowed with competitive global environment, buffalo milk should be pushed as 'buffalo milk' to hold its own in terms of its advantage in fat, protein quality. Mozzarella cheese and other products manufactured nationally and globally from the buffalo milk need to be marketed as buffalo milk products. A technology supported marketing strategy will ensure in much higher remuneration to the producer and profit to the industry. The coagulated milk products have a ready world market. The conflicting situation needs to be resolved with policy intervention to promote buffalo milk for its excellence without compromising on the milk volume and quality advantage.

The Male Wastage Dilemma

Due to the limited use of buffalo, male calf in the agricultural power cycle has led to its existence as an abject picture of neglect resulting into male wasting. Over 57 million buffaloes in milk annually (total population 109 million) are supposed to have approximately 30 million calves annually out of which male-to-female birth ratio would give a frequency of approximately 15.0 million to each of these sexes. In fact, the available census analysis (2007) shows over 17.5 million female buffalo calves less than one year of age and only 9 million males in the comparative age group implying systematic neglect of the male calves. There is further erosion of numbers, and the number of buffalo male calves is drastically reduced to half by three years.

- This clearly indicates that over 15 million calves are annually wasted due to shear neglect and/or allowed to die, which is a colossus national loss in terms of energy loss, economic loss and loss to environment sustainability.
- The conflict is driven primarily by no tradition for fatting of buffalo calves for meat. No industrial production has been encouraged by either private or public funded industry to harvest the dividends from this animal.
- Even if we can bring up just 50% of the 15 million animals (male calves), say 8–9 million, in 16 months post birth, we would have excellent buffalo meat available with great potential for profits and export earnings. The added 4.0 million tonne of quality meat for export, will propel the demand growth of this sector to an almost 80% increase in growth and export earnings. The available models of male buffalo calf rearing, in recent years has taken the economic returns from this animal beyond all previous calculations over and above the returns from milk alone. Thus, mindset conflict in the gender rearing is severely eroding the animals' economy. Farmers in several states have caught on the idea and we see that more and more people opting for buffalo production.
- Intensive male fattening programme on an integrated basis using poultry production model to serve as export units linked to central slaughtering facilities should involve a feeding regime, health care, disease prophylaxis and control. This could generate fattening units with the farmers linked to a meat processing plant for export. This would have facility for recycling/rendering of slaughter waste.
- There is also a need for breeding buffaloes for quality and quantity of meat. Selection should be to ensure a better weight gain and better body mass without affecting the milk potential. This will also address traceability and quality control issues. This will serve as a reference for quality germplasm to be inducted in male selection and rearing

programme, which is essential for breed improvement in any eco-regional breed situation.

Buffalo Meat Advantage

Buffalo being backbone of Indian dairying and meat exports is often referred as 'Black Gold'. Out of the total buffalo population of 109 million, approximately 13.86 million buffaloes are slaughtered annually, producing buffalo meat (carcass wt.) of 1.50 million tonnes. Buffalo meat production potential (boneless meat) is approximately 2.00 million tonnes. There are approximately 12 modern abattoirs-cum-meat processing plants leading to an export of 462,780 metric tonnes of meat. Buffalo meat is exported to 75 countries, and the value of the meat exported in 2009 was INR 48,400 million. Buffalo meat with lower intramuscular fat, cholesterol and calories; higher units of essential amino acids, biological value and iron content and residue-free status would make buffalo meat one of the healthiest for consumption. Buffalo meat is excellent for producing various meat products such as emulsion products, smoked and cured products, restructured products and traditional meat products. The demand of buffalo meat is fast escalating in India and abroad and giving better revenue for male calves.

Draught Animal Power

Though there are large populations of draught breeds such as Hallikar, Ongole and Khillari found in Southern and Western parts of the country, their use over past decade has drastically reduced in agriculture operations as a draught power. The trend witnessed in 2007 Census has sharply dipped further as evidenced in the reports from several states limiting the dependence of these draught animals as a work animal for ploughing the land.

Although role of draught animals in rural economy of the country needs not and cannot be under emphasised, the fact remains that the contribution of animal energy has significantly decreased from 71.6% in 1961 to 23.3% in 1991 and human energy from 23.3% in 1961 to 10.1% in 1991. Of the 5.8% in the animal energy; cattle, buffaloes, equines and camels contribute to it; suggesting a very low contribution of buffalo to this pool.

As per the 2007 Census, out of approximately 38 million female calves belong to up to the 1-year age group. The figure for the male calves available every year is approximately 23.0 million (50% sex ratio for every calving should give us a figure of same number of males born). The Census further revealed that the number gets almost halved by 2.5 years to 17.5 million. Most of these calves are lost from our production system because the farmer has no use for same; especially the annual crop of approximately 4 million buffalo male calves is lost. We need to take urgent steps to convert the large population of male buffalo calves as the backbone of the meat industry.

Past Animal Breeding Strategies and Programmes

Historically setting up of the RCA in 1926 and the National Commission on Agriculture in 1970 were landmark decisions that gave impetus to planning and execution of various cattle and buffalo breeding programmes for improving the genetic potential of specified breeds and improving the non-descript cattle and buffaloes to increase the productivity of cattle and buffaloes as well as increase the milk and meat production in India.

Since the beginning of planned development in India (1950), undertaking genetic improvement of Indian cattle and buffalo breeds have been at the centre of the objective of all the programmes. The Key Village Scheme, the Intensive Cattle Development Programme have been major programmes covering/ intending to cover a very large population of cattle and buffaloes. Their failure to achieve the genetic improvement has many causes and deficiencies. None of these programmes have been found faulting as far as planning was concerned. It is the deficiencies in execution of programmes at the field level that hindered the achievement of aims and objectives. Among many deficiencies important ones have been: lack of field organisational structures to implement the model programme; lack of trained manpower to reach farmers and organise them into either co-operatives or self-help groups; lack of infrastructure for generating technical inputs required to be provided to the farmers such as good quality bulls and semen, facilities for artificial insemination; lack of facilities to produce feed and fodder; lack of facilities to procure milk, process and market it. Above all, the major deficiency has been the lack of facilities to record the various production parameters that would ultimately help in determining whether genetic improvement is taking place.

Cross-breeding among Murrah and Kundi with Surti led to production of Mehsana breed having improved reproductive efficiency and high milk yield. Cross-breeding of Murrah with local nondescript buffaloes led to the production of Godavari buffaloes in coastal Andhra Pradesh.

Method of Buffalo Genetic Improvement

Within breed improvement through selection and use of between breed diversity through cross-breeding or breed substitution are the two main methods of livestock genetic improvement. Selection within breeds or strains is intended to increase the average level of genetic merit of the population for economic production of final marketed products, for example, live buffalo young for veal/meat. Objective of within-breed selection usually involves measuring and selecting on the basis of productivity (for example, litter size, milk yield, growth of the young and size at maturity). Within breed improvement through selection is necessarily a fairly long term (at least 10–15 years to see appreciable results) because the achievable rate of genetic gain is low. Even with high accuracies of selection and reset about resulting from new methods for managing field data and suitable models of analysis, the annual rate of genetic gain achieved in output per animal in dairy cattle in western free market countries was 1% (Cunningham 2010).

In developing countries, with small herd sizes, large fluctuations in rearing conditions and management between herds and over time within a herd, lack of systematic livestock identification, inadequate recording of livestock performances, pedigrees and constraints related to the subsistence nature of livestock rearing (where monetary profit is not the most important consideration), the accuracy of selection will be much lower, resulting in even lower rates of genetic gain.

However, locally adapted breeds are likely to be highly variable and the highest performing animals of such breeds can have great productive potential. Therefore, the screening of livestock populations previously not subjected to systematic selection is likely to give quicker results to provide high genetic merit foundation stock for nucleus herds. This is because identifying animals of high production performance from a large (>5,000) population and bringing them together in one herd lifts the flock straightway to a higher genetic level compared to the general population. This presumes accurate identification of superior animals in the field, based on their production performance, which may not always be possible. But the lower accuracy of selection can be compensated for by a higher selection intensity, through selecting only a small proportion of a large population for further breeding. Population screening for traits such as milk yield in buffaloes or weight-for-age in buffaloes (in the absence of birth date records, only after the age of 15–18 months, which can be estimated from the number of permanent teeth) can be effectively done by holding competitions and offering attractive prizes. The prize winning, high performance animals then must be identified with ear tags and they or their progeny purchased for use for further breeding. Systematic recording and selection must be continued in the nucleus so generated to achieve continuous genetic gain. Genetic improvement so achieved in the nucleus is disseminated to flocks of livestock keepers by way of distribution of breeding males or semen. Improvement in production due to such genetic improvement can contribute to improved income and the livelihood of livestock keepers who depend on low-input systems.

Cross-breeding with a more productive breed can yield faster improvement. However, an appropriate improver breed must be available, which will adapt to the conditions where it is to be introduced. Since it would be a completely new introduction, it is better to conduct a trial and monitor the performance of crosses for a few years before undertaking large scale cross-breeding. There are many other considerations such as ensuring supply of good quality animals of the improver breed, livestock keepers' acceptance of the improver breed, a distribution strategy of breeding males or semen and a strategy for maintenance of the local breed so that it is not displaced completely by the cross-breds. Financial consideration is also important because it could be expensive to import the improver breed.

Genetic improvement to be successful, it usually needs to be accompanied by improvements in nutrition, health and management. However, while improvements in these other components of livestock production give short-term benefits and incur ongoing expenses, genetic improvement is permanent and is passed on from one generation to the next automatically so long as the improved animals are used for breeding and their selected progeny retained for further breeding. It is also true that improvement in components of livestock production other than genetics, are more effective and successful in improving incomes and livelihoods if they are accompanied by genetic improvement. Genetic and environmental improvements thus have synergistic relationship.

Criteria for Prioritizing Production Systems for Genetic Improvement Interventions

Production systems where genetic improvement interventions are more likely to succeed are those:

(i) that have a high level of market orientation;

(ii) that have a reasonably good resource base so that buffaloes are not ordinarily in danger of starvation;

(iii) in which livestock keepers care about the quality of their livestock or about their distinctive features and make special effort to procure such livestock and maintain them reasonably well;

(iv) in which livestock keepers have reasonably large flocks or are relatively well-off so that they are not likely to sell good breeding animals and are willing to incur a small amount of expenditure for inputs for the livestock; and

(v) in which livestock keepers are literate and have high level of awareness.

Genetic improvement of buffaloes has not made any major impact due to lack of government support for an improvement programme, although the background methodology was well recognised. There is no record of any specific Plan of Improvement of Buffaloes on records of the government except as adjunct to cattle, as all schemes of cattle included buffalo also. Thereby, all the problems, issues real or imaginary identified for cattle have been accepted as a part and parcel of buffaloes and the stress has been on bovines (which generally means cattle and buffaloes). In view of this, all programmes discussed and problems identified and solutions adopted have been discussed in the chapter on cattle. Discussing the areas of genetics, disease prevalence etc., this para is briefly summarised the factual position from 1889 to 1986 when the ICAR decided to establish the Central Institute of Research on Buffaloes at Hissar in Haryana.

2.2.2 Buffalo Breeding

The first attempt to progeny test a Murrah buffalo bull was undertaken during 1960s at the Progeny-testing Farm, Hissar in Haryana, and at Babugarh and Madhurikund farms in Uttar Pradesh. However, the projects failed to achieve the objectives due to non-availability of the facilities for deep freezing of semen. By the time the sires were ranked they were lost, due to old age and non-availability of facilities for freezing semen. Thus, top-ranked bulls could not be mated to bull mothers, which would have led to production of superior young sires. Another drawback of the earlier programme was testing of sire within a single herd which led to reducing the accuracy of selection *vis-a-vis* multiple herds testing.

Progeny testing in small institutional herds does not provide effective accuracy and intensity of selection. It does not provide the superior male germplasm required for breeding available buffalo female population in the country. Significant progress made in researches in nutrition, health and management has mainly contributed to growth in overall development of milk production from buffaloes.

The importance of the buffalo as the major source of milk in India was recognised as early as 1928 by the RCA. The Commission in its Report stated that buffalo milk with high fat content had a good market. It observed that while seeking an index of milk production of a province it is the number of she buffaloes and not the number of cows that should be taken into consideration. Compared to the cow, the she buffalo was tended with greater care by the farmer's family and quite frequently careful selection was exercised in breeding the buffalo. The RCA emphasised that there should be no relaxation in the efforts to improve the buffalo.

A network programme on buffalo breeding is currently in progress and 16 sires have been put under progeny testing. Six organised farm herds are involved in this programme. Progeny testing has been undertaken at the CIRB in Hissar. Sire index based on daughters' performance on nine sets of buffalo bulls have so far been evaluated. In addition, nine sets of Murrah buffalo bulls have been already evaluated at the NDRI, Karnal and the PAU, Ludhiana centres. Frozen semen of seven top proven sires are available and are being used for mating elite dams to produce future young sires.

2.2.3 Buffalo Physiology and Reproduction

Major developments in reproductive biology have made it possible to develop a package technology of freezing of buffalo semen. This is in wide use in the field. A germplasm bank containing 55,000 semen doses of bulls having high genetic merit is available. A major breakthrough has been obtained in embryo transfer technology. Procedures have been standardised for super ovulation, embryo collection and embryo transfer to obtain successful pregnancies. Embryo sexing has been standardised using a recombinant probe with 98% efficiency. Upstream research in *in vitro* fertilisation is currently directed to produce embryos *in vitro* using slaughterhouse material.

Frontline research on embryo biotechnologies including embryo transfer, cloning and *in vitro* fertilisation and cutting edge technologies on stem cell research, genomics and metagenomics, proteomics and bioinformatics are being pursued in several institutions. Buffalo cloning which has application for faster multiplication of elite individuals has resulted in the birth of first cloned calf 'Garima' produced from skin fibroblast using the Hand-Guided cloning technique. Later, 'Garima 2' and 'Shreshta' (the first cloned male buffalo) were also produced through cloning using fetal fibroblast and embryonic stem cells of buffaloes at NDRI. Further effort to get cloned male calves from semen has also been rewarding.

2.2.4 Feed Efficiency

Feed-conversion efficiency studied in a high-yielding cross-bred cow, a high-yielding Indian dairy cow, a high and low-yielding buffalo and a goat at the National Dairy Research Institute, Karnal gave interesting results. The high-yielding cow, averaging 3,000–3,500 litres of milk per lactation, has feed-conversion efficiency of 25%– 30 % for milk nutrients for energy and of 25% for protein. A buffalo averaging 1,500–2,000 litres of milk per lactation has feed-conversion efficiency of approximately 20% for energy and 15% for protein. A buffalo producing approximately 1,000 litres per lactation but maintained on coarse forage and grain by-products gave 35% efficiency for energy and 26% for protein. A local cow yielding approximately 400 litres of milk per lactation had conversion efficiency for energy and proteins of only 4%–6%. Thus, high-yielding cow with greater efficiency will need less land when put to quality animal feed, whereas a low-yielding animal with low efficiency in feed conversion will need greater amount of land. A medium-yielding buffalo (1,000 kg/L) on coarse feed gave the best efficiency when raised on crop residues and agricultural by-products. This would be ideal in a situation when land cannot be allotted for raising high-quality forages for high-yielding cows.

Buffalo husbandry in India does not contribute to ecological degradation because these animals are mostly stall-fed. In future, more animals of superior type are likely to be kept that too on stall feeding system rather than on grazing due to shrinking of land area meant for cultivation. Their population will start declining but the average productivity is likely to increase. The demand for milk and milk products will continue to rise. India should make special effort to promote the export of buffalo germplasm as well as dairy products and meat. Excellent quality white mozzarella cheese is produced from buffalo milk and this may be exported to fetch greater returns.

Biotechnologies

Frontline research on embryo biotechnologies including embryo transfer, cloning and *in vitro* fertilisation and other cutting-edge technologies on stem cell research, genomics, meta genomics, proteomics and bioinformatics are being pursued in several institutions with encouraging results.

2.3 SHEEP

Sheep play an important role in Indian economy by providing food, fibre, manure and rural employment in hilly, tribal, arid and semi-arid regions of the country. They are important source of livelihood for small and marginal farmers, landless labourer, migratory farmers and farmers who follow the transhumance system for raising sheep. There were 65 million sheep in 2012 which produced 47.9 million kg of wool in 2013-14 (Fig. 3).

There is large genetic variability in sheep as is reflected by 40 diverse breeds found in varying agro-climatic areas which have been classified into four major zones, *viz.* (i) Northern temperate; (ii) North-Western; (iii) Southern Peninsular and (iv) Eastern zone. Sheep in the Northern temperate region produce wool of good apparel quality. In climatically similar areas of southern hills, Nilgiri sheep also produce wool of apparel quality. In North-Western region sheep produce wool of superior carpet quality. The sheep of Southern peninsular region either produce no wool or little wool of 36S quality. They are primarily used for mutton production. In the eastern region because of very high humidity sheep are found in only small valleys with low rainfall such as Sikkim.

The sheep population at the time of first Census 1951 was 39 million. It has now decreased to 65 million registering an increase of 66.7%. The sheep meat (mutton) has however registered an increase of 42.7%.

With the constitution of the Imperial (now Indian) Council of Agricultural Research (ICAR) in 1929, pilot experiment for breeding superior sheep were undertaken from 1938 to 1949 in different parts of the country primarily to focus attention on improvement in wool production. Based on the available knowledge and experience at that time, it was decided that the improvement of indigenous breeds in plains through selection should receive priority, and that cross-breeding with European breeds should be restricted to temperate Himalayan region and to sheep-rearing areas in the western parts of Deccan plateau. Grading up with superior indigenous breeds (Bikaneri) types was undertaken in Andhra Pradesh, Karnataka, Tamil Nadu and Uttar Pradesh. The results were not very encouraging in the Southern states. There was, however, an improvement in wool production and quality as reflected by increase in average fleece weight and decrease in average fibre diameter in crosses.

Cross-breeding trials with exotic breeds, both Merino and British were undertaken in the mid 30s, resulting in marked improvement (2–2.5 fold) in wool production and wool quality. Merino crosses thrived better than British crosses, and Rambouillet crosses were better than Merino crosses. The men responsible for these introductions were Dr. P.N. Nanda and Dr. G.S. Mahal. A new strain called 'Hissardale' was evolved through the interbreeding of Rambouillet half-breeds with Nali by Dr. Mahal. This strain, however due to small numbers was highly inbred and deteriorated due to problems of infertility. However, the results obtained in Kashmir were spectacular, where Merino/Rambouillet x Gaddi local lead to development of Kashmir Merino, a new synthetic.

In 1952, a comprehensive plan to deal with the problems of evolving superior breeds and studying different aspects of sheep husbandry was prepared by the ICAR. Sheep breeding farms were established under the Plan at Banihal and Riasi in Kashmir, Gwaldom in Uttar Pradesh, Sarhan in Himachal Pradesh, Darjeeling in West Bengal, Jorbir (Bikaner) in Rajasthan, Patan in Gujarat, Pune in Maharashtra, Nilgiris (Ootacamund) in Tamil Nadu and Gaya in Bihar. Breeding programmes included selective breeding within indigenous breeds and crossing of exotic fine wool breeds with local breeds of temperate region. Data generated under the selection experiments at the Government Livestock Farm (GLF), Hissar were analysed. Selection within indigenous breeds did in general show a significant improvement in wool quality and correlated decrease in average fibre diameter. This was also associated with decline in body size and fertility.

The Central Sheep and Wool Research Institute (CSWRI) was established by the ICAR in 1962. The Institute started functioning with Dr. M.V. Krishna Rao as Officer on Special Duty with 1,620 ha of land acquired near Malpura, approximately 84 km from Jaipur. This Campus was given the name Avikanagar. The Institute established two regional stations one in Himalayan region at Garsa (Kulu) and the other in Southern region at Mannavanur (Tamil Nadu) with broad objectives of genetic improvement of sheep.

The general principles of improvement remain the same—within breed improvement through selection and the use of between-breed diversity through cross-breeding. These are the two main methods of livestock genetic improvement. Selection within breeds or strains is intended to increase the average level of genetic merit of the population for the economic production of the final marketed product/s, e.g. live lambs or meat. Objective of within-breed selection usually involves measuring and selecting on the basis of productivity (for example, litter size, milk yield, growth of the young and size at maturity). Within-breed improvement through selection is necessarily a fairly long-term process (at least 10–15 years) to see appreciable results in sheep because the achievable rate of genetic gain is low. Even with high accuracies of selection and new methods for managing field data and suitable models of analysis, the annual rate of genetic gain achieved in output per animal in dairy cattle in western free market countries was 1% (Cunningham 2010). In our situation, it could be even less considering the fact of operative management of small flocks and poor quality of the field data on pedigree identification.

Major thrust to sheep breeding research was the development of multi-disciplinary co-ordinated projects on sheep for wool and mutton initiated by the ICAR in 1990. These projects involved 8,000 breedable females of nine Indian breeds and four exotic breeds in a scheme of crossing such that several genotypes could be developed suitable to varying ecologies.

Methodology and Basic Genetic Principles in Livestock Selection in general and specific to Sheep and Goats

In developing countries, the annual genetic gains would be low with small flock sizes, large fluctuations in rearing conditions and management between flocks and over time within a flock, lack of systematic livestock identification, inadequate recording of livestock performances and appropriate pedigree data and feed constraints related to the subsistence nature of livestock rearing (where monetary profit is not the most important consideration), the accuracy of selection will be much lower, resulting in even lower rates of genetic gain.

Locally adapted breeds are likely to be highly variable and the highest performing animals of such breeds can have great productive potential. Therefore, a new model proposed by Nimbaker (2014) involving the screening of livestock populations previously not subjected to systematic selection is likely to give quicker results to provide high genetic merit foundation stock to develop nucleus flocks. This is because identifying animals of high production performance from a large (>5,000) base population and bringing them together in one flock lifts the flock straightaway to a higher genetic level compared to the general population. This presumes accurate identification of superior animals in the field, based on their production performance, which may not always be possible. The lower accuracy of selection can be compensated for higher selection intensity, through selecting only a small proportion of a large population for further breeding. Population screening for traits such as milk yield in goats or weight-for-age in goats and sheep (in the absence of birth date records, only after the age of 15–18 months, which can be estimated from the number of permanent teeth) can be done effectively by holding competitions and offering attractive prizes. The prize-winning, high performance animals then must be identified with ear tags and they or their progeny purchased for use for further breeding. Systematic recording and selection must be continued in the nucleus flocks, to achieve continuous genetic gain. Genetic improvement so achieved in the nucleus flocks is disseminated to the flocks of livestock keepers by way of distribution of breeding males or semen. Improvement in production due to such genetic improvement can contribute to improved income and better livelihood of livestock keepers, who depend on low-input systems.

Cross-breeding may yield faster improvement but an appropriate improver breed must be available. Genetic improvement, to be successful, usually needs to be accompanied by improvements in nutrition, health and management. However, while improvements in these other components of livestock production give short-term benefits and incur ongoing expenses, genetic improvement is permanent and is passed on from one generation to the next automatically so long as the improved animals are used for breeding and their selected progeny retained for further breeding. It is also true that improvements in components of livestock production other than genetics are more effective and successful in improving incomes and livelihoods if they are accompanied by genetic improvement. Genetic and environmental improvements thus have a synergistic relationship.

To successfully use the system of breeding alone specifically, criteria for prioritizing production systems for genetic improvement, interventions are discussed below. However, production systems where genetic improvement interventions are more likely to succeed are those:

(i) that have market orientation (Remunerative Market):

(ii) that have a reasonably good feed resource base so that sheep are not ordinarily in danger of starvation (Good Nutrition).

(iii) in which livestock keepers care(Good Management and Health Care)

(iv) in which livestock keepers are not likely to sell good breeding animals(Good Care of Breeding Studs)

(v) in which livestock keepers are literate (High literacy and skill).

2.3.1 Cross-breeding for Fine Wool

The first major experiment spread over 12 years at the CSWRI involved testing the effectiveness of the three important fleece type breeds, *viz.* fine carpet wool (Chokla), medium carpet wool (Jaisalmeri) and coarse carpet wool (Malpura) in crosses with Rambouillet at three levels of inheritance, *viz.* ½, 5/8 and ¾, and subsequent interbreeding at ½ and ¾ levels. Estimates of heterosis, both from comparison between F_1 and average of the contemporary pure-breed, and F_2 and F_1 were obtained. Although Chokla crosses were inferior in body weights and survivability to Malpura crosses, they produced heavier fleece of superior quality. With the increase in Rambouillet inheritance there was a small but non-significant increase in greasy-fleece production, body weights at weaning and yearling age and survivability; however, fleece quality improved substantially as reflected by decrease in medullation percentage and average fibre diameter. There was a large component of heterosis in weaning and six-monthly body weights. There was significant improvement in all the three crosses in greasy fleece weight and medullation percentage. Increasing Rambouillet inheritance beyond ¾ did not show any improvement in these traits over ½ breds. Interbreeding of F_1 Chokla and Malpura crosses resulted in further improvement in growth, survivability and fleece-quality traits, and slight decline in greasy-fleece weights (Acharya and Mani Mohan 1978). These three groups, *viz.* Malpura half-breeds, Chokla half-breeds and ¾ breds of Malpura and Chokla, formed the base population from which a superior carpet wool strain 'Avikalin' and the apparel wool strain 'Avivastra' were developed by selection and interbreeding. Though, no economic analysis is available, farmers interest in fine wool production was lacking.

Bharat Merino is a strain of fine wool sheep evolved by crossing the half-bred ewes of Chokla, Nali, Malpura and Jaisalmeri with Rambouillet and/or Russian Merino rams and stabilizing the population at 75% exotic fine wool inheritance. Bharat Merino in temperate climate of Kodai hills produced 2.6–2.8 kg of annual Greasy Fleece Yield (GFY) with fibre diameter of 19–20 micron, staple length of 7–8 cm and medullation of <1.0%. Six-monthly body weight of 25.0 kg on an average was achieved in temperate climate. This breed has become very popular among the farmers particularly of Kolar district of Karnataka state where they are using it as improver breed for higher mutton production.

One of the findings in cross-breeding experiments for improving sheep for fine wool was that there were little non-significant differences among the exotic fine wool breeds in crosses with native breeds. There was however, substantial improvement in greasy wool production especially in low fleece yielding breeds and in fleece quality as reflected by a decline in average fibre diameter and medullation percentage. In the second finding, however, there was a slight increase in greasy wool production beyond 50% exotic inheritance in fleece weight. There was, however, some improvement in fleece quality at ¾ level. New strains, *viz.* Avivastra, Nali and Chokla Synthetics and Niligiri Synthetic, have been evolved through inter-breeding and selection of half-breeds and ¾ cross-breds. 'Bharat Merino has been evolved through the interbreeding and selection of ¾ Rambuillet/Merino cross-breds with Chokla and Nali. This strain produces 2.8 kg of annual greasy fleece with 17.6–20.5 m fibre diameter and 1.1%–1.61% medullation. No studies on economics of these strains have been conducted, unfortunately neither the government, nor the Farming Communities or the wool Industry have taken up follow-up studies to make use of these new breeds and innovations to help the farmers and the industry.

2.3.2 Improving Carpet Wool

Avikalin: A new synthetic carpet wool strain: A cross-breeding experiment for improving wool production was initiated at CSWRI, Avikanagar in 1964–65 involving Rambouillet and Malpura. The new strain Avikalin was developed and established at 50% Rambouillet and 50% Malpura inheritance. The quality of wool produced by Avikalin sheep is of carpet quality. Least square means of first six monthly greasy fleece yield (GFY) and adult annual GFY was 1.034 and 1.560 kg, respectively, with average fibre diameter, medullation and staple length of 30μ, 30% and 4.5 cm, respectively.

The first six-month GFY and adult annual GFY produced by Chokla were 1.438 and 2.386 kg. Being a Best Carpet wool breed, Chokla produces wool with average fibre diameter and medullation of approximately 30μ and 30% with staple length of more than 6.0 cm suitable for all kind of carpet preparation.

Marwari: Marwari is one of important carpet wool producing sheep breeds of North Western arid and semi-arid region of India. The adult annual GFY was improved from 1,209 g to 1,484 g (22.8%). The superior breeding rams produced in the project were supplied to the government/farmers/NGO/other developmental agencies for the genetic improvement of the animals in the farmers' flock.

Magra: Magra is an important carpet wool breed of Rajasthan. The sheep produces extremely white and lustrous fleece. The wool therefore produced by Magra is most suitable for carpet production and is in great demand due to its lustre. Magra sheep produce 950g GFY at first six months, 1050 g GFY at the second six months and 2200 g GFY annually with fibre diameter of 32μ, staple length of 5.15 cm and 40%–45% medullation. No economic studies are available.

Conclusion

Crossing indigenous carpet wool breeds with exotic fine wool, mutton and dual-purpose breeds resulted in improvement not only in greasy wool production but also in quality traits such as decrease in average fibre diameter and medullation percentage. The medullated fibres in the half-breeds were not only finer but mostly heterotypic. Kemp was also reduced. A new carpet wool strain, *viz.* Avikalin, was evolved, which adapted well to the hot arid conditions of Rajasthan, and showed better performance of annual greasy-fleece yield (1.9 kg), fibre diameter (27 m), medullation (22.4%) and lambing percentage (73%) (There has however been no follow up by either the State Wool Boards or the Industry).

For improving production and quality of carpet wool breeds, selection within superior carpet wool breeds or cross-breeding inferior carpet wool breeds with exotic fine-wool breeds stabilizing at 50% is a possible breeding solution.

Mutton Production

There is no specialised sheep breeds for mutton production. Any well adapted breed which produces higher body weight is good mutton producing animal.

There is no tradition of fat lamb production as is known and practised in the west. Only some special feeding protocols are in vogue for both sheep and goat for Qurbani by Muslims at the time of ID as an important sacrifice which are made as a part of religious function.

Malpura, an indigenous sheep, is well adapted to harsh climatic conditions of semi-arid region. Lambs attained body weight of 15.58 kg at 3 month, 25.74 kg at 6 month and 31.90 kg at 12 month of age. Tupping and lambing rates of 93.64% and 85.75% were attained in natural mating system.

2.3.3 Cross-breeding for Mutton Production

For specialised mutton or fat lamb production, crossing of coarse wool/meat type breeds with Dorset and fattening the half-breeds will provide a body weight of 30 kg at 6 month of age with 3:1 feed conversion and 50% carcass yield. However, for creating new strains attention must be given to the improvement of reproduction in half-breeds. 'Avimans', a new mutton synthetic sheep, was evolved through cross-breeding of Malpura/Sonadi with Dorset/Suffolk. Its body weight was 20 kg (Arora, 1983). The 'Avimans' has an average survivability of 95.6%, tupping 86.4% and lambing 66.6%.

The average pre-weaning daily gain was highest in mutton synthetic (134 g), followed by those in Malpura selected (130 g) and Malpura control (121 g). The survivability ranged from 94% to 99% in different age groups.

Nellore synthetics at the APAU, Palamner were superior to indigenous Nellore lambs in body weight at birth, weaning and 6, 9 and 12 month of age by 5.70%, 40.42%, 31.85%, 39.40% and 36.10%, respectively. The average daily gain (19.85%), efficiency of feed conversion (32.37%) and dressing (1.93%) were more in Nellore lambs. At village level, the quarter bred lambs showed improvement at all ages over Nellore lambs.

Muzaffarnagari sheep, attained 25.4 kg body weight at the age of 124 days (in feed lot after 60 days of age) with an average daily gain of 171 g. It had a feed-conversion efficiency of 20.02% and gave a carcass yield of 50% on fasting weight basis. Fertility was 98.4% on Ewes available basis and 104.4% on tupped basis. The survivability at different ages was over 94%.

Although good experimental evidence is available that a fat lamb industry can be developed in this country so far there is no well-developed structure available to make use of these Research Innovations.

2.3.4. Lamb-pelt Production

Karakul sheep from the Soviet Union were introduced into India in 1975 to study its performance as pure-breed and when crossed with Indian coarse carpet wool breeds for lamb-pelt production under hot arid and cold arid ecologies in India. The performance of Karakuls with respect to growth, survivability, greasy-fleece production, reproductive performance and pelt quality was highly satisfactory. Their crosses showed a significant improvement in pelt quality over the native pelts when tested in the ecology of Ladhak and Rajasthan. There was however a lot of variability in wool pattern and shape in crosses. Jacket type pelt was found in 46% lambs. There appears to be a great possibility of developing Indian pelt sheep through grading of and selection of Karakul pelt types.

The research technologies generated have not been adopted by the State Animal Husbandry/Sheep Husbandry Departments in the field except in parts of Jammu and Kashmir, Himachal Pradesh and parts of Rajasthan because of problems of being still trans-humane (migratory and semi-migratory). The farmers are illiterate, harbour their own prejudices and convictions, and are satisfied with their meagre income.

There have been many problems with transfer of these technologies. Cheap availability of apparel wool from abroad has been discouraging investment in machines suitable to short staple lengths, leading to problems of non-disposal of cross-bred wool.

The changes in the pattern of the life style of sheep breeders and their ecological environment needs radical interventions if these technologies are to be of some use. There are, however, success stories in Jammu and Kashmir.

Kashmir Merino replaced native sheep population in the state by 1992. This was brought about by Dr. G.N. Kaul and Dr. G.A. Bandey. Cross-breeding of local Gaddi sheep in Himachal Pradesh with Rambouillet and Merino has also been rewarding. Here more than 60% of sheep population is of cross-breds. The credit goes to the Late Dr. G.C. Negi, Director. Experience in Rajasthan has also been encouraging but adoption on large scale is very poor.

(i) The All India Coordinated Research Projects (AICRP) on Sheep Breeding for fine wool and mutton:

The objective of this project was 'cross-breeding of genetically low yielding indigenous sheep breeds with high yielding exotic sheep breeds such as Rambouillet (wool and meat), Dorset and Suffolk (meat)' (Singh 1986). This project was started during the Fourth Plan (1969–74) and continued up to the Seventh Plan (1985–90). The exotic fine wool breeds used for crossing were Rambouillet and the Soviet Merino. The objective was to develop new fine wool breeds suitable for different agro-climatic regions of the country, capable of producing 2.5 kg annual greasy fleece yield and weighing 35 kg at 2 years of age. Some of the new breeds/strains that were developed were Avikalin, Avivastra, Sandyno and Bharat Merino (CSWRI, 1998). Wool quality improved appreciably (fibre diameter and medullation percentage reduced substantially) but wool yield increased only by 2–3% in Chokla and Nali crosses. F1 crosses of Nali and Chokla breeds had 22%–51% lower average fibre diameter but approximately 18% lower staple length. The crosses also had lower reproductive performance than the indigenous breeds (Singh, 1986). There is no economic analysis available of the extent to which the improvement in wool quality could translate into higher prices for the wool, raising the total income from sheep rearing. The effect of these developments on smallholder sheep rearers was negligible.

Wool prices in India have fallen considerably in recent years, due to the large-scale import of better quality wool whereas meat prices are increasing fast. Therefore, the importance of wool from the point of view of breeding for smallholders has declined sharply. In fact, in coarse wool breeds such as the Deccani, it costs more to shear the sheep than the income earned from the sale of wool.

(ii) The exotic mutton breeds selected for cross-breeding to increase mutton production were Suffolk and Dorset. These were crossed with the indigenous breeds, Malpura and Sonadi, in Avikanagar and Rajasthan, and with the Deccani in Rahuri, Maharashtra. A composite mutton breed called 'Mutton synthetic' was developed by pooling together cross-breds of Suffolk and Dorset with Malpura and Sonadi and stabilizing the exotic inheritance at 50%. Cross-breeding was found to be advantageous for improving mutton production under intensive feeding and management, as indicated by a 10–20% increase in 3- to 12- month weights, except for the Muzaffarnagari breed (Singh, 1986). Intensive fattening is, however, not the norm in field conditions. Kandasamy (2009) states that the success of cross-breeding with exotics for mutton was modest at best.

(iii) The Network Project on Sheep Improvement (NWPSI) (Arora and Prince, 2011):

This project was started by the ICAR in 1990 by changing the objectives of the AICRP on sheep breeding. According to Kandasamy (2009), this project appears to be a tacit admission of the failure of cross-breeding for fine wool and mutton in the arid and semi-arid and hot and humid agro-climatic regions of India. The duration of the project was until 31 March 2012, which was the end of the 11th Five-Year Plan but the project has probably been continued into the 12th Plan. Its objectives were to carry out surveys to find out the status and performance of indigenous sheep breeds, genetic evaluation and genetic improvement of indigenous sheep breeds in the local environment. There are six farm-based units:

- Chokla sheep for carpet wool at the CSWRI in Avikanagar, Rajasthan
- Marwari sheep for carpet wool at the CSWRI Arid Region Campus in Bikaner, Rajasthan
- Muzaffarnagari sheep for mutton and wool at the CIRG in Makhdoom (Mathura), Uttar Pradesh
- Deccani sheep for mutton and wool at the Mahatma Phule Krishi Vidyapeeth in Rahuri, Maharashtra
- Nellore sheep for mutton at the Livestock Research Station of Sri Venkateswara Veterinary University, Tirupati, in Chittoor district, Andhra Pradesh
- Patanwadi sheep for mutton and carpet wool at the Sardarkrushinagar Dantiwada Agricultural University, District Banaskantha, Gujarat

Prolific Sheep

Realizing the importance of FecB mutation, extensive work has been done in India for improving prolificacy and its effects and economics in sheep production traits, introgression of the FecB gene from 'Garole sheep' into 'Malpura' to produce the cross-bred carrying FecB gene. The viable benefits accrued by gene infusion are reflected by 45.7% increase in the prolificacy and 35.05% in weaning rate of Garole/Malpura (G/M) compared to that of Malpura. A single mutated copy of FecB gene produced 0.70 and weaned 0.58 extra lambs in GM sheep. Patanwadi inheritance was introduced in the GMM to increase the milk yield of dam. In this process, GMM and Patanwadi were crossed to produce three breed crosses which result in higher prolificacy, higher birth weight along with more milk yield of dam to feed twins/triplets.

Table 2.9: Comparison of Growth performance in different genotypes w.r.t. Triple Cross

Particulars	Birth Weight	3-m Weight	6-m Weight	12-m Weight
Malpura	3.08±0.02 (520)	15.58±0.17 (369)	25.74±0.24 (237)	31.90±0.31 (181)
GMM	2.54±0.03 (338)	13.65±0.18 (265)	19.60±0.21 (143)	26.39±0.28 (114)
GMMxP	3.51±0.08 (57)	20.87±0.55 (44)	31.82±0.93 (29)	39.35± 1.24 (19)

GMM= Garole x Malpura x Malpura and

GMMP = Garole x Malpura x Malpura x Patanwadi

Physiology and Biochemistry Researches

Research in areas of adaptation and stress physiology, semen preservation and artificial insemination, estrus synchronisation, super ovulation, *in vivo* and *in vitro* embryo production, embryo freezing and transfer to enhance reproductive efficiency, advanced reproductive technologies pertaining to cryopreservation of ram semen and embryos, trans-cervical and laparoscopic artificial insemination, embryo technology and control breeding in ewes were developed and standardised.

First time in the country, lamb born following transfer of frozen thawed embryo in 1998 and IVM-IVF-IVC embryo in 1999, Embryo Transfer Technique (ETT) was perfected and 240 elite lambs produced through ETT in the institute (CSWRI). Twenty-four *in vivo* embryos were collected in single flushing in Garole sheep. Cost effective indigenous progesterone impregnated intra-vaginal sponges for induction and synchronisation of estrus developed and made available to numerous farmers and other organisations across the country. Heat and nutritional stress module was also developed in sheep. An accelerated lambing system successfully developed to enhance more lambs from ewes in their life time. A total of 76.7% Malpura ewes produced 1.5 lambs per year up to consecutive 4 lambing. Ovine kisspeptin was synthesised and purified.

Artificial insemination of sheep using liquid ram semen and 'Estrus induction and synchronisation in sheep' techniques were developed and successfully used in the field. A lambing rate of 63.26% (916/1448) by AI with liquid chilled semen in farmer's flock was achieved. A commercial kit (Avikesil-S) for estrus synchronisation in sheep and goats developed and more than 20,000 vaginal sponges (Avikesil-S) sold to progressive farmers, SVUs, ICAR institutes, KVKs and NGOs.

Sheep Health Management

Region specific worm management programme in sheep flocks (based on disease epidemiology) was developed. *Haemonchus contortus* (>80%) followed by *Oesophago stomum, Trichostrongylus axei* and *T. colubriformis* were major species recorded in the region. Warm moist season with adequate rainfall is more suitable for survival of *Haemonchus contortus* larvae. The annual incidence of strongyle worms in sheep flocks was 55.0% and 65.0% in field and farm flocks, respectively. The other gastrointestinal parasites recorded were *Trichuris* spp. (<2.0%, higher incidence in winter), *Strongyloides papillosus* (10%–15%, maximum during monsoon), *Fasciola gigantica* (<0.2%, higher prevalence during summer in semi-arid region and during winter in arid region), *Paramphistomes* (1.6% in arid region in a single peak during August to 7.5% in semi-arid Rajasthan with two peaks/year in June and August), *Schistosoma indicum* (0.5%; higher incidence during May to September), *Moneizia* spp (5.0%; higher incidence during May to August) and *Eimeria* spp (40.0%; maximum during monsoon). Worms cause estimated losses of INR 973.715 million/annum in adult sheep and INR 217.993 million/annum in yearling sheep in Rajasthan.

The emergence of resistant strains of parasites frustrates the worm control measures and limits the available anthelmintic drugs to combat the primary problem of worms. Benzimidazole (BZ) and Tetrmisole (TEM) resistance in *H. contortus* was observed in 85.7% and 56.6% of flocks, respectively, in Rajasthan. For the first time, a new RFLP-PCR technique was developed for detection of the benzamidazole drug resistance in adult and larvae of *Haemonchus contortus* parasite.

Based on epidemiological investigations, a regional Modified Worm Management Programme (MWMP) for sheep flocks of Rajasthan has been developed and tested for its effectiveness. It involves single targeted anthelmintic treatment of flocks during mid to late monsoon with tactical treatment (if required) in November. Following successful implementation of single drench/annum schedule during mid to late monsoon, efforts were made to further reduce the use of anthelmintics in flock to harvest the benefits of over-dispersion phenomenon in Fecal Egg Counts (FECs) through Targeted Selective Treatment (TST) or breeding for resistance/resilience. A conjunctiva colour chart (graded in 5 categories, correlated in farm flocks since the last six years. The implementation of TST resulted in anthelmintic intervention to only 20%–30% of sheep without hampering the flock productivity.

The studies on breeding for resistance/resilience to *H. contortus* through selection resulted in creation of divergent lines with moderate (0.2–0.3) heritability for faecal egg counts. The animals of resistance line (in Malpura and Avikalin breed) have been successfully kept without anthelmintic intervention since last eight years at CSWRI. Efforts are underway to identify genetic markers for disease resistance using biotechnological tools.

Forecasting systems for gastrointestinal nematodes in sheep have been formulated in the form of bio-climatographs and mathematical modelling. Mathematical simulation model (FROGIN: Forecasting for Rajasthan on Ovine Gastrointestinal Nematodiasis) based on development, mortality and establishment rate, climate data, translation of larvae on pasture, fecundity of parasite, stocking density and migratory behaviour of larvae was used to predict the level of pasture contamination and subsequent possible intensity of infection in sheep. It is a computer based mathematical modelling of *H. contortus* population tailored for different zones of Rajasthan. It gives result as predicted FEC on start of the month, intensity of FEC for next 60 days and pasture larval burden for that month. Testing and validation of FROGIN exhibited >70% agreement in all the locations and management systems studied.

2.4 GOAT

2.4.1 Breeds

The goat is perhaps one of the very useful domestic animals in the arid deserts of India - dry areas of Rajasthan and Gujarat. India ranks first among the countries in goat population. According to the 2012 Livestock census, there were 135 million goats in the country. They contribute 36% of the total meat and 4% of total milk in India. In addition, they produce pashmina, skins, casings and hair. Their droppings enrich the soil. Goats are completely misconstrued animals and are often blamed for desertification, because of their browsing habit and hardy nature. Its mouth parts are such that it can feed on leaves of plants without touching the thorns. They can consume oddest types of plants and produce milk and meat under highly harsh conditions. They are the last to leave the desert because they can consume even the twigs when the other animals and the man who are responsible for desertification have left.

The Indian breeds have mostly evolved through natural selection for adaptation to specific agro-ecological conditions. Very little directed effort has been made for developing these breeds through artificial selection to meet any socio-economic goal. Most of the breeds are very well adapted to harsh climate, long migration, tropical diseases, poor quality of nutrition, shortage of drinking and poor quality of water.

Goat breeds can be grouped on the basis of size, *viz.* large sized: Jamunapari, Beetal and Jakhrana; medium sized: Barbari, Sirohi, Marwari, Gohilwadi, Zalawadi, Kutchi, Surti, Sangamneri, Osmanabadi, Gaddi, Kannai Adu, Malabari, Mehsana, Ganjam, Changthangi and Chegu; small sized: Black Bengal and other Bengal types. They are also classified on the basis of major production characteristics, *viz.* dairy and meat, fibre and meat, and meat types.

There has not been any organised effort with respect to research and development work on goat production in Indian subcontinent, except for establishing some goat farms in the home tracts of some of the breeds, e.g. Beetal at GLF, Hissar and Jamunapari at U.P. State Farm, Etahwa. Kaura (1943) gave the first comprehensive description of goat population.

For the development of long-haired native goats such as Mehsana and Marwari having some *Capra falconeri* blood, introduction of *Capra aegagrus* blood, through Jamunapari in the first stage to increase appetite and displace the long-haired characteristics was necessary. This is preferable than to use highly developed dairy goats from Britain and America to improve appetite and milk yield in such a way and to such an extent as permitted by the environment. For the introduction of milk traits, cross-breeding with Saanen or British Alpine breeds was of choice.

2.4.2 Angora goats

Angora goats are the ideal choice for improving mohair yield and good quality meat in hill goats (Lall, 1968). Kashmiri and Gaddi goats of hill areas and Mehsana and Marwari of Gujarat were useful for hair production because of their long staple and yield of hair. Kashmiri goats yield 85–113 g of cashmere fibre (pashmina) per annum; Gaddi and Chamba breeds also yield a small quantity of pashmina.

In India, mohair production is limited to a few introductions of Angora goats. They are found in small numbers in the hills of Himachal Pradesh and Uttar Pradesh.

Developmental programmes in Jamunapari and Barbari breeds were initiated in the Gorakhpur Division of Uttar Pradesh by maintaining bucks of these breeds at the veterinary hospitals.

2.4.3 All India Coordinated Project on Goats

In 1970–71 research work was initiated on goat breeding in an organised way for bringing about improvement in milk, meat and fibre in indigenous goats through cross-breeding with exotic milk/meat/fibre breeds and selection within indigenous breeds. Creation of research facilities at several Institutions in different agro-climatic regions of the country helped in collecting useful data on the performance of indigenous breeds and their crosses with exotic dairy/fibre breeds under semi and intensive management conditions.

The work undertaken under the AICRP on Goat, led to the establishment of a Goat Research Centre during the Fifth Plan at Makhdoom, Mathura (Uttar Pradesh) by the ICAR. During the Sixth Plan period, apart from intensifying the All-India Co-ordinated Research Project, the ICAR decided to upgrade this existing Research Centre to a National Research Institute (Central Institute for Research on Goat in 1979) with the main responsibility of developing suitable technology for goat development in the country and in co-operation with the All-India Co-ordinated Research Project, evolving strains/breeds suited for different agro-climatic zones of the country [ICAR-K.L. Sahni (OSD), 1981].

The AICRP on Goat also carried out studies relating to milk, meat and fibre spread over eight different research centres with varying agro-climatic conditions. Important data on native breeds generated, indicated their genetic potential for development. Heritability and repeatability of multiple births has been estimated as 15% and 9%, respectively. Heritability estimate for gain in body weight at weaning (6 month) is 15%–20% and for adult body weight (one year) 30%. There is apparently considerable scope of selection for body weight.

The milk yield is low in Indian goats but there is wide range in milk yield. Lactation length shows a large potential for development. The half-breeds of Saanen and Alpine with Beetal and Malabari produce twice as much milk as their contemporary pure-breeds. In improving the indigenous goats for milk production, cross-breeding in selected areas with adequate feed availability is recommended. Cross-breds of Black Bengal with Jamunapari and Beetal were superior to Black Bengal. Out of the two crosses, Beetal x Black Bengal performed better than Jamunapari x Black Bengal (51.6% improvement in pre-slaughter body weight, 87% in hot carcass weight and 17.4% in dressing percentage). Crosses of Beetal with Assam goat and Sirohi have also been encouraging. The half-breeds can attain 16 and 25 kg, respectively, at six months of age.

Jamunapari goats showed 15.6% improvement in milk yield in 90-day lactation period at the CIRG, Makhdoom, over the last eight years.

Jamunapari and Barbari goats attained 29.6 and 23.2 kg live weight at the age of 12 month. Jamunapari x Black Bengal and Beetal x Black Bengal crosses showed substantial improvement in live weight compared with Black Bengal (11.2 ± 0.3, 8.2 ± 0.2 kg).

At the IVRI, Mukteshwar, Chegu goats weighed more than the Changthangis, but Changthangis were superior to Chegu in pashmina production at maturity. The differences were negligible in fibre length and diameter. Heritability was low for fibre yield and fibre length in general, and was moderately high for fibre diameter.

The research results obtained so far showed that there was enough variability in the 17 breeds of goats spread over the Indian sub-continent that could be utilised for a selection programme. The twinning and triplet percentage varied from 100% to 70% from small to large breeds. They are adaptable to stall feeding and can be developed to meet the product specification as needed.

Major Small Ruminant Breed Improvement Interventions in India in the Past 30 Years

The Indo-Swiss Goat Development and Fodder Production Project (ISGP) was started in 1981 in Rajasthan, with the objective of improving goat production through genetic improvement and increasing fodder production. The project was formalised through an agreement between the governments of India and Switzerland and was implemented through the Rajasthan State Department of Animal Husbandry, with technical support from Inter-cooperation, Switzerland. Its main objective was to develop strategies for sustainable improvement of goat production in the semi-arid farming systems of Rajasthan to improve the income-generating capacity and nutrition of families belonging to the weaker sections of the rural community.

The 180-day milk yield of 178 Alpine x Sirohi and 92 Toggenburg x Sirohi goats at the Ramsar farm (160.3±4.0 and 160.9±4.8 kg, respectively) was only approximately 23% higher than the Sirohi at the Ramsar farm, and the kidding rate of the Sirohi (69%) was higher than the crosses (48%) (Kropf et al 1992). Based on these results, the project decided that the 23% increase in milk yield of crosses was not enough to justify the investment and running cost of a large-scale cross-breeding programme and started to concentrate on selective breeding within the Sirohi breed. Following this, the project carried out the first ever large-scale performance study of Sirohi goats at the field level. Sirohi goats belonging to farmers were identified individually with ear tags. From 1988–91, milk records were measured monthly in 461 flocks. The average of 3,406 records of 60-day milk yield was 90±29.7 kg and the average of 2,860 records of 180-day lactation was 245.3±74.2 kg (de Groot et al 1992). This milk production performance of the Sirohi measured in the field was 88.5% higher than the least squares mean of 180-day lactation yield (130.1±3.1 kg) of 362 Sirohi goats kept at the Ramsar farm and was considerably higher than previous published reports on the Sirohi breed from small research station trials carried out by agricultural universities and research institutes. This is also a common observation in other programmes, in which the performance of animals kept at research stations when compared with those in livestock keepers' flocks. The reasons for better performance of animals in smallholder flocks may be the genetic differences but are likely to be differences in management between owners managed flocks compared to flocks managed by paid employees at organisational farms because owners often graze the animals for longer hours and make efforts to graze them on better pastures. Sirohi goat owners are particularly known for taking good care of their animals and providing the goats with ample concentrate feed. In another ISGP publication (ISGP 1990b), an analysis of 433 lactations of 180 days, each of Sirohi goats concluded that 35% of the lactations produced more than 300 kg and 9% produced more than 400 kg, whereas the average was 269 kg. Goats with such high levels of milk production can be considered 'elite' animals. These findings confirmed the importance of the Sirohi as a well-adapted and well-performing, dual-purpose breed for semi-arid conditions.

The ISGP, thus, proved for the first time in India that individual identification of goats and milk recording in the field was feasible, thus fulfilling an important condition for the establishment of a breeding programme.

A milk-recording and buck-rearing scheme was introduced in 1989. Buck kids from does with 180-day lactation yields of >300 kg were identified at 2–3 months age as potential breeding bucks and were purchased by ISGP straight away or when they were one year old. In 1989–90, 139 such bucks were distributed for breeding. The project operated in 146 villages, with 677 goat keepers and 564 poorest of the poor goat keepers owning approximately 18,000 breedable does (ISGP 1990a).

However, the entire sector of small ruminants played only a marginal role in the state government livestock policy. This resulted in an almost total lack of ownership of ISGP by the Rajasthan Animal Husbandry Department. Finally, when concrete results and recommendations for the extension of the project were becoming available, the department had lost interest in the project and did not fulfil its obligations. By mutual consent, the project was prematurely terminated in 1992. Fortunately, the essential part of the know-how gained could be transferred to interested partners in the newly started NGO Programme Rajasthan (Wieser et al 2000).

Milk component: Misra (1988) analysed the data on milk component of the Kerala centre (Thrissur, Kerala) of the All India Coordinated Projects on Goat. Malabari goats of Kerala were cross-bred with the European dairy goat breeds Saanen and Alpine in Thrissur, Kerala, with the objective of developing a milch goat suitable for the agro-climatic conditions of Kerala. The optimum level of exotic breed inheritance was found to be 50%. Saanen x Malabari and Alpine x Malabari crosses were found to be superior to pure-bred Malabari in growth, survival, milk yield, fecundity and feed conversion efficiency. The Saanenx Malabari had an average lactation yield of 211.5 kg in 200 days, which was 147% higher than contemporary pure-breed Malabari goats. Saanenx Malabari required 2.47 kg dry matter per kg milk produced, as compared to 2.71 kg for Alpine x Malabari and 3.54 kg for pure-breed Malabari goats. Thirty-nine cross-bred bucks were distributed to local farmers and development agencies. Artificial insemination with cross-bred buck semen was also made available and approximately 500 does were inseminated. However, these did not appear to have been any recording of the cross-bred animals with farmers. Nor was a suitable synthetic goat breed developed.

The unit had problems with inbreeding because live animals or frozen semen of the exotic breeds could not be imported and presumably inbreeding was not controlled by maintaining an appropriate population structure. The difficulty with importing germplasm was due to a ban imposed by the government on the import of live animals or semen of exotic dairy goat breeds since these were suspected to be carriers of the disease caprine arthritis encephalitis (CAE). In 1988, it was recommended that the unit be phased out, because it had generated enough basic data for the purpose with which the unit was setup.

Meat component

The meat production performance of three north-western Indian goat breeds, Sirohi, Marwari and Kutchi, was evaluated at the Central Sheep and Wool Research Institute (CSWRI) in Avikanagar, Rajasthan. One of the major objectives was to compare the pure-bred performance of the three breeds under semi-intensive and intensive management systems, although intensive feeding is not the norm in the field. Perhaps this was to encourage intensive commercial goat rearing enterprises. The three breeds are, however, from different geographical areas with varying climate and other conditions and each breed is adapted to its distinctive sub-environment.

Under intensive feeding, the six-month body weight of Sirohi, Marwari and Kutchi male kids was 26.4, 24.7 and 25.4 kg, respectively. The feed conversion ratio of the Marwari was 7.24 kg dry matter per kg gain in body weight, which was lower than the other two breeds.

Black Bengal goats were cross-bred with the Beetal breed in Ranchi (now in the state of Jharkhand), with the objective of developing a composite meat breed suitable to the local agro-climatic conditions. Beetal x Black Bengal male kids achieved an average body weight of 14.7 kg at six months as against the target of 15 kg and this was 61% higher than pure-breed Black Bengal. The crosses also produced approximately 60% more milk than the Black Bengal and had 45.6% greater feed conversion efficiency than pure-breeds. In farmers' flocks, the crosses were significantly heavier till three months of age. However, the difference between pure-breeds and crosses was not significant at six months of age. The hot carcass weight of crosses was 28.7% higher than pure-breeds; however, the dressing percentage and percentage of bone in the carcass were the same. The age at first kidding of the crosses was approximately 100 days higher than that of pure-breeds. The conclusion was that the Beetal could be used as an improver breed in the Chhotanagpur plateau region and that a farmer could earn 70% more by rearing a cross-bred kid rather than a local pure-breed kid (Singh et al 1989). No details of the economics are, however, given in the report.

Two programmes were carried out for Genetic improvement of goats for meat production in farmers' flocks-one among Marwari goats in Bikaner district in Rajasthan and one among Malwa goats in Mhow in the Indore district of Madhya Pradesh. These programmes were started in 1987, and were termed 'On-farm Client oriented Research'. The average body weight of male Malwa kids under conventional range management conditions was reported to be 21.1 kg at 9 months in the Gavali Palasia village, District Indore. This was significantly higher than the 9-month weight of Malwa male kids (15 kg) in Simrol village, District Indore. This indicated the genetic variability available for selection.

Marwari is a widely distributed breed, covering nearly the whole of Western Rajasthan. According to the programme document, superior germplasm was to be identified in farmers' flocks and improver flocks were to be formed. Male kids were to be selected from improver flocks on the basis of body weight at six months and milk yield of the doe, based on at least three consecutive milkings in the first month of lactation, and on carcass traits, if possible, giving preference to multiple-born kids over singles. Selected males were to be used for breeding in farmers' flocks. The meat production performance of the progeny of improved sires was to be evaluated.

(i) Body weight and measurements of 1,522 goats were recorded for breed description.

(ii) Performance recording of 263 kids of 15 sires distributed the year before revealed that their weights were higher than their contemporaries but not significantly. It is not clear whether the comparisons accounted for the probable environment and management differences between flocks.

(iii) Twenty-six bucks were purchased at the age of six month; their body weight was 7.1 kg above the population average. In 1996, the six-month weight of selected bucks was 1.4 standard deviations above the population mean.

(iv) More than 6,000 goats were dewormed and protected against ectoparasites, and free treatment was given to approximately 4,000 goats.

It appears that the programme was well planned and had ambitious targets for the selection of male kids. It however seems that improver flocks were not formed and the selection of male kids was done only on the basis of 6-month body weight. It is also not clear how the males were to be evaluated for carcass traits if they were to be used for breeding. It also seems that proper genetic analyses of the data were not done. Therefore, the extent of genetic progress was unclear even after 10 years. The programme also suffered from other constraints such as inadequate funds to purchase superior bucks and inability to recruit adequate and suitable staff.

Fibre component

Cross-breeding of Angora goats was carried out at the Mahatma Phule Krishi Vidyapeeth in Rahuri, Maharashtra, to develop a composite Angora suitable for mohair production under local agro-climatic conditions. The 7/8 Angora had superior mohair yield and quality compared to other genotypes although its reproductive efficiency was poor. Reciprocal cross-breeding of ¾ Angora and 7/8 Angora was, therefore, done. The technology of rearing Angora goats for mohair production was not accepted by farmers in the vicinity of the university. Commercialisation of mohair production was not found to be a practical proposition because the quantity of mohair produced was too little.

Pashmina

With the experience gained and information generated, it was decided to abandon cross-breeding in the Eighth Five-Year Plan (1992–97) and improve meat and milk production of indigenous breeds through within-breed selection (Swarup and Singh 2011). Institutional flocks (farm units) of Jamunapari, Barbari and Sirohi breeds were established in 1993: Jamunapari and Barbari flocks at the Central Institute for Research on Goats (CIRG), Makhdoom (Mathura), Uttar Pradesh, and the Sirohi flock at the CSWRI, Avikanagar (Tonk), Rajasthan. A field unit for Jamunapari goats was also established in 1993 in the Chakarnagar block of Etawah district of Uttar Pradesh, which is considered the home tract of this breed. The main objective of the farm-based units was to conduct selective breeding, and to improve the production and reproduction performance of the breed. The sub-objectives were to:

(i) estimate genetic variance of economically important traits in goats (for example, body weight at birth, 3, 6, 9 and 12 month of age, milk yield, lactation duration);

(ii) estimate the breeding value of male and female animals;

(iii) facilitate *in situ/ex situ* conservation of elite germ-plasm and its effective utilisation;

(iv) estimate production economics of goats under farm conditions and

(v) validate farm-based goat production technologies under field conditions.

The farm-based flock of Barbari goats at the CIRG had 342 adult females and 91 adult males on 31 March 2011, whereas the Jamunapari flock had 305 breeding females. The Sirohi unit at the CSWRI had 361 adult females and 52 adult males. These animals generally had good production performance. The Sirohi unit reported that the 6-month weight achieved was 20.8 kg against a target of 17.5 kg and 150-days milk yield was 112.2 kg against a target of 87.5 kg (CSWRI, 2011). The kidding percentage of the flock was 89.9% and the survivability was 97%. It is, however, not mentioned over what period the targets were achieved; nor was there an economic analysis. None of these units reported the breeding values of selected animal vis-à-vis the average breeding value, standardised selection differentials, selection intensities or generation intervals. It appears that proper genetic analyses of the data were not done and, therefore, genetic progress was not estimated. This is inexplicable because each unit had qualified scientists (quantitative geneticists) with PhDs in animal breeding. A proper assessment of the success of these genetic improvement programmes cannot, therefore, be made.

Bucks are sold from the farm units to various agencies for breed improvement. Preference is given to various State government agencies (Swarup and Singh, 2011). Only the Sirohi unit report has given a list of the places where bucks were supplied during 2010–11 (CSWRI, 2011). Twenty-two of the 63 bucks supplied were given to Uttarakhand, where the terrain, climate and other conditions are vastly different from Rajasthan, the home-state of the Sirohi breed. When providing bucks from such government-funded flocks established for genetic improvement, priority should be given to goat rearers in the areas where each breed is traditionally reared. There also needs to be some kind of a multiplier flock structure for distribution of bucks and a follow-up mechanism in place, to evaluate the performance of the bucks and their progeny. There are apparently no guidelines on where the breeding animals should be distributed. Efforts are not made to restrict distribution only to the traditional tracts of the breed and other suitable areas and production systems. Since 2006, 175–280 Barbari breeding bucks and 55–165 Jamunapari breeding bucks were supplied for breed improvement every year. Moreover, farm flocks are not open to the introduction of good quality animals from village flocks. Therefore, there is no opportunity of accessing genetically superior animals with smallholders to raise the genetic merit of the organised flock and reduce inbreeding.

Six field units for goats were established in 2001, based on flocks owned by smallholders and maintained under a village management system in the specific home tract of each breed. The objectives of the field-based units were to:

(i) assess the production performance of goat breeds in farmers' flocks and improve the germ-plasm through selection;

(ii) evaluate the socio-economic status of goat breeders and the economics of goat production in farmers' flocks; and

(iii) disseminate pro-poorgoat-based technologies under field conditions and assess their impact on goat production.

The field-based units were of the Black Bengal goat at Kolkata; the Ganjam in Bhubaneshwar (Odisha); the Jamunapari in Chakarnagar, Etawah district (Uttar Pradesh); the Marwari at Bikaner and Sirohi in Udaipur (Rajasthan); the Sangamneri in Rahuri (Maharashtra); the Surti in Navsari (Gujarat); and the Malabari at Thrissur in Kerala (Swarup and Singh, 2011). The results of this project indicated that there was substantial improvement in meat and milk production of goats in farmers' flocks of different breeds (Swarup and Singh, 2011). However, there did not appear to be enough evidence for this statement. It was found that genetic and environmental influences on animal performance such as body weight or milk yield were not separated in the data analyses. Perhaps, this could not be done as full pedigree records were not available; in fact, individual animal identification was not done in some units. Trends of estimated genetic progress in important traits over time were not reported. The annual reports of individual field units were found to have reported phenotypic superiority of bucks selected for breeding, in body weight at a certain age compared to all available bucks (for example, 3–4 kg in 3-month body weight and 4–9 kg in doe's 90-day milk yield in the Sirohi breed unit). These should have been reported in standard deviation units for proper comparison.

The farm and field-based units of various goat breeds have, thus, been in operation for 10–20 years (except the new field units that were established three years ago). Most field-based units have been operating in the same villages and with the same selected households over this period; although with fairly large numbers of goats. For example, the Sirohi field unit is recording the data of 787 goats belonging to 66 households in 15 villages in Udaipur; the Marwari field unit is recording the data of 3,427 goats in 119 households in four villages in Bikaner; and the 3-year-old Osmanabadi unit is recording the data of 814 goats belonging to 280 households in six villages in Maharashtra. However, because of the limitations of the technical programme, they are not able to identify genetically superior genotypes from a wider area. This is because a lot of detailed recording such as body measurements of kids at birth, and at 3, 6, 9 and 12 month is required. However, these records are not used for genetic evaluation. The Osmanabadi unit has established goat keepers' groups in the villages and has started giving them training in one to two-hour sessions in the village once a week although the project does not include a training component (NARI, 2011).

The economics of meat production, fleece development and pelt production have also been commercially validated by the special experts of the ICAR and Central Leather Research Institute.

The time is ripe that a two pronged efforts be made for commercial dividends which will make goat as central to the system which will integrate feed mill → farm co-operative (small and marginal farmers and financially non-viable farmers into a financially viable farmers/entrepreneurships based on a single village, several villages and a larger framers who will get facility of credit/ animal health/hygiene slaughter and processing of skins → mean processing → better processing in useful products using meat research done so far into a wholly industrialised cheaper partnership and create the necessary projects at the ground level. This will create jobs and wealth together.

Goat Health

Goats suffer from several diseases of bacterial, viral, fungal and parasitic origin, which result into significant economic losses due to the morbidity, mortality, retarded growth, reproduction and production losses. Among the bacterial diseases, enterotoxaemia, brucellosis (Brucella *melitensis*), pasteurellosis, Johne's disease *(Mycobacterium paratuberculosis),* colibacillosis and mycoplasmosis, are the major diseases. Development of various vaccines not only mitigated the losses due to mortality but also improved their performance. The combat contagious caprine pleuro pneumonia (CCPP), a notifiable disease, a live attenuated vaccine was developed at IVRI, followed by killed Al(Oh) 3gel M. capri vaccine (1993) and saponified vaccine in 2007–08 (Anon 2012).

After the eradication of Rinderpest in 2006 from India (Yadav 2011) among the viral diseases, goat pox, peste des petits ruminants (PPR), foot and mouth disease, and blue tongue are of major concern being responsible for morbidity, mortality, besides productivity and reproduction losses in goat population. Dependable vaccines have been developed against FMD, PPR, and goat pox indigenously. A killed goat pox vaccine was developed in 1986 (Yadav et al, 1986). Subsequently, in 2005–06 a live attenuated vaccine, which gives protection for two years and safe for use in pregnant animals was developed (Anon, 2012).

The development of Monoclonal antibody (Mab) based PPR-ELISA kitas at IVRI (Singh et al, 2000) and a safe potent cell-culture attenuated vaccine (Sreenivasa et al, 2000) helped in providing baseline data on the actual prevalence of the disease in goats and sheep in India but also made it possible to launch a nationwide control programme. Due to the availability of potent vaccine, incidence of PPR reduced significantly in the vaccinated flocks and states using the vaccine at large scale. This encouraged the government of India to launch a nation-wide mass vaccination program against PPR in the 12th Plan (2012–17). Although currently three indigenous live attenuated vaccines—namely, Sungri 96, Arasur 97 and Coimbatore 97—are available, the Sungri strain vaccine developed by IVRI, which has undergone extensive field trials is being used in most of the states. Recently, a nano-scale biosensor chip surface for detection of PPR was developed using monoclonal antibody parasitic infections/ infestations are other major causes of mortality in kids and loss of productivity in adult animals. Among the parasitic diseases coccidiosisx, fasciolosis, strongyloidosis constitute the major parasitic burden. Effective drugs developed are now available to combat these diseases. Herbal ectoparasiticidal product Alquit and anti-diarrhoeal and anti-coccidial formulations developed have been commercialised. Feeding of phyto-chemicals such as condensed tannins @ 1.5% of diet through leaf meal mixture (*Ficus infectoria* and *Pswidium guajava*) resulted in faecal egg count reduction, improved growth in kids and decreased gastrointestinal parasitic load in kids and decreased gastrointestinal parasitic load in kids (Anon 2012), thus demonstrating the potential of secondary plant nutrients as an alternative environment—friendly sustainable novel strategy of worm control. Genetic parameters of indicator traits of gastrointestinal infection were elucidated and the relationships between indicator traits with growth parameter at the genetic, phenotypic and environment level quantified.

2.5. EQUINES

In India, most animals in the equine species belong to equines are owned by socio-economically weaker sections of the society. They use them for transport of goods and passengers in villages and remote towns. There is also a class of people who use them for tourism and sport, the most important being horse racing. There is little scope of earnings through exports although we have more indigenous breeds for trade.

Like other livestock, equines enrich soil with organic and mineral constituents through their urine and faeces. There use for transport can reduce the load on fossil-fuel-driven vehicles which are the major cause of environmental pollution. If alternate pollution-free, renewable and eco-friendly energy does not replace the fossil fuels in coming 50 years or so, we may have to revert to the utilisation of animal power where equines will play very dominating role. Therefore, without affecting the macro and micro-environment preservation and development of equines is a sustainable and eco-friendly proposition. We do have environmentally adapted six breeds of horses, *viz.* Marwari, Kathiawari, Spiti, Manipuri, Bhutia and Zanskari (Siver 1986). Presently India possesses 1.13 million equines including some thoroughbred and wild asses.

2.5.1. Equines - a Cavalry Force: Breed Characteristics and husbandry

Before independence the horse had a prominent role in battles and warfare of the country and states. Largely the security of states was dependent on equines, large cavalry forces with excellent quality animals were maintained by many rulers. Excellent quality horses were being bred in the country, before the advent of the British rule. This has been mentioned by Maj. Gen. Tweedie in his book *The*

Arabian Horse. He served in India from the middle till the end of the 19th century. A century ago Hindustan yielded horses second to none in stamina and 'we have often perceived among Rajputs and Marathas the same anxiety as distinguished the Arabs to keep pure-breed horses'. 'The mares in Marwari bred by Rajputs are most tenacious, display the clean muscle, leaned head, thin nostrils and large dark eye of Nejd. This breed owed nothing to crossing, but on the contrary had been handed down unaltered in the pure-blooded Rajput families through centuries of warfare'. It is unfortunate that following the establishment of British rule the quality of indigenous breeds of horses declined badly due to bad management but there are still some remnants left of the excellent quality animals of indigenous breeds (NCA 1976).

Soon after the establishment of East India Company, the two studs, *viz.* Ganjam Stud Depot in Madras Presidency by Lieut Evans and Pusa Stud in Bengal Presidency by Lieutenant William Frazer, were established in 1795 based on the breeding policies of Tipu Sultan. The studs started with 20 Arab stallions with a target covering of 400 mares. By 1799, for breeding purposes 1,972 mares were covered. Stables were managed by Bhola Thakur, who had vast experience in breeding. To begin with, foaling efficiency was approximately 66%. Since the number of barren mares was increasing with no solution in sight Dr. Moorcroft, a renowned veterinary surgeon and expert in horse breeding was appointed in place of Lt. Frazer. Later, for the supply of horses, stud depots were created at Bareilly (1803), Hissar (1814), Hapur (1820) (the present equine breeding studs), Aligarh (1828), Saharanpur (1843) and Karnal (1846). It was in April 1876 that Army Remount and Horse Breeding Department was created. The Horse Breeding Department was under the control and direction of Col. J.H.B. Hallen, who was also the Chairman of Indian Cattle Plague Commission. He was responsible for conducting horse and mule breeding operations in North-West Province, Punjab and Rajputhana. Later, donkey stallions were imported from Italy, France, Spain, Cyprus, Persia and Arab countries for mule production.

2.5.2. Army Remount and Horse Breeding Department

In April 1876, the Stud Department was abolished and many studs were closed including at Pune and an Army Remount and Horse Breeding Department was created. Three different methods, *viz.* Circle system, Zamindary system and Home system, were adopted to produce stock. During the period of Maharaja Ranjit Singh in Punjab, the government provided stallion to selected breeding districts. The objective of horse-breeding department was to establish and improve a native breed of horses. Similarly, various state departments provided stallions from their studs for free service. Selected mares were branded to prove their right to free service by Government stallions and to prevent their purchase by cavalry or police horse buyers. A system of prize giving at the fairs and shows with some slight advantages to the progeny of branded mares was instituted. All young stocks fit for army service were bought by Government at 18 month of age and reared in Remount Depots before issue to units (NCA, 1976).

In addition to these, pony stallions were distributed in suitable districts under the control of District Committees, to cover small and unbranded mares. Grave objections to this so called 'Diffused System' of breeding were voiced in knowledgeable quarters. It was in 1900, the Government of India constituted a *Horse and Mule Commission* under the Presidency of Major General T.B. Tylar. The Committee reviewed the entire question of *Horse and Mule Breeding* and gave recommendation about the breeding policy adopting the unbound and bound system of breeding. It was further noticed that due to unplanned breeding policy, followed over the years, well-known indigenous breeds of horses suffered badly. The Commission noticed that the unbound system, in reality, was similar to the 'Diffused System' except that it was confined to favourable districts where there was natural advantage of climate and soil, and the people were interested in the breeding of horses. The bound system or the controlled system of breeding envisaged issue of service condition grants of land in canal colonies in the Punjab. The condition was the provision of brood mares for breeding horses and mules, and the Government lien on the young stock up to 18 months of age. Animals rejected as unsuitable were at the absolute disposal of their owners. The unit of grant was 2 squares of land per brood mare approved by Government. The services of Government stallions were provided free of charge but all coverings were strictly regulated by the Government officers. The guarantees were confined to members of those professions with hereditary taste for, or practical experience of horses and mules. The scheme under competent administration and arrangement was expected to put the country and the Government in possession of an extensive and valuable remount reserve. Consequently, a breed of horse was evolved, which is known as Country Bred or Indian Bred. It is a type of horse rather than breed (NCA, 1976).

Another most important point the Commission brought out was the recognition of the fact that in certain states of India there existed pure-breeds of horses, which the local breeders strived to maintain pure and were, in the judgment of the Commission, well worth preserved in their purity. They particularly referred to Kathiawari and Marwari breeds. The Commission recommended that the Indian Government should purchase the most suitable stallions of these breeds and encourage the people of Kathiawar and Marwar to breed these horses both for army and civilian requirements.

Moreover, the Army Remount Department was made responsible for supply of horses and mules only to designated circles: Circle I: Shahpur, Rawalpindi, Hazara, Jhelum and Attock; Circle II: Jhelum canal colony, Jhang, Multan and Dera Ghazi Khan; Circle III: Lahore, Gujranwala, Amritsar and Ferozepur; Circle IV: Baluchistan and Sind; Circle V: Bulandshahr, Meerut, Aligarh and Muzaffarnagar; and Deccan Circle: Pune and Ahmednagar. In the above circles, there were approximately 14,012 mares, 246 horse stallions and 303 donkey stallions. Total breeding was under the command of Army Remount Department. The Commission further recommended that Kathiawari and Marwari breeds need to be preserved for their purity. This resulted in the establishment of Kathwari and Marwari horses stud farms especially at Ahmednagar, which was closed in 1930.

Recorded information revealed that majority of thoroughbred horses were imported from UK, approximately 200 annually until 1930. It was the effort of Sir Victor Sasson who in early 1930 imported excellent thoroughbred horses and presented them to the existing studs. By 1938 these studs produced nearly 120 broodmares. However, the import of foreign horses was stopped in 1941 due to World War II, and the ban was lifted in 1945. The number of broodmares had increased from 385 in 1940 to over 1,000 by 1947.

Racing as a sport

Racing in India was conducted during British Raj in every cantonment in India. It dates as far back as 18th century according to records extracts from G.O. No. 2054 of the Government of Madras dated 17 November 1922. The race course of Madras (now Chennai) was formed in April 1777. Records of racing in Calcutta (now Kolkata) at Akra Farm on the bank of Hooghly River dates back to 1794. Mr W.F. Ferguson was responsible for the establishment of Calcutta Turf Club in February 1847. Bombay Turf Club was established in 1880 by Sir Charles Forbes, G. Hall, A. Combell and P. Hadow. Madras Race Club was founded in 1896 (Asian Conference 1995).

In ancient times, the outcome of wars between nations and communities often depended on the horse. Similarly, efficient Governance depended on faster transmission of information and messages through them. In the changed scenario, after Second World War and independence, the equine use for information transmissions has declined in all these aspects and replaced by motor transport which lead to their decline in population, health and production.

2.5.3 Equine Health

The need to protect equines against diseases was felt with the growth of equine population and its use in Army in early 1900s and later part of 18th century, and Troupe Committee and Special Stud Commission under the chairmanship of General Gilberts were established in 1872. Due to recommendations of this committee, the Imperial Bacteriologist, Dr. A. Lingard and his colleagues H.T. Pease, V.R. Rajgopalan, Mohamad Israil and F.C. Minette made special contribution on various diseases of equines, e.g. Dourine Glanders, Strangles, and on problems associated with abortion and pulmonary disorders. Most of these researches were conducted at Imperial Bacteriological Laboratory of the Imperial Veterinary Research Institute (IVRI) from 1889–1946. Dr. Lingard and his team were responsible for developing mallein, and its use for the diagnosis of Glanders in equines. This development was followed (1880–91) by the control of this disease. The IVRI had been the single producer of this product. Soon after the availability of mallein, disease monitoring system became effective. For the control of glanders, Farcy Act was framed (The Glanders and Farcy Act, 1889) and was implemented authorising the destruction of positive reactors (Lingard, 1903, 1906).

Dr. Lingard considered dourine as an outstanding problem among equines in India. He recorded the first case at Babugarah in 1901 in Quarter Arab Stallion (Lingard 1903, 1905). According to Major Pease working at Lahore Veterinary College, dourine was described in a Persian book *Kitab-ul-vaitert* which was the original translation of Sanskrit text which was also discrilled by Major Pease of Rahuore vet. Col. *Trypsanosoma* in Equine Blood was also demonstrated by Pease (1903, 1905a). Lingard referred that *Trypanosoma* multiplies in serosanguineus exudate of the plague. He also advocated the use of drugs *viz.* arsenic and trypan blue preparation. Application of strict quarantine measures after dourine became notifiable disease under the Dourine Act of 1910 reduced its incidence. The Act provided power for castration of dourine-affected stallion and the destruction of branded mares. Systematic researches and implementation of control programme was responsible for the decline of dourine after 1906. The last case was detected in 1920–21 (Gill 1991).

Epizootic lymphangitis which warranted mandatory destruction of horses was detected by Iyer (1936). He isolated the causative agents and described the lesions which were confined to the intestine and mesenteric lymph glands.

Abortion and sterility were affecting the breeding programme of equine establishments. Intensive studies on the above problem were done at the IVRI, Mukteswar. In 1919–20 *Salmonella abortus equi* infection was reported from the Government Livestock Farm, Hissar (Edward 1919–20). Subsequently, serotypes *of Salmonella abortus equi* strains were identified and vaccine development was initiated in 1930. A vaccine containing 12 strains of *Salmonella abortus equi* and 2 strains of *E. coli* was prepared.

African horse sickness (AHS), a viral disease of equines, was identified among ponies at 61 Cavalry Unit of Indian Army, Jaipur in April 1960 by Dr. G L Sharma, Dr. N S Datt and S.B.V. Rao. In all, 22,977 horses were affected; out of these 20,822 horses died. These scientists worked on the isolation, identification of the virus, adaptation in mice and embryonated eggs, and subsequent development of mice brain vaccine (Shah 1964). Later, Sinha and Madhusudan (1963–64) produced vaccine comprising seven AHS virus strains as recommended by the FAO under the consultancy of Dr. Gamuldin Zaharan (1961). This finally paved way for the control of the disease. Later, the (strain Karen) was removed from the vaccine due to reports of unfavourable reactions. A safe and potent monovalent African horse sickness vaccine of chick embryo origin was also developed. The disease was finally controlled by the strenuous scientific efforts made by the scientists at the IVRI along with the co-operative efforts of the Indian Army which also included vaccination practices. The last case of outbreak of AHS was noticed in 1963. India stopped vaccination of equines against AHS for more than last 30 years. Later, S Kumar, B B L Mathur, R N Sharma and Ramakant (1971–75) formulated immunological methods such as complement fixation test, HI, AGID and Conglutinating complement absorption tests to conform the virus of AHS (Kumar 1976). These were used for monitoring and surveillance of the disease.

Histoplasma farciminosum, the cause of epizootic lymphangitis, was a menace in the army horses and was formidable to cultivate in dimorphic phase. Tej Singh and Vermani (1960–63) worked at the IVRI, Mukteswar and isolated the fungi and evolved a suitable medium for its growth. Further studies led to the development of successful chemotherapy. Sikdar (1966) developed immunological methods, *viz.* complement-fixation, allergic and passive haemagglutination tests for the diagnosis of the disease.

Monitoring and surveillance programme on glanders would have not been possible without the refinement on the production of mallein and development of diagnostics. Sen and Ganguli (1960–64) developed complement-fixation conglutinin absorption and indirect haemagglutination inhibition tests and recorded the superiority of intra-dermal palpebral test over the routine mallein testing. Further, a quicker diagnostic test, *viz.* counter-immuno-electrophoresis, was developed by Uppal and Singh (1986–88). Subsequently they employed micro-complement fixation and other tests in the diagnosis and control of glanders in Haryana and Punjab. This resulted in limiting the occurrence of this disease. Mishra and Verma (1989) developed Dot Enzyme Linked Immuno-sorbent assay (dot ELISA) test. They were responsible for controlling of glanders among equines in army units.

Capt. Vishwanathan (1981–84) worked on strangles under the ICAR scheme at CMVL, Meerut. He isolated *Streptococcius equi* and *Str. zooepidemicus* and developed vaccines and Treatment regime for them.

Keeping in view the importance of equines, the ICAR established the National Research Centre on Equines in 1986 with the mandate to conduct research on health and production of equines; impart training, provide diagnostic and advisory services, and transfer technology to the beneficiaries.

2.5.4. Equine Production

Whatever little success was achieved through the bound and unbound system of breeding before independence was abruptly brought to naught as all the main breeding establishments located in West Punjab went to Pakistan and the studs which existed in India could not provide even 200 young stocks to the Indian Army or thoroughbred industry. This resulted into importation of horses to the tune of INR 70 million between 1949 and 1954. Later, the Government of India, Ministry of Defence streamlined infrastructure facilities for breeding and rearing of equines. Accordingly, the status of Babugrah was raised to equine breeding stud in 1957 and another breeding stud in 1965 was established at Hissar. Besides these remount depots at Saharanpur and Hempur met the demand of trained horses to other animal holding units. Indian Army imported heavily both warm blood (Thoroughbred, Arab, Anglo/Arabs, Half-breeds, Cheval De Selle Francais, Cheval De Selle, Netherlands and Belgium Saddle, German

Hannoverian, Oldenburg, and Halflinger) and cold blood (Polish, Comtois, Brittany, Yugoslavian, Polish Arab, Trait Cobs) horses (Bhalla 1988).

The police departments also carry out small-scale breeding operations at their horse lines where the animals are kept for police duty and ceremonial purposes. Mention may be made of the horse lines at Madras (Chennai), Bhopal and Ahmedabad, where several foals have been born to mares mated to outstanding stallions. Such breeding operations by the police departments are of a casual nature and there are no specific plans for breeding horses to meet their own requirements. The Defence Department provides some horses to police departments, Border Security Force and riding clubs, and for ceremonial purposes at Rashtrapati Bhavan and other occasions. However, police departments have been buying horses from private breeders at organised fairs held in different parts of the country from time to time. The support of the Central Government extended through the then Ministry of Food and Agriculture and the Indian Council of Agricultural Research for horse development remained confined to providing health cover to the animals. Important studies on diseases have been conducted under the schemes for disease investigation in some States and the Indian Veterinary Research Institute. In most of the states no scheme was undertaken on horse breeding for improvement of indigenous stock, perhaps mainly because very little information was available on the requirements of the horse breeding industry. Moreover, the need for developing other species of livestock was so pressing that the development of horses, mules and donkeys did not receive priority in allocation of funds which had great limitations. The activities in most of the states were, therefore, restricted to provision of health cover (NCA, 1976).

The Government of Himachal Pradesh also had undertaken development of horse breeds during the first three Plan periods. In addition to establishment of several stud centres, a small nucleus herd of imported Connemera horses was established for this purpose in the State. Efforts were made to grade up the local horses with the introduction of Connemera inheritance by stallions stationed at veterinary hospitals. During the Third Plan, the progress of the breeding activities was reviewed. It was concluded that this breed was not quite suitable for the hilly areas, and the limited resource available was reserved for the development of other types of livestock. The stud centres and the Connemera horse farms were therefore wind up.

India had the distinction for exporting two horses, *viz.* Astonish and Adler who won in races at Hong Kong and the USA in 1994. The former horse belonged to Cyrus and Zavary Poonawalla and the latter to Vijay Mallya. It is believed that Indian Breeding Industry is now looking for exciting future since the Government of India has introduced further economic reforms.

Non-availability of proven donkey stallion and jennies was a major constraint to produce quality mule in the private sector and improvement of local donkeys for ages. National Research Centre on Equines, Hissar (ICAR) imported a nucleus herd of 20 jennies and 5 jacks of superior exotic Poitou donkey germplasm from France in October 1990. This was successfully adopted and maintained at the Centre for mule production.

The Centre has been instrumental in phenotypic and genotypic characterisation of various Indian breeds of equines including horses, ponies and donkeys besides playing important roles in the conservation of various breeds, improvement of production potential of equines through semen cryopreservation and artificial insemination, establishment of nucleus herd of Marwari horses, Zanskari ponies, Manipuri ponies and indigenous and Poitou donkeys, and the use of sustainable mule power for agricultural and defence operations.

National Research Centre on Equines, Hissar

Equines play an important role in the economy of small and marginal farmers through supplementing their livelihood, besides their role in military and paramilitary forces, responsible for maintaining external and internal security of the nation. Realizing the unique role of equines both for civilian and army establishments, the National Research Centre on Equines (NRCE) was established on 26 November 1985 at Hissar (Haryana). The Centre has been taking up most sustainable high-quality research, training and extension activities that have resulted in the improvement of the equine health and production in India through control and eradication of several infectious diseases and genetic improvement of equines. A new sub-campus—the Equine Production Centre—was established in 1989 at Bikaner for undertaking breeding and production research.

The research activities of the Centre continue to bridge the gap between basic biology and clinical applications thereby providing cutting-edge technologies and products for improving the health of the Indian equines. The ongoing research areas include equine virology, bacteriology, parasitology, immunology, pathology, medicine, biochemistry, biotechnology, genetics, breeding, physiology, reproduction, nutrition, and extension. The institution is disseminating information at national and international level to equine keepers through publications and direct contact.

The NRCE has contributed significantly in the field of equine health through cutting edge research on development of diagnostics, vaccines and immunobiologicals, molecular characterisation of equine pathogens and surveillance and monitoring of equine diseases in India. The Centre has been recognised as a national referral laboratory for the diagnosis of 10 important equine infectious diseases by the Government of India, Ministry of Agriculture, Department of Animal Husbandry, Dairying & Fisheries, New Delhi.

A National Veterinary Type Culture Collection (VTCC) was established in the year 2005 at NRCE for collection and storage of microbes with a basic mandate of acting as a repository of microbes of animal origin for education, teaching and training, research and development, and microbial diversity conservation, which will serve as a biological resource not only for the present generation but also for the posterity.

Salient achievements

Diagnostics

Development of immunoassays and PCR-based diagnostics: Various immunoassays and diagnostic tests developed include equiherpes B-ELISA for diagnosis of EHV-1; an EHV-1/4 recombinant glycoprotein G antigens-based type-specific ELISA for differentiation of EHV-1 and EHV-4 infections; anti-ERV VP6 monoclonal antibody-based sELISA for detection of equine rotavirus (ERV) from stool samples (also detects rotaviruses from cattle, buffalo, sheep, human), haemagglutination inhibition (HI) assays for the diagnosis of equine influenza, Japanese encephalitis and West Nile; indirect haemagglutination (IHA) test and dot ELISA for the diagnosis of glanders; EMA-2 recombinant protein antigen based-ELISA for serodiagnosis of *Theileria equi,* indirect ELISA for serodiagnosis of *Trypanosoma evansi*; EIAVgP26 synthetic recombinant protein antigen-based Coggins test and indirect ELISA for sero-diagnosis of equine infections anaemia (EIA), field-oriented immunostick ELISA for EHV-1/4 detection, and virus neutralisation (VN) test for serodiagnosis of equine viral arteritis (EVA). In the field of molecular diagnostics, the centre has developed qRT PCR for equine Influenza, RT PCR for diagnosis and subtyping of equine influenza (EI) virus, qPCR for differentiation of neurogenic and non-neurogenic strains of EHV-1, multiplex PCR targeting glycoprotein C for differentiation of EHV-1 and EHV-4; PCR for RoTat 1.2 and EMA-1/2 gene-specific PCR for sensitive detection of *T. evansi* and *T.equi* infection, respectively; *Cho*E gene and species specific genes-based PCRs for detection of *Rhodococcus equi*. Centre has developed tissue based diagnostic facilities using indirect immunoperoxidase technique (IPT) for the diagnosis of EHV-1, EI, *R. equi,* etc.

Vaccines and Immuno-biologicals: Vaccines and immuno-biologicals developed include, cost effective updated inactivated equine influenza vaccine (A/equi-2/Katra/2008), inactivated EHV-1 vaccine, improved *Salmonella abortusequi* (bacterin and outer membrane protein-based) vaccine; monoclonal antibodies and recombinant antigen(s)-based diagnostics for EHV-1, equine rotavirus, equine influenza virus, Japanese encephalitis virus, EIA, glanders, *Trypanosoma evansi* and *Theileria equi* infection.

Surveillance and monitoring of equine diseases: The Centre regularly carries nation-wide monitoring and sero surveillance for important equine infectious diseases and acts as a referral laboratory for almost all the important equine diseases. The NRCE has been instrumental in reporting and controlling equine influenza, equine infectious anaemia, equine coital exthema and glanders, etc. for which P.K. Uppal, M.P. Yadav, B.K. Singh, A.K. Gupta, B. Rattan, P. Malik and R.K. Singh made important contributions through it's a proactive strong sero-monitoring program.

Phenotypic and genotypic diversity of indigenous equines: Populations of six equine breeds registered by the Indian National Bureau of Animal Genetic Resources have drastically decreased due to indiscriminate breeding. These breeds namely Marwari, Kathiawari, Spiti, Zanskari, Bhutia and Manipuri were characterised phenotypically on the basis of their biometric indices and coat colour. Significant differences among different biometric indices were observed due to breed as well as sex. Genetic diversity analysis, population structure and relationship among six Indian horse (Kathiawari, Marwari) and pony breeds (Manipuri, Spiti, Zanskari and Bhutia), along with English Thoroughbred horses as an out group were carried out, which indicated a high genetic diversity in all Indian breeds.

Artificial insemination using frozen semen for production of superior quality Marwari horses, superior mules and donkeys has been perfected. The pure germplasm of endangered indigenous breeds of horses is being conserved using this technology.

Equine development related activities undertaken, such as (i) establishment of nucleus herd of equines (exotic donkeys, indigenous large white and small grey donkeys, Marwari horses, Zannskari and Manipuri ponies), (ii) generating baseline data on different biochemical, physiological and haematological indices in equines; (iii) work on equine work efficiency, (iv) collection and cryopreservation of semen of elite horses at farmers' door, (v) early pregnancy diagnosis, and developing technology for utilisation of animal energy and bio waste have immensely benefitted the stakeholders.

The Veterinary Type Culture Centre (VTCC) – Repository: The Veterinary Type Culture Centre (VTCC) established in 2005 at NRCE, Hissar focuses on the acquisition, authentication, production, preservation, development and distribution of standard reference microorganisms, cell lines and other microbial resources for research in veterinary and life sciences. The VTCC repository includes a variety of bacteria (760), viruses (135), bacteriophages (21), recombinant clones (399), genomic DNA (176), phage library (27), rumen microbes (including 82 bacteria and 111 fungi), and dairy microbes (432 bacteria).

2.6. CAMEL

The role of the camel as domestic animal is undergoing fundamental changes as subsistence nomadism shifts towards semi-sedentary cash demanding system. The camel in the arid and semi-arid areas must become more productive and competitive. Its unique qualities to survive under harsh climate and to utilise low quality feed resources and convert it into animal protein, power and other products must be fully exploited and improved upon. Camel can thrive for several days with little or no water.

Camels in India are bred and managed under extensive system by camel rearing communities mostly in 11 arid districts of Rajasthan and parts of Gujarat. Small to large camel herds (50–500 camels) are managed on the range land for breeding purposes. Camels are also kept in small numbers under intensive system, particularly for the routine work such as camel cart traction, transport of man and material, ploughing and other agricultural operations.

In the livestock species, it produces the major sustenance of livelihood—milk, meat, skin, hide manure, bone—and continues to provide these for a survival system to resource poor farmers.

The camel *(Camelus dromedarius)* is an important component of the desert ecosystem. It provides transport and subsistence to the rural societies in habitating marginal, arid and semi-arid zones. In the Indian dry land farming system, the camel is a major source of power for transport and agricultural activities. In the livestock species, it produces the major sustenance of livelihood—milk, meat, skin, hide, manure, bone—and continues to provide the access for a survival system to research programmes and other minor subsistence, *viz.* milk, hair, hide, manure, bones and meat to the societies residing in the poorly resourced arid and semi-arid zones. It is a single livestock species which carries the load of human species single handed of human species in dried zone.

The camel possesses many unique qualities which make it distinctly superior to other domesticated animals in the hot and arid desert ecosystem. Most of the areas where camels play a traditional role are subject to continuous social and economic changes. Different production systems are now emerging which tend to be less extensive and more commercially oriented. In this process the role of camel apart from providing source of bioenergy will assume additional dimensions, particularly as a source of milk and game animal. Other by-products such as hide, hair and bones are gaining prominence in the small-scale handicraft sector. Camel meat is also popular as a delicacy in some countries but in India it is not eaten.

Seasonal movements of camel herds to the areas having comparatively better pasture facilities had been a common practice. The Indian camel keepers have quite extensive knowledge about camel health care and management.

The camel breeding was practised by the local breeders using need-based characteristics and sires within a strain/bred in a locality. Private breeders use mostly sires which were either herd bred or purchased from nearby areas. Due to the popularity of Bikaneri breed, there is tendency among private breeders to use Bikaneri sires. Thus, continuous grading up of local strains or breeds to Bikaneri characteristics is taking place. It is necessary to evaluate different breeds and to establish separate farms maintaining different pure breeds so that genetically improved sires can be distributed for breed improvement of strains/breed which have been developed in different camel breeding tracts.

Limited numbers of Bactrian camels are found in the Leh district of Jammu and Kashmir state. It is believed that these double-humped camels were prevalent in India during pre-Christ era. The Indian double humped camels have well developed muscular body with soft fine long hair coat. The under-coat is woolly. The colour varies from light brown to dark brown. The height at withers varies from 190 to 220 cm, body length from 130 to 150 cm, heart girth from 200 to 235 cm and neck length from 110 to 120 cm.

Except for one camel breeding farm each in the Rajasthan and Gujarat states, there are no organised breeding farms to provide genetically improved animals. These two farms have been maintaining only Bikaneri and Kachchhi breeds. The other important camel breeds/strains are being propagated only by private breeders. Therefore, it is likely that unless timely action is taken, the country might lose very important camel genetic resource. There is more urgent need in the case of Bactrian camels where only 50–55 animals are left in the cold desert zone of Jammu and Kashmir state.

The ICAR sanctioned the establishment of a camel breeding farm and 10 stud centres which were to operate broadly on the pattern of the All India Key Village Scheme. The object was to collect scientific data on the camel and to improve various breeds by selective breeding. Due to recurrent famine conditions in Bikaner where the farm was established, the farm along with stud centre was handed over to the Department of Animal Husbandry, Government of Rajasthan, for being continued as a non-research development activity. The Rajasthan government handed over the farm to the College of Veterinary and Animal Sciences of the University of Udaipur for teaching, research and development.

The ICAR also sponsored a visit to India by an eminent foreign expert in anatomy for a detailed anatomical study of the camel. After the visit of this expert, schemes were sanctioned for detailed studies of the anatomy and physiology of the camel. In addition, the College of Veterinary and Animal Sciences, Bikaner, carried out some studies on the physiology, nutrition, husbandry, disease incidence and parasitism of the camel.

Considering the importance of the camels in the socio-economic development of arid and semi-arid zones of the country, the National Commission on Agriculture (NCA) in 1976 recommended establishment of a research institution in the country specifically dedicated for camel improvement research programmes. Consequently, the Government of India approved establishment of National Research Centre on Camel (NRCC), at Bikaner during the Fourth Plan with a total outlay of INR 37 million. The NRC on Camel came into existence on 5 July 1984. The physical facilities available at the erstwhile Camel Breeding Farm, Bikaner, consisting of 149 Bikaneri camels and approximately 2,060 acres of land were, therefore, transferred by the Govt. of Rajasthan to the ICAR.

The centre has developed a camel herd for research purposes. It consists of 218 animals belonging to three breeds *viz.* Bikaneri, Jaisalmeri and Kachchhi. Good useful information has now become available on these breeds and their cross-breds.

A nucleus of cross-bred camels between Arbi and Indian breeds has been developed for producing camels for race competitions. The progeny is showing promising results in growth and endurance.

Base line data on production and reproduction traits have been collected. Useful data on camel genetics and breedings, cytogenetics, haematology, physiology, nutrition, hormone estimations during breeding and non-breeding seasons and reproduction have been collected. Studies at the farm revealed that 40% of subabool fodder can be incorporated in camel feeding during crisis period.

The centre also provided free of charge services of genetically improved sires to female camels brought by the villagers during breeding season.

Among the livestock species, camel known as the 'Ship of Desert' is endowed with unique anatomical, physiological, biological and behavioural qualities enabling it to survive and thrive under the extreme harsh climatic conditions, inadequate feed, fodder and scares water resources of both hot dry and cold dry deserts of India. The value of camel *(Camelus dromedaries)* was realised long ago, particularly by the pastoral communities. Camels had special role and place in Indian Army while confronting the enemy in the Thar Desert due to its speed, stamina and adaptation to critical conditions for survival. Camels also used to be the source of communication both in peace and war times. This versatile animal is well adapted to arid and semi-arid regions with their unique physiological and metabolic pathways to survive without food and water for many days. The usefulness of camel in terms of draught ability, milk and commercial eco-tourism is well documented. In India, camels are mostly found in the Thar Desert, particularly Rajasthan and adjoining areas of Gujarat, Madhya Pradesh, Punjab, Haryana and Uttar Pradesh.

Focused research conducted on various aspects of camel husbandry since 1984 through the Project Directorate on Camel at Bikaner and subsequently after its up-gradation to National Research Centre on Camel (NRCC), Bikaner in 1995 has immensely benefitted the nation in generating scientific and technical information with emphasis on restoration of economic and aesthetic cultural value of camel in rural life of Indian desert.

The NRCC has contributed enormously to breed improvement through genetic characterisation, up-gradation and evaluation; strategically designed interventions to enhance optimum production and reproductive efficiency; development of nutritional packages for camel farmers; evaluation of camel milk for its therapeutic value and its exploitation as functional food; value-addition of camel milk and exploiting the nutraceutical value of camel milk. A very useful database containing information, *viz.* inventory of central camel herd, pedigree, growth, body measurement, milk production, reproduction, mortality, culling of Jaisalmeri, Kachchhi and Bikaneri breed, etc. was developed. A good initiative of *in situ* conservation of Jaisalmeri, Bikaneri and Kachchhi breeds of camel is in hand.

Camel diseases and the common health disorders such as surra, pox, mange, abortions, mastitis, skin dermatoses, tick parasitoses, worm burden, respiratory, metabolic and digestive disturbances, etc. were documented and package of practices for health management were developed. Several diagnostic techniques for *surra* were evaluated including antigen-antibody and nucleic acid based diagnostic assay (Pathak et al 1997, Omanwar et al 2001). Using shot-gun proteomic approach over 160 proteins of *T. evansi*were identified (Roy et al 2010). Proteins identified belonged to various functional categories including metabolic enzymes; DNA metabolism; transcription, translation as well as cell-cell communication and signal transduction.

The NRCC carried out a wide variety of value-addition experiments on camel dairy products, *viz.* cheese, *paneer*, *kulfi*, milk chocolate, chocolate *barfi*, milk powder, milk *barfi*, *khoa/mawa*, *gulab jamun*, milk *peda*, *rasogolla*, *sandesh*, *rabri*, *lassi*, different types of yoghurt and a camel whey-drink. Products such as *kulfi*, flavoured milk, *lassi*, tea and coffee have been commercialised through the NRCC's camel milk parlour (Anon 2012).

Camel milk was evaluated for its therapeutic value and its exploitation as functional food. Extensive studies were carried out on identification, characterisation and structural studies of proteins from camel milk, etc. The medicinal value of camel milk for use in a skin cream preparation to compare its anti-wrinkling effect with commercially available skin creams has also been tested with encouraging results (Anon 2012).

Over the years, there has been a steady decline in the camel population in the country, which as per the 19th Census (2012) was 0.4 million, decreasing by 22.48% compared to 18th Census (2007).

2.7. PIG

Pig is one of the most efficient feed converters among domesticated livestock. It is the only litter-bearing animal among meat-producing livestock having the shortest generation interval and high feed-conversion efficiency.

In India, pig farming has a special significance as it plays an important role in improving the socio-economic status of a sizeable section of the weaker sections of rural community. For a long time, pig rearing has, by and large, remained confined to this class of people. Big population of 10.29 million pigs in India in 2012 Census decreased by 7.54% as compared to 2007 Census. Pig husbandry is an important ovation in North Eastern Hill (NEH) states and tribal belts of the country providing livelihood to the poorest of the poor landless farmers. Various infectious diseases are major hurdle towards profitable pig production.

A co-ordinated Pig Development Programme was initiated in the country in 1955. Under this programme, schemes were drawn up for the establishment of bacon factories, regional pig breeding stations, pig-breeding farms/units and piggery development blocks. These schemes were further extended during the subsequent Plan periods. Piggery Development Scheme was strengthened by a special development programme taken up at the end of the Third Five-Year Plan. In preparing a model scheme for this programme, the Central Government took the decision of attaching the regional pig breeding stations to the bacon factories. The primary objective of the Regional Pig Breeding Stations was to acclimatise and further improve the breeding stock of exotic pigs imported to the country through selective breeding for distribution to pig-breeding farms/units and farmers for further multiplication.

A secondary objective was to supply pigs to the bacon factories from surplus stock. The pigs produced at the breeding farms/units were to be distributed in compact rural areas for grading up of the local stock. The bacon factories were to provide ready and remunerative markets for pigs raised and to produce pork and pork products for sale under ideal hygienic conditions. Approximately 12,000 improved pigs were distributed to the farmers till 1972. The major policy initiative was initially to replace the local population by improved exotic pigs.

The organisation of pig production in rural areas has been concentrated particularly around the bacon factories.

Establishment of eight bacon factories during the Second and Third Plan periods was the beginning of an attempt made for making wholesome pork and pork products available to the consumer.

An All-India Co-ordinated Research Project on Pigs was set up during the Fourth Five-Year Plan. The project, sanctioned in October 1970 started functioning early in 1971. The major objective of the project was to develop strains using imported breed stock suitable to regional ecology of India.

It has been also realised that in the north-eastern and other states, several local genotypes are available, which are preferred by the breeders. Pigmy hog is the most well-known breed. Similarly, in parts of Kerala, Ankamali pigs are prolific producers. These stocks have been given up in preference to producing new strains of pigs testing exotic genes for improved growth and feed conversion efficiency.

Piggery among the various animal farming models offers the most potential source of meat production as one of the efficient feed converters after the broiler birds besides contributing to the lowest socio-economic strata of this country. The creation of National Research Centre on Pig (NRCP) at Guwahati in 2002 during the 9th Plan was an outcome of these imperatives.

One of the major findings of the AICRP on pigs was that the replacement of local pigs by exotics should be two stage process and all the three genotypes; desi pigs, graded pigs and exotics need to be concurrently raised under different ecologies.

The NRCP and AICRP on Pig through selective breeding, improved AI practices and distribution of improved piglets to farmers and marketing avenues offer sustainable solutions to livelihood improvement of the stakeholders. Pig village concept has been popularised among local farmers.

Two phenotypically different strains of pigs identified have been registered as first indigenous breeds of India based on phenotypic, cytogenetic and DNA profile as per FAO recommended micro-satellite loci. An improved variety of cross-bred pig was developed by crossing Ghungroo (indigenous) with Hampshire (exotic breed). AI technique was standardised and transferred under field conditions to farmer communities for testing.

Unconventional feed resources for economic pig production were standardised by incorporation of azolla and colocassia were standardised. Feeding standards were developed and standardised for Indian raw materials.

Classical swine fever (CSF) outbreak was reported in India in early 1960 in Uttar Pradesh. Subsequently, this dreadful disease was reported from many states which necessitated development and refinement of sensitive and specific diagnostics. This included isolation of the virus from buffy coat cells in PK-15 and SK-6 cell-lines. Detection of viral antigen from tissue samples by direct immune-fluorescence antibody technique (FAT) as well as by immuno-peroxidase method was done. Recently, RT-PCR was developed at IVRI, which is more sensitive. At Madras Veterinary College, Chennai, Mabs specific to E2 glycoprotein and Erns were produced and tested for application in easy-to-use diagnostics for detection of CSF virus antigen. A potent swine fever virus vaccine was developed (2006–07) at IVRI after serial passages of virus in PK-15 cell culture and tested in field (Anon 2012). Japanese encephalitis virus was isolated from pigs in Kolar region in 1987.

The HSADL, Bhopal developed an enzyme immunoassay and nucleic acid based diagnostic tests for detection of porcine reproductive and respiratory syndrome and a diagnostic technique for the detection of transmissible gastro-enteritis virus infection in pigs in India. Swine influenza (H_1N_1) reported in pigs is a threat both for pigs and humans. In recent years, new diseases/infections reported include porcine circovirus-2, porcine parvo virus, swine hepatitis E, Porcine epidemic diarrhoea and Torque Tenosus virus.

2.8. YAK

The yak, also called the grunting ox *Bos (Poephagus) grunniens* Linn (Hindi—Ban chour for wild; Chour-gau for domesticated) is an animal adapted to high altitudes. The yak is a native of the Tibetan Plateau and the surrounding countries in Central Asia. They are found at altitudes of 4,300–6,100 metres above sea level. Wild yaks live in large herds in cold, high-altitude, desolate regions where other livestock do not survive. Grown up bulls make small groups of their own. Winter is the breeding season and calves are born during the following autumn season.

In India, yaks are found in Ladakh Valley of Jammu and Kashmir; Pangi, Sangla, Lahaul and Spiti valleys of Himachal Pradesh, Uttarakhand and in small numbers in Arunachal Pradesh. These animals generally live along the snow line moving up during summer and down during winter. They can live on grasslands at high altitudes which other species of livestock are unable to utilise. The present population of yaks in India as per 2012 census is 0.07 million, a reduction by −7.6% over the 2007 Census.

Yaks are used to a considerable extent for cultivation, riding and as pack animals. They are also used for pulling small carts over small distances. Yaks are particularly valued as a source of supply of meat, fibre and milk. Female yaks in India yield 1–2.7 litres of milk testing 5%–6% fat. Butter and cheese are also made of yak milk, particularly during summer when the animals are usually taken to higher pastures. Cross-bred females born out of ox female by yak male produce 3–4 litres of milk and these constitute a source of commercial milk production in some areas of Ladhak and *Lanhal* of Spiti. Yak skin made into rough leather is used for various purposes. Hair, with or without other animal fibres such as goat hair, is made into ropes or woven into bags and blankets. The tail tufts, particularly if white are valuable items of commerce and are greatly prized for ceremonial purposes in temples and monasteries. Dung of yaks is used as manure, and in treeless plateaus this is the only source of fuel. Crossing of yak males with cows is quite common. Females born out of such crossing are fertile but males are infertile and hence can be used only for work and meat. Cross-bred males are known as *dzo* in Tibet and *choru* in Himachal Pradesh. Females are called *dzomo* and *chaori*, respectively, in the two places. Different names are, however, used for such crosses in different areas. The dzomo is supposedly a higher yielder of milk than yak female or the female oxen and is popular as a dairy animal in areas where yaks abound. There is a popular belief that *dzos* and *dzomos* are as well adapted to high altitude areas as yaks but are better adapted to lower altitudes. It is possible to cross the dzomo with either yak or cattle. The males out of such crosses are infertile like dzomos. Female crosses are not as easy to maintain as *dzomos*. They are also poorer milk producers.

The yak is one of the least studied domestic animals. The authenticity of the information obtained by oral enquiry is questionable. The information reported in this book is based on several resources including All India Coordinated Project on Yaks. Selection of male breeding yaks was based entirely on phenotypic body characteristics of the animals. There was no proper record-keeping to assess the results.

The Himachal Pradesh Government has established a yak-breeding farm near Chitkool in Sangla valley. It has recently been handed over to the Agricultural University, Himachal Pradesh for research and development.

Improvement of Yak (Poephagus grunniens L)

The yak *(Poephagus grunniens* or *Bos grunniens*) has long documented history closely associated with the culture and social life of the pastoral people of the high-mountainous cold regions and best grazing lands of Asia. Yak is regarded as one of the world's most remarkable domestic animals that thrive in conditions of extreme harshness and deprivation while providing livelihood support and nutritional security for highlanders, especially farmers (*Brokpas*) on remote hills of Arunachal Pradesh (Fig 2.1). Yaks are reared on pastoral system by the tribal people on high hills where no other livestock or crop production is possible. They can sustain severe cold (up to −60°C) and can travel comfortably on snow bound steep hills under hypoxic conditions. They are multipurpose animals which provide milk, meat, fibre, dung and are also useful for transportation of household goods. The milk yield of yak varies between 1–2 kg/day. Yak husbandry is highly remunerative with a high cost-benefit ratio up to 4.1, when yaks are reared on highland pasture during summer on alpine pasture (above 13,000 ft msl) (known as alpine pasture).

Fig. 2.1: Typical yak rearer (Brokpa) in Arunachal Pradesh

Progressive decline of yak population in the 60s and 70s became a concern, which led the ICAR to establish the National Research Centre on Yak (NRCY) in the year 1989 with the mandate '*Research in conservation and improvement of yak with the aim of making the institute a global leader in yak science and generating cost effective technologies for socio-economic uplift of yak husbandry centric highlanders'*. Some of the important projects on genetics, breeding behaviour and reproduction, health management and value addition are currently in progress. This species-specific institute has in the first phase made in-depth study on traditional yak rearing and developed strategies and programs of intervention using science and technology for future plans for overall development of sustainable yak husbandry.

The institute conducts focused research in the areas of yak genetics and breeding, nutrition, physiology, reproduction, health, management, product technology and extension for its composite development in the Indian Himalayan region.

Breed Characterisation and Conservation

The physical appearance of yak was found to be polymorphic and a total of nine different horn types were discovered in yak tracts of Arunachal Pradesh alone. Yak types were differentiated for the first time based on its physical conformation. The karyotype of both male and female were produced (Fig 2.2). On the basis of physical characteristics and genetic mapping, these were classified into various categories. Five categories of yak—namely, the Common type (used for milk and meat), Bisonian type (used for draught), Bare back yak (used for milk and draught), Hairy forehead yak (used for milk and draught) and White yak (used for milk)—have been identified (Fig 2.3). The common types of yaks are docile, predominantly milch type and constitute approximately 60% of yak population in Arunachal Pradesh. The Bare back type is a dual type having good milking and draught characteristics.

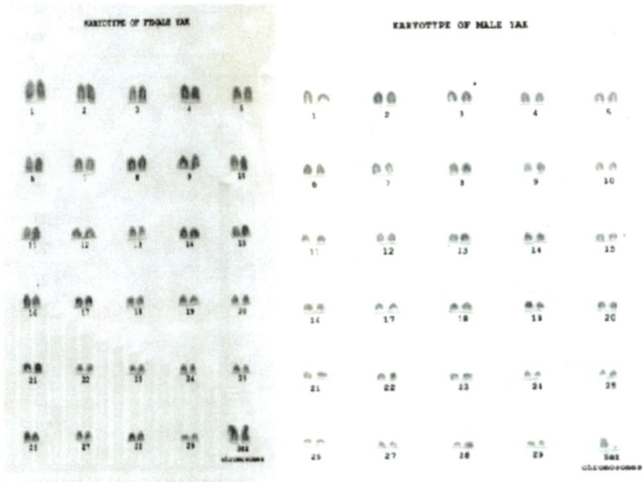

Fig 2.2. Karyotypes of female and male yak

Fig. 2.3: Five categories of yaks in the population of Arunachal Pradesh (Clockwise: White, Hairy forehead, Bareback, Bisonian and Common types)

Conservation of superior yak germplasm

To address the issue of conservation and multiplication of germplasm of yak, artificial insemination (AI), embryo transfer technologies (ETT) with *in vitro* fertilisation (IVF) technique were developed and are in use (Fig 2.4) in breeding of yak population.

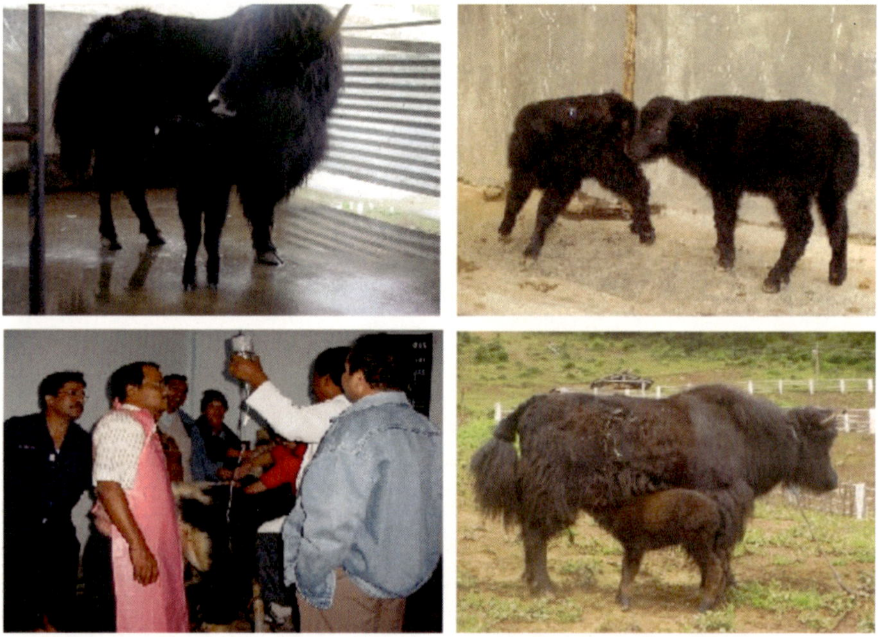

Fig. 2.4: Reproductive technologies developed at NRCY (Clockwise: First yak calf through AI, Calves born through estrus synchronisation, Non-surgical ETT, First ETT calf-MISMO)

Genetic interventions

Toll-like receptors (TLRs), a group of trans-membrane proteins play an important role in the recognition of components of pathogens and subsequent activation of the innate immune response, which leads to development of adaptive immune responses. Presence of polymorphism in TLR2 gene located at relatively conserved domains of yak genome is significant. Genetic variation in these loci may be useful in selection of yaks for disease resistance, which needs in-depth study and confirmation.

Yak Nutrition

The centre has taken up mapping up of feeds and feed materials used by yaks. The mapping of nutrition type trees and bushes have been done and besides these several technologies and best practices were developed by the centre for rejuvenation of degraded high-altitude pastures by promoting temperate grasses and legumes. Awareness was created among the yak husbandry centric high-landers about poisonous plants for safe grazing of their animals.

Nutritional intervention in winter

During summer yaks migrate to the high altitude of 8,500–14,000 ft above msl or even upto 16,000 ft above msl). In winter, yak herders come down to lower altitude and during this period yaks lose 25%–30% of their body weight due to shortage of feed. To overcome inadequate fodder resources during winter, different complete feed blocks (CFB) were developed utilizing locally available crop residues *viz.* corn stover, finger millet straw along with edible tree leaves, concentrate mixture and molasses (Fig 2.5).

Fig. 2.5: CFBs prepared NRCY using locally available resources of Arunachal Pradesh

Area-specific mineral mixture was prepared when soil samples from 14 yak inhabiting pockets revealed the deficiency of some micronutrients. The mixture made of zinc, copper, cobalt and manganese in the ratio of 40:20:2:1 was found to be essential for yak production, health and immunity.

Yak husbandry is facing several challenges due to transhumance system of farming, grazing tax and degradation of the natural grasslands. To mitigate the issues and to enhance production potential of temperate grasses and legumes a package was developed for high land pasture management. Technology for conservation (ensiling) of surplus forages (grass and tree leaves) during rainy season for use during winter was developed and popularised.

Yak Health

Identification of infectious agents/toxicants causing loss to yak husbandry

Virulent genes of enteric pathogens isolated from yaks were identified using molecular techniques and compared with other organisms. Molecular signature of *Cutaneous papillomatosis* of yaks was done for the first time in the world. The first reports of *Mammomonogamus laryngeus* and *Cutaneous papillomatosis* in yak were identified for the first time in Indian yak. Cryptic stage of *Babesia bigemina* was detected by standardised molecular approach, which could detect 10 pg of parasite DNA.

Alkaloid containing poisonous plant such as *Senecio crysanthomoides* was identified having fatal effect on yak. The pathway of *senecio* poisoning was explored through pro-inflammatory, oxidative and pro-apoptotic signalling methods. Using herbal preparations lead to innovative research in yak health includes development of poly-herbal wound healer, fly repellent, herbal eye drops for treating pink eye disease, etc.

Yak Reproduction

Notable research accomplishments include standardisation of AI, ETT and 'ovsynch', protocol for synchronisation of oestrus with high conception rate. Both AI and ETT were standardised and the first female yak-calf was produced through ETT in 2005 and the first calf through AI in 2006.

The first test tube yak calf was born at NRCY on 15 July 2013 (Fig 2.6). The male yak calf, weighing 19 kg, was named **Norgyal**, is the result of transfer of cryo-preserved yak embryo produced through *in vitro* fertilisation of oocytes retrieved through ultrasound guided ovum pick up (OPU) technique from donor yak. The achievement is a breakthrough towards conservation and multiplication of elite yaks in the country and this advanced reproductive biotechnology will help towards *in situ* and *ex situ* conservation of yaks, the most precious animal of highlanders.

Fig. 2.6: The first test tube yak calf, NORGYAL born at NRCY on 15th July 2013

Technology developed and refined

Technologies were developed for yak production and health and for meeting the nutritional requirements of resource poor farmers of North Eastern states. Yak production was improved by use of CFB. Intervention of pink eye disease in yak was accomplished by in house preparations of herbal drugs. Value added milk, meat and wool products having good market value (Fig 2.7) were developed.

Fig. 2.7: Technologies developed for value addition of milk, meat and fibre products

Value Added Product Development

Recipes for value-added yak products such as 'churpi', a wet soft cheese, and butter, both made from fermented milk were standardised. The other important value additions made are 'low-fat paneer'(1% fat and 2% fibre), meat sausages (60%) and pork (20%) and pork fat (20%), milk whey beverages incorporated with kiwi fruit pulp, improved yak fibre technologies for making 'chawar', cap and carpets/wall hangings, etc.

The yak farmers are trained in modern husbandry practices and skill development for better marketing of milk, meat and fibre products. Efforts are being made for alternate use of yaks as pack animals for use in military and paramilitary forces and riding for encouraging tourism. Development of modules for combating winter crisis of nutrition, disease containment, value addition of yak products, ensuring more production and ensuring food safety measures and attracting more farmers in yak husbandry to make it more profitable and understanding unique traits using molecular biological techniques are some of the priority research areas of this institute with the aim to uplift economy and livelihood of the poorest of the poor highlanders.

Research on characterisation of yak breeds, AI and other services are being used to improve the yak production which has helped the yak keepers in reaping rich benefits.

Broad Future Research Projections - Focus 2030

Flagship programme: Strategic feed supplementation for growth and milk production in yak during winter (Basis: The yak herdsmen starts upward migration along with their yaks to alpine pastures for summer grazing and during winter move down to mid altitude and graze the yaks in adjoining pasture lands, simultaneously grazed by science interventions with to reduce the body weight of 25% to 30% as at present. Other livestock and yaks lose body weight 25%–30%). Mostly genetic improvement and nutrition takes priority followed by diagnosis and treatment.

Fig. 2.8: Indian Yak

2.9. MITHUN (*Bos frontalis*)

Origin and History

Mithun (*Bos frontalis*), a unique bovine species is mainly found in four different states of India, *viz.* Arunachal Pradesh, Nagaland, Manipur and Mizoram. This heavily built bovine species is mostly reared under free-range system in the tropical evergreen rain forests with a natural living approach. This geographically distinct woodland animal is today regarded as a socio-cultural emblem and plays a vital role in the overall economy of the tribal people of this region. It is well adapted to humid climate and hilly terrain at an altitude of 300–3000 m above mean sea level (MSL). Mithun is extremely efficient browser on steep hilly slopes compared to other animals. This unique species is believed to have been domesticated for more than 8,000 years. Mithun is the semi-domesticated form of wild gaur (*Bos gaurus*). Nevertheless, at present both mithun and gaurs are taxonomically grouped into the species *Bos frontalis.*

Mithun is a large ruminant. It is presently confined in North-Eastern Hill Region, *viz.* Arunachal Pradesh, Nagaland, Manipur and Mizoram. According to the All-India Livestock Census 2012 their population was estimated at 0.29 million. Mithun has got great potentiality for meat and milk under special topographic and climatic conditions of north east.

Mithuns are integral component of agriculture in upper hills of North-Eastern hill states, particularly of tribal communities. They have an important place in the economic, socio-cultural and religious life of the local tribes particularly of Arunachal Pradesh, Manipur, Mizoram and Nagaland. The animal is semi-wild in nature but fully domesticated and owned by the farmers.

Fig. 2.9: A mithun bull

Mithun R&D

National Research Centre on Mithun (NRCM), Jharnapani, Nagaland, established in the year 1988, is engaged in identification, evaluation and characterisation of mithun germplasm, its conservation and improvement for meat and milk, and to act as a repository of germplasm and information centre for this unique woodland bovid, which is regarded as a socio-cultural emblem of the tribal population in the NEH region of the country at an altitude ranging from 300 to 3000 m above sea level.

At present, the animal is reared by the tribal community mainly for meat purpose, besides other religious reasons. It is considered as a sign of prosperity. The status of an individual in his community is determined on the basis of the number of mithun heads he/she possesses.

The peculiar characteristic of the animal is its ability to graze on slopes of hills where other animals cannot reach. Mithun prefers cool forest ranging from 300 to 2,700 m altitudes above mean sea level and is well adapted on the steep slope of the forest.

Through mithun production biomass of inaccessible deep forest can be harvested into the best quality meat and milk. Mithuns are also closely associated with the preservation of forest, conservation of soil and water resources. These animals enrich the soil fertility with dung and urine. The soil is otherwise acidic and deficient in micronutrients.

Mithun Nutrition

Mithun keeping is practised under free range livestock production system at higher altitudes between 300 and 3,000m (MSL). They are quite hardy and thrive well on tree leaves, jungle grasses etc. A survey was conducted to identify the elite browse and tree species, biomass production and nutritive characteristics of forage resource development programme (Rajkhowa 2008). During this survey 79 types of shrubs, 128 types of trees, 107 types of grasses and 18 types of herbs were collected. Of these, only 29 shrubs, 74 trees, 55 grasses and 8 herbs could be identified by the Botanical Survey of India (BSI), Shillong. The samples were collected during monsoon. Of the total number of collected samples, only 70.37% shrubs, 80.4% tree leaves, 91.58% grasses and 83.30% herbs belonged to edible species. Mithuns are comparatively hardy and thrive well on coarse fodder of the jungle.

Useful nutritional interventions developed include development of area-specific mineral mixture, efficient drying of thick stemmed fodders, and feed block preparation by incorporating high moisture content agro-industrial by-products.

A total of 260 different forest based foliages were collected from different pockets of mithuns rearing states Nagaland, Arunachal Pradesh, Manipur and Mizoram. All the fodder samples were evaluated for nutritional quality. Voluntary feed intake and nutrient utilisation on feeding locally available tree leaves based total mixed ration were investigated. Comparative nutritional evaluation of different tree lopping and concentrate based total mixed ration was done. Effect of different fodder and concentrate based total mixed rations on nutrient utilisation and growth performances in calves were evaluated. In sacco degradability of forest based foliages available in Nagaland was studied. Effect of feeding tree leaves based total mixed ration on nutrient utilisation, growth performances and rumen fermentation patterns was investigated. Nutrient utilisation and nitrogen balance in adult animal fed on urea treated rice straw and concentrate based feed blocks was carried out. Important digestive enzymes in rumen fluid were estimated. The contents of different macro and micro-minerals in important locally available fodders were estimated.

Fig. 2.10: *Lagerstroemia* spp. (Crude Protein 23.1%)

Scarcity of fodders is a general scenario in the NEH region during winter. However, during rainy season availability of naturally grown fodders is abundant and underutilised. Underutilisation of fodder during rainy season is due to its abundance as well as non-existence of preservation practices in this region. Besides silage making, innovative drying method for fodders, especially the thick stemmed ones, which are very difficult to dry during rainy season were developed. Processing of high moisture content thick stemmed fodders mixing with low moisture content paddy straw reduces duration of drying time by 33.33% and transformed its good ingredients for preparation of feed blocks.

Breed Characterisation and Genetic Studies

The origin of this species has been studied using latest methodologies. Scientific characterisation of the mithun strains was done using traditional and molecular genetic approaches. Molecular markers were identified, which could be subsequently used for genetic improvement for meat/milk through MAS. They could be used for estimating the genetic diversity among strains of mithun. Four strains of mithun, namely, Arunachal, Mizoram, Nagaland and Manipur were identified using RAPD analysis. The gene sequencing of the exotic region of leptin gene of mithun was submitted to the Gene Bank (Accession No GQ 411537) (Anupama Mukherjee et al, Chromosome Research 22(3):427).

Genetic Studies so far reported have been concentrated on Chromosome Research.

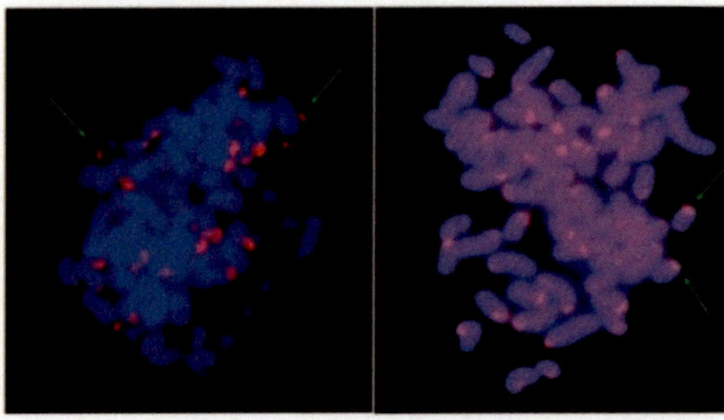

Mithun metaphase chromosome showing centromeric signals with rhodamine as a flurochrome in bovine probe

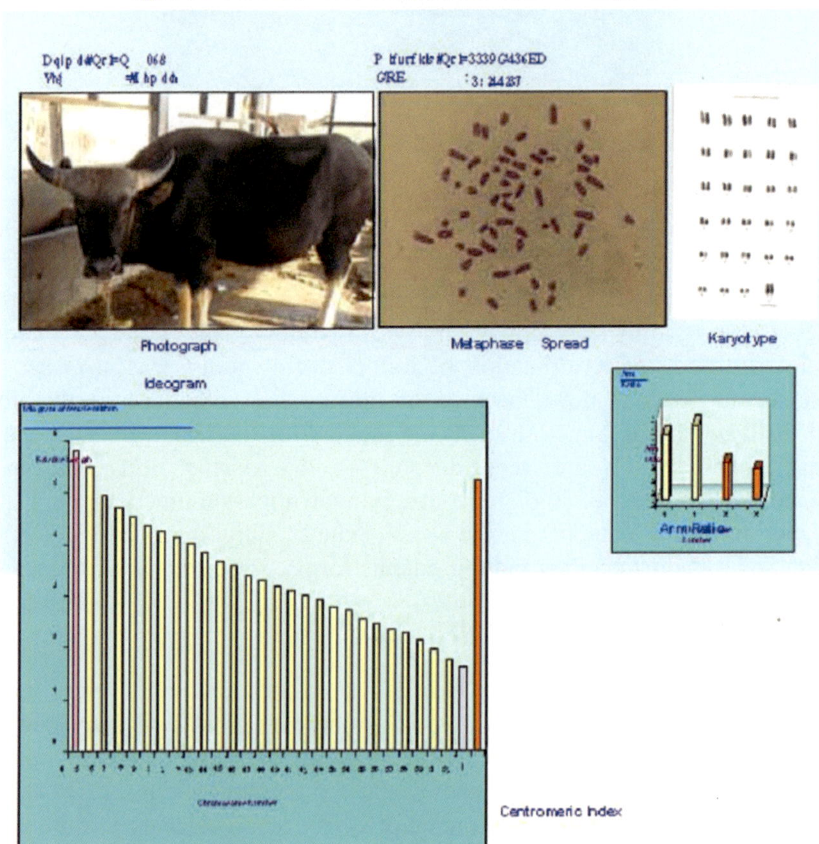

Fig 2.11 Cytogenetic Digital Album of a Mithun

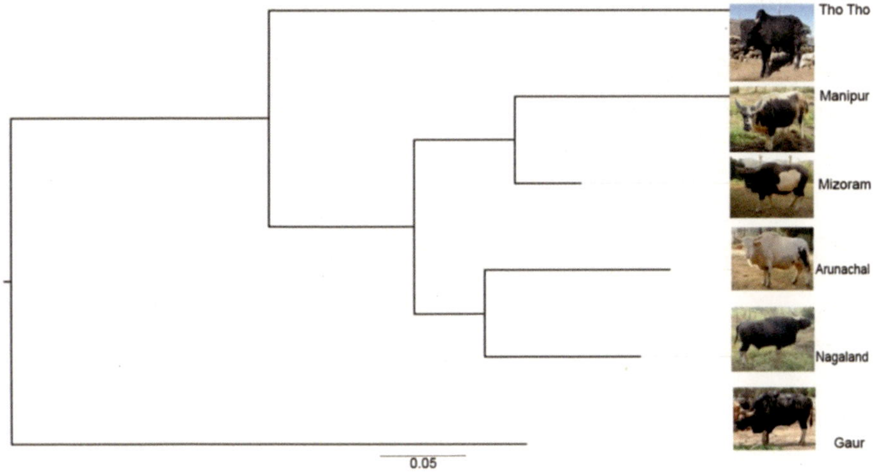

Fig. 2.12: Phylogenetic tree of Mithun

DNA Investigation with RAPD markers

Genetic studies were taken up using RAPD markers for estimating intergenic diversity. A total of 19 random primers were screened against all four genetic groups of mithun. Most of the primers (17 out of 19 primers) detected polymorphism between the genetic groups of mithun. The average number of bands per primer ranged from 8.1 to 9.3 in different genetic groups. The result also revealed that different genetic groups of mithun differed genetically. Arunachalee mithuns generally showed higher genetic differences with other genetic groups, particularly with Mizoram genetic group compared to others. The results are suggestive of genotypic differences among the strains (Mukherjee et al 2014).

Measurement of Genetic Relation of Mithun at DNA level with different Livestock Species using RAPD and MASA Markers

A total of 21 random primers were screened against five livestock species. The RAPD profiles generated by the polymorphic primers among various livestock species were utilised for estimating the inter-population diversity. The estimates of genetic distance between the livestock species ranged from 0.58–1.90. The results revealed maximum genetic similarities between sheep and goat (0.56), while cattle and sheep (0.15) showed least genetic similarity. The mithun showed maximum genetic similarities with buffalo (0.45) followed by cattle (0.41) and sheep (0.32). Least genetic similarity of mithun was observed with goat (0.25). One can safely guess that mithun's point of origin is related to cattle and buffalo linkage tree.

Three MASA primers were also screened against five livestock species. Out of them, two primers (OAT 18.2 and 033.15) produced highly reproducible amplification. The value of band sharing between livestock species, pooled over primers ranged from 0.35 to 0.72. The genetic distances were estimated as −1 n (BS) and pooled over the primers ranged from 0.31 to 1.04. The results revealed maximum genetic similarity between sheep and goat (0.72), while goat and mithun showed least genetic similarities (0.35). Mithun had almost same genetic similarity with cattle (0.66) and buffalo (0.63) as already shown with RAPD analysis (Mukherjee et al 2014).

Genetic characterisation of mithun using cytogenetics and microsatellites

The chromosome number in mithun is 58 (2n=58), which includes 28 pairs of autosomes and one pair of sex chromosomes.

The normal karyotype of mithun showed that the first pair of autosome is the largest sub-metacentric chromosome and the rest autosomes are acrocentric. Among the sex chromosomes, the X-chromosome is sub-metacentric. Y-chromosome is the smallest metacentric chromosome such as*Bos taurus*. Cytogenetic analysis was carried out for both male and female mithun based on the morphometric parameters, *viz.* relative length, range, centromeric index, arm ratio and morphological index.

The band patterns of chromosomes were also studied by conventional G-, R- and C- banding technique. The karyotype generated based on these band patterns was first reported in mithun (Mukherjee et al 2014). The R- banding confirmed that the X chromosome is the second largest sub-metacentric chromosome, whereas the largest is the first pair of autosome. The C- banding highlighted the centromere position in the metaphase chromosome. Based on all these analysis, a digital cytogenetic album of mithun was constructed.

Cattle micro satellite markers were utilised for analysis of genetic diversity of mithuns, gaur and Tho-tho cattle and 19 (63%) microsatellite markers successfully amplified the genomic DNA of mithun and wild gaur; where 15 (79%) of these loci were highly polymorphic in mithun and gaur, respectively, with polymorphic information content (PIC) greater than 0.50. Phylogenetic tree was prepared to show the genetic status of four mithun strains with wild gaur and Tho-tho cattle.

Phylogenetic tree does not show any large genetic relationship with Gaur while it seems closer to Tho-tho cattle.

Mithun Reproduction and Physiology

Data were generated on the productive and reproductive traits, the Arunachalee strain has comparatively better performance record especially in terms of growth parameters, while calving interval was lowest in Mizoram strain (349 days), suggesting that the Arunachalee and Nagaland strains of mithun could be explored for meat production and the Mizoram strain for both meat and milk production as it has excelled other three strains in terms of milk yield and calving interval. This needs further verification based on farm records.

The problem of detection of silent oestrus was resolved by adopting 'Ovsynch' protocol and controlled intra-vaginal slow drug-releasing device. The techniques developed include protocols for oestrus synchronisation for fixed time inseminations, standardisation of AI method, super-ovulation and ETT protocols. Embryo transfer technology was successfully used and the first mithun calf was produced on 12 May 2012 through ETT.

Oestrus synchronisation protocols for mithun were developed using single and double injection of prostaglandin F2α, co-synch protocol (GnRH-prostaglandin F2α combination), OvSynch protocol (GnRH- prostaglandin F2α-GnRH), was tried using controlled intra-vaginal drug releasing device (CIDR), OvSynch ± CIDR and heat sync (using oestradiol cypionate; ECP). Ovsynch protocol of oestrus synchronisation was effective for heifers, anoestrous and cyclic mithun cows to synchronise heat and fixed time insemination. CIDR and OvSynch ± CIDR protocols were effective in post-partum anoestrous in mithun cows. Co-synch and double injection of prostaglandin F2α protocols are effective in cyclic cows to synchronise oestrus. Heat synch protocol of oestrus synchronisation made it possible for the mithun cows to manifest behavioural signs of oestrus prominently and was the best method for oestrus synchronisation in mithun.

Methods were standardised for preservation of mithun semen at refrigeration temperature using tris-egg yolk diluents. Cryopreservation of semen was also standardised using tris-egg yolk and citrate-egg yolk diluents with graded levels of glycerol. Seasonal variation in quality of semen was also studied.

Artificial insemination protocol was standardised and successfully used in the farm and field condition. In 2010 the first mithun calf was born in field condition (Khunoma village) through AI.

Super-ovulation protocol was standardised in mithun using FSH. Protocol standardised for super-ovulation and embryo transfer resulted in the birth of Mohan, the first mithun calf born through embryo transfer technology (ETT) from cryopreserved embryo at the institute on 12 May 2012. The embryo was preserved in liquid nitrogen (−196°C) for 100 days. This is the third successive birth of mithun calf using embryo transfer technique. The first calf named as Bharat was born in 27 March 2012 and the second calf (Prithvi), which was born on 11 May 2012.

Mortality: The problem of calf mortality due to weak mother-neonate bonding/attachment (called mismothering behaviour) was addressed with encouraging results by intranasal spray of oxytocin that can induce mother-neonate bonding/attachment in mithun and it is also acclaimed to be the 'world-first' report.

Antimicrobial effect of mithun lactoferrin (mLF) was determined on different pathogenic bacteria. Mithun lactoferrin could inhibit the growth of *Staphylococcus aureus* and *Escherichia coli* by 53% and 28%, respectively, and was comparable to ampicillin and kannamycin therapy.

Livestock Production and management: Biochemical properties of mithun rumen fluid were studied. Total and differential counts of mithun rumen ciliates were established and fourteen different species of rumen ciliates were identified. Changes in milk production and different milk constituents during different lactation stages were established. Levels of different macro and micro-elements in mithun plasma and locally available fodders were estimated, Absorption efficiency of colostral immunoglobulins in neonatal mithun calves was investigated. Profiles of different immunoglobulins in serum during first three months of age in mithun calves were established.

Endocrine control of natural estrous cycle and estrous behaviour was established. Methods for preserving mithun semen at 4°C and ultra-low temperature (−196°C in liquid nitrogen) were developed.

Artificial insemination (AI) technique using cryopreserved semen was standardised and birth of the AI mithun calves could be achieved for the first time in our country during 2006.

Draught power: Draught capabilities of mithun bull were evaluated. Preliminary investigation indicated that the average draught power of mithun is 0.60 KW/h.

Value added product development: These include preparation of cheese from milk, development of leather processing technology and leather products from hides. Other technologies developed and provided to mithun farmers include determination of age by dentition pattern.

Mithun health: Mithuns reared under intensive system of rearing were surveyed and sampled for IBR, TB and brucellosis using ELISA and virus neutralisation test. A total of 650 animals were surveyed in eight villages in Nagaland and Mizoram and approximately 250 serum/blood samples were collected form mithuns reared under free-range condition for sero-monitoring of different diseases; 150 mithuns reared under intensive system of rearing were also included in the study. Bacterial flora of the lower reproductive tract of mithuns was studied and 16 different bacteria were isolated by culture. These were studied for their anitbiogram.

Antibiotic sensitivity pattern of clinically important bacteria isolated from mithun calves was established. Various sero groups of *E. coli* have been isolated from different types of samples of mithun calves. Virulent genes were detected from *E. coli* (isolated from diarrhoeic faecal samples as well as cases of fibrinous pericarditis) using PCR technique. Screening of faecal sample of mithun from free-range and farm condition revealed *Trichostrongylus* sp., *Haemonchus contortus*, *Oesophago stomum* sp. *Trichostrongylus* sp., Moniezia expansa, M. benedeni, Cooperia sp. and *Trichuris* sp. The species of *Toxocara vitulorum*, *Strongyloides papillosus* and *Bunostmum phlebotomum* and *Haemonchus contortus* were the major parasites causing anaemia and mortality. Herbal and chemical drugs have been identified. Standardisation of dose of these formulations was done for the prevention and treatment of GI parasites as well as tick infestation.

The animal is highly susceptible to foot and mouth disease and Broncho-pneumonia due to *Psedumonas*. Approximately 15 animals from two herds in Nagaland (out of a total of approximately 450 animals) died in the outbreak.

Other studies of interest

Genetic characterisation of *Fasciola gigantica* was done based on two nuclear spacer sequences (ITS-1 and -2), mitochondrial gene (NadH1) and ribosomal genes 18S and 28S. Likewise, molecular identification of *Mecistocirrus digitatus* was carried out based on two marker genes namely of beta tubulin gene iso type 1 and internal transcribed spacer 2.

Technologies ready for commercialisation

Area-specific mineral mixture for mithun: In an analytical study of mineral content of fodder, tree leaves showed that the levels of calcium, magnesium and iron contents were higher whereas, concentration of phosphorous, copper and zinc were lower than the standard levels. Based on this study, an area specific mineral mixture was formulated. Field study revealed enhancement of growth and reproductive performance in mithun.

Low-cost high-nutritious feed block for mithun: Scarcity of fodders is a general scenario in the NEH region during winter. However, during rainy season availability of naturally grown fodders is abundant and underutilised. Underutilisation of fodder during rainy season is due to its abundance as well as non-existence of preservation practices in this region.

Besides silage making, the institute has developed innovative drying method of fodders especially the thick stemmed ones which are very difficult to dry during rainy season. Processing of high moisture content thick stemmed fodders mixing with low moisture content paddy straw reduces duration of drying time by 33.33% and transforms it good ingredients for preparation of feed blocks.

The feed blocks were made from locally available feed ingredients. Tree leaves, straw and concentrate feeds were mixed together and pressed by using feed block making machine. Similarly, feed blocks were also made from paddy straw and spent grains (a breweries industry by-product) with high moisture content (75%–80%). Simultaneously, methods were also developed for drying high moisture content by products (spent grain) more efficiently during rainy season. Performance of mithun was good with average daily growth (ADG) of 500 g on tree leaves based feed blocks and 553.6 g from spent grain based feed blocks as against ADG of 396.2 g in control feeding.

Impact of Researches in the Field

Due to the technologies and best practices developed/adopted and transferred by NRCM to the farmers, the declining trend in mithun population has been reversed. As per the latest livestock census conducted in 2012, the mithun population registered a growth rate of 12.98 % over the previous census in 2007, i.e. a growth rate of 2.6% every year. The estimated population stands at 0.29 million today. Eight patents filed for the technologies developed include method for processing rabbit pelts and fur, an energy efficient post tanning process for mithun hides, a novel method of chrome tannage of mithun hides with glyoxalic acid, an improved method for chrome tanning mithun hides, a process of making leather without any large machinery, a method of removal of chromium from chrome liquor obtained after processing of mithun hides, and a method of processing rabbit fur on leather.

Fig. 2.13: Mithun Nutrition

Fig. 2.14: Bharat, first Mithun calf born through ETT

The harness fabricated for mithun bulls for pulling load/cart

Mithun bulls are pulling load on earthen path

Fig. 2.15: Mithun Bulls for Pulling Load

Fig. 2.16: Area-specific mineral mixture developed in the Institute

Fig. 2.17: Feed block machine with feed mixer

Fig. 2.18: Preparation of feed block and a Mithun feeding on a feed block prepared in the Institute.

2.10 RABBIT

Rabbits are found in India both as wild animals and as domestic pets. Until recently, rabbits were mainly raised for laboratory purposes, but are now being kept as a domestic farm animal for production of wool, meat and fur. Rabbit population in India was 0.593 million in 2012, showing an increase of 39.55 % over 2007 figures.

Rabbits have a high feed-conversion ratio, consume a large amount of forage from diverse origins and so can be reared on small amounts of concentrates. They can be adapted to any set of circumstances, right from a kitchen garden to a large intensive commercial enterprise. Because of the high reproduction rate, rabbits produce meat all round the year. The meat is pearly white, is low in fat and cholesterol, can be eaten in all seasons, and does not have any religious implications. Its fur has good market value. Angora rabbits produce wool of superfine quality, suitable for producing highly valued woollen apparel such as shawls and mufflers.

Angora rabbits are of many types. The differences are mainly in the amount of wool produced and the percentage of guard hair in the wool. German Angora is the best and annually yield 1,000–1,200 g of wool with a very low amount guard hair. Besides the popular White Angora, there are several colour Angoras which produce wool of a variety of natural colours. Such wool, however, has a limited market.

There are approximately 35 internationally recognised meat- and fur-type breeds of rabbit. The main differences are in the final or adult body weights and type of fur they produce. Rabbits have 22 pairs of chromosomes.

Areas having 15–20°C temperature and 55%–65% relative humidity ranging throughout the year are ideal for rearing rabbits, particularly Angora. Meat rabbit is more adaptable to stress and can tolerate temperature as high as 35–40°C. However, under such conditions the indoor temperature must be artificially lowered to 30–32°C.

Hilly regions are suitable for raising both Angora and meat rabbits but plains are suitable for growing meat rabbits only. Angora rabbits were introduced in Kangra Valley in 1974 by a private breeder for increasing livelihood opportunities of small holder hill farm women. To support this activity, Garsa station of the CSWRI started a rabbit unit.

Research work on breeding rabbits for meat and fur was initiated in 1978 with the import of New Zealand white rabbits. Subsequently, addition of four strains, three of meat type, *viz.* Soviet Chinchilla, White Giant and Grey Giant from USSR in 1979, several fibre types (German white, British and Russian were also introduced) and several research projects on breeding rabbit for meat were taken up under the AP Cess Fund Scheme of the ICAR. This was done under the leadership of Dr. B.K. Soni and implemented by Dr. R.M. Acharya, Dr. C.L. Arora and Dr. J.M. Mahajan.

The CSWRI institute under the ICAR and its centre at Garsa adopted the technologies from Europe and Soviet Union for rabbit raising and perfected a management and health care system. Several crosses were also made to generate new genotypes most suited to Indian climatic conditions, ecology and way of life.

At Garsa, Angora crosses of German (GA), British (BA), Russian (RA), and their cross-breds GBR and GRA in reproduction flocks produced 524, 390, 392, 257 and 420 g hair in five clips except GBR which was shorn four times. Males of GA, BA, RA and GRA yielded 546, 430, 381 and 417g wool, and females produced 620, 430, 417 and 487g, respectively.

A feed efficiency experiment was conducted with 35 male and 33 female rabbits belonging to 16 genetic groups from 5th to 24th week of age. The average feed intake per kg body weight gain from 5 to 12 and 13 to 24 weeks of age was 4.8 and 9.7 kg, respectively. Thus, rearing of bunnies beyond 12 weeks of age appeared to result in higher feed costs.

2.11 POULTRY

Poultry keeping in India is probably as old as its civilisation. Red Jungle fowl found in India is believed to be the progenitor of all modern domestic breeds of poultry. Despite this, poultry farming remained a backyard venture until recent years. India possesses the largest diversity in poultry germplasm. This is yet to be exploited by scientists. They have genes for survival under unfavourable environments, disease resistance, heat tolerance and adaptability to diverse systems of life support. One breed has a melanin gene which produces black muscle protein as against red, found normally. Naked neck gene has been used to develop nude chicks. Indigenous poultry does have all these advantages but is not commercially viable. Lack of interest in poultry improvement was due to low productivity of indigenous birds, low prices for eggs and meat and inadequate knowledge of husbandry practices to make it a financially worthwhile enterprise.

During the last three and a half decades, the entire scenario of poultry farming in the country has changed. It is now recognised as an organised and scientifically managed industry with tremendous employment potential to fight poverty and malnutrition. Thousands of poultry farms, both small and large, have come up all over the country and indigenous desi birds have given way to highly specialised layers and broilers. A network of hatcheries was established both in public and private sector throughout the length and breadth of the country to supply commercial hybrid chicks. Most of the private hatcheries receive their grandparent/parent stocks from internationally reputed breeders, thus ensuring a wide choice in the availability of commercial layer and broiler stocks for commercial exploitation. Basic breeding programmes were initiated to achieve self-sufficiency in the availability of pure line stocks for commercial use.

Indian poultry industry today is a mixture of several systems of management: starting from traditional backyard system, keeping through semi-intensive, intensive and intensive to modern space age technology. The latter has evolved from the former and developed itself into a scientifically based agro-industry. The modern sector of the industry has achieved significant growth during this short period. The growth of the industry has not only been in size but also in sophistication, quality and productivity. Egg production has recorded an average growth of 7%–8% during 1971–1995. Although late to come, broiler industry averaged a growth rate of 15%–20% per year during the same period. Consequent to increase in productivity, the per capita availability increased to 30 eggs and 566 g of poultry meat in 1994 compared to only seven eggs and 155 g of poultry meat in 1961 with a record 74.75 billion eggs in 2013–14 per capita/year/egg availability is 61 against the ideal requirement of 180 eggs. With 2.39 MMT broiler meat production in 2013–14, per person/year availability of poultry meat is 2.8 kg against the requirement of 11 kg. There are about 60,000 poultry farms under intensive production system using latest technology and inputs either developed locally or imported. Their production coefficients have been as high as that in developed countries. Of the total egg and meat produced at present, 60%–65% comes from intensive modern production system. Approximately one lakh poultry farms with 25–250 birds exist in rural areas. These are either raised on intensive or semi-intensive scale. The indigenous native fowl and their crosses continue to be the mainstay of rural poultry farms, whereas modern hybrid layers and broilers are exclusively used under intensive system of poultry production.

The primary business of poultry keeping has given rise to several supporting and allied industries such as egg and meat processing, compounded feed plants, units manufacturing, equipment and machinery, pharmaceuticals and biologicals enterprises.

All this started with the establishment of a Poultry Research Section at the Indian Veterinary Research Institute in 1939 on the recommendations of the Royal Commission of Agriculture. This was upgraded to the status of a Division in 1951. This was the first major step for promotion of Poultry Research and Development in the Indian subcontinent. The pioneering work done by the Scientists of this section under leadership of A.J. McDonald and Dr. S.G. Iyer, officers in-charge of the Poultry Research Section, not only helped to make poultry farming a viable enterprise but also paved the way for further research and development. Establishment of this Section also provided moral and scientific support to Indian Poultry Club and its mouthpiece 'Indian Poultry Gazette', the only poultry periodical being published in the country at that time. Establishment of the Poultry section at the IVRI also served as a catalytic agent for establishment of several military and civil poultry farms for multiplication and supply of exotic pure-line stocks for promoting poultry production in the country. Poultry research and education infrastructure was further strengthened with the establishment of agricultural universities in different states.

The indigenous *desi*-fowls were the mainstay of Indian poultry farming prior to 1960. Genetic improvement of these stocks utilizing appropriate breeding technology, therefore, received major emphasis in early stages of poultry development in the country. Although encouraging results were reported from conventional methods of selection and upgrading by Iyer and his co-workers from the IVRI, large-scale importation of elite poultry stocks in the late 1950's and early 1960's in the form of grandparents by the private sector and in the form of purelines by Government institutions to accelerate the growth and development of the industry caused a setback to the improvement of indigenous poultry stocks.

Broadly the decision to develop poultry industry was implemented by introducing the basic improved hybrid stock into the country as a replacement for new highly developed birds from Europe and USA.

Commercial poultry farming in the country began in early 60s when at the Government poultry farms, particularly Dr. J.N. Panda, and Mr Marcadale, FAO. Consultant demonstrated the efficacy of modern methods of poultry rearing. This finding received wide recognition by the poultry scientists and extension specialists of the newly formed agricultural universities and put this into practice to promote poultry production in the Indian sub-continent. American Peace Corps volunteers helped to transfer this technology to the villages to popularise poultry production in rural areas. It also created enough interest among the private entrepreneurs leading to establishment of franchise commercial hatcheries for production and supply of commercial hybrid chicks. The Government of India established four regional poultry farms at Bombay, Bengaluru, Bhubaneshwar and Chandigarh for multiplication and supply of the exotic stocks received under USAID programme. Several farms were also established in different states for similar purpose.

During the 60s, the ICAR sanctioned several poultry improvement programmes such as popularisation of baby chicks scheme, random sample units, improvement of poultry through family selection, diallel crossing for broiler production etc. to accelerate poultry research and development. Family selection programme was also initiated at central poultry breeding farms of Government of India. While the results realised from these projects did not measure upto the expectation, they did provide the confidence and paved the way for undertaking large scale poultry improvement programmes thereafter.

The ICAR sanctioned two co-ordinated projects during the Fourth Plan for the development of superior broiler and layer genetic stocks. Similar breeding programmes were also undertaken at four regional poultry breeding farms of the Government of India and in some agricultural universities such as the PAU, Ludhiana.

Under the programme of the Government of India, four large breeding farms were established at Chandigarh, Bombay, Bhubaneswar and Hesaraghatta, Bangalore. Establishment of chick sexing training school at the Central Poultry Training Institute, Bengaluru and random sample test units at Mumbai, Bengaluru, Bhubaneshwar, and Gurugram by the Government of India are also significant milestones in the history of poultry breeding research and development in the country. Results reported by random sample tests were of limited value in choosing the commercial stocks by the poultry farmers. It has however helped considerably the pureline breeders to develop competitive stocks to remain in the business of poultry breeding.

The import of grandparent (GP) stocks received a serious setback when the recommendations of Narayana Committee on import of poultry stocks were accepted by the Government of India in early 80s. The recommendations of the Committee, however, helped in liberalised import of pureline poultry stocks by the private breeders, for establishment of breeding programmes in the subcontinent. More than a dozen of pureline breeding programmes are currently under progress in the private sector for developing genetically improved layer and broiler germplasms for commercial exploitation. Due to this policy, India is not only self-sufficient in improved poultry germplasm; however, a net exporter of this germplasm to other countries. This effort was crystallised by Dr. B.V. Rao who was the architect of Venkateshwara Hatcheries group of companies. To him, the credit must go for the establishment of this industry in India.

The Central Avian Research Institute (CARI), the premier institute of poultry research in the country, was established in 1979 at Izatnagar by the ICAR to provide research, education and extension support to growing Indian poultry industry. Besides undertaking research in all disciplines of poultry science, the Institute imparts post-graduate education leading to M.V.Sc. and Ph.D. degrees in Poultry Science. It also offers National Diploma in Poultry Husbandry. The CARI has also actively been engaged in conducting specialised training programmes of various durations on different aspects of poultry production. Recently, the ICAR has established the 'Project Directorate on Poultry' (PDP) at Hyderabad to conduct and monitor research programmes for poultry breeding. A germplasm centre of elite layers and broilers has been established at PDP, Hyderabad for their improvement, multiplication and supply to commercial users. The maintenance, multiplication and supply of control populations for egg and meat have also been entrusted to PDP, Hyderabad. The Central Institute on Poultry Production and Management, Bengaluru, functioning under the aegis of the Department of Animal Husbandry and Dairying, Ministry of Agriculture, Government of India also provides training on various aspects of poultry production for various officers of state governments.

Improved commercial genetic stocks developed so far owing to these efforts in the public sector (ICAR-DAH-SAU) include ILI-80, ILM-90, HH-260, KALINGA Brown, CARI Gold and Punjab White of commercial layers and B77, IBL-80, IBB-83 and IBI-91 of commercial broilers. Most of these layer stocks have topped random sample tests and are comparable in their performance to the commercial layer stocks available in the country from private hatchery sources. The commercial broiler stocks developed and released have also given comparable performance with those of industrial broilers with respect to feed conversion efficiency and viability to the age of marketing. With respect to body weight, however, they are 200–300 g poorer than industrial broilers at the age of marketing as expected since colour broilers do not grow as fast as industrial broilers, though they are more efficient to withstand sub-optimal conditions.

As feed alone accounts for 60%–75% of the total cost of poultry production, efficiency in feeding is the major economic and management concern for raising poultry in the country. The research efforts in the area of poultry nutrition, therefore, were primarily directed to (i) determine the nutrient requirements of different classes and age groups of poultry to optimise performance in tropics, (ii) identify alternate feed resources and their evaluation for preparation of economic poultry rations, (iii) devise suitable processing techniques for improving the nutrition value of promising non-conventional poultry feeds, (iv) determine the effect of aflatoxins on the performance of layers and broilers and (v) develop methods for detoxifying the toxins present in the feed to improve the availability and utilisation of nutrients. Efforts were also directed to study the interaction of management and nutrient requirements with particular reference to interaction of housing systems with energy, protein and amino acids. Among the systems of feeding, restricted feeding both quantitative and qualitative received greater attention.

The country has not only been able to develop her own standards on the nutrient requirement of poultry and systems of feeding but also identify several agro-industrial by-products, forest wastes, animal by-products and slaughter house wastes which can suitably be incorporated in poultry rations in different proportions to economise on the cost of compounded poultry rations. Methods were developed and standardised for assay of aflatoxins in the feed and to improve the feeding value of some of the non-conventional poultry feeds for their inclusion in higher proportion. Feed values of some of the by-products such as rice bran and clusterbean meal were also improved by use of solid state fermentation technology and other biotechnological methods developed by Dr. S.V.S. Verma and co-workers at the CARI. Results obtained so far indicated that most of the by-products investigated can be incorporated up to 10% in the ration of broilers and layers without affecting their performance.

Effects of aflatoxins on production and reproduction in birds, determination of their safety levels for different classes and age groups of poultry, and development of methods to detoxify toxins in feed were established (Dr. P.V. Rao). A detoxification plant was set up at PTI, Hessaraghetta (Dr. G. Devegowda). Studies on vitamins are limited. Dr. B. Panda reported the requirement of vitamin A in poultry rations; Dr. Ramakrishnan and his group undertook investigations for establishing the nutrient requirement of ducks; Dr. B. Panda and his group for determining nutrient requirement of quails; and Dr. S.V.S. Verma, CARI and Dr. N.N. Pathak of IVRI determined nutrient requirement of guinea fowls for optimising their performance in tropics.

The poultry processing sector is still in infancy. Only 10% of chicken are dressed in mechanised poultry processing plants including small poultry dressing units operating in different parts of the country. The bulk of eggs produced are marketed as table eggs and only a very negligible fraction of them is converted into processed egg products. Several mechanised poultry egg and meat processing units are being set up in the country. These will be of great value in developing processed poultry egg and meat products for the domestic market and boosting the export prospects of poultry products.

The research in poultry products technology has primarily been directed to (i) develop and standardise the technology for increasing the shelf life of poultry meat and egg, (ii) identify the factors affecting meat yield and meat quality and (iii) standardise the processing techniques for manufacture of fast foods. The work carried out on the above aspects by Dr. B. Panda, Izatnagar; Dr. A.K. Chaterjee, Mysore; Dr. Nagendra Sharma, Dr. N. Kondaiah and Dr. P.C. Panda has resulted in the development of effective methods for preservation of shell eggs at ambient temperatures and to reduce the damage to eggs during transport and distribution. To market clean and graded eggs a cost-effective egg-washing machine was fabricated and egg washing powder developed. An egg coating oil for the preservation of shelled eggs was identified. Processing techniques for the manufacture of several further processed egg products such as egg powder, albumen flakes, scrambled egg mix, omelette mix and pickled eggs were standardised and keeping quality of products evaluated. Techniques were developed to control gelatin in frozen egg yolk. Suitability and efficiency of some of the economic thermoplastic mono-layer laminated packaging materials for modified atmosphere packaging were evaluated.

Meat yield and meat quality in chicken broilers, spent hens and other species of poultry were evaluated for their marketing as dressed whole carcass, cut-up parts or boneless meat and their utilisation in the manufacture of comminuted/ restructured meat products. Enzymatic methods were evolved for the tenderisation of tough meat of spent hens and culled breeding stock. Apart from evolving techniques for preserving poultry meat a wide range of value-added further processed meat products such as chicken sausages, patties, nuggets, loaves, steaks, kababs, chicken essence, canned chicken, chicken soup, cured and smoked chicken as well as gizzard pickles were developed for catering to the growing demand of such products. Processes have been developed for the effective utilisation of inedible poultry by-products resulting from poultry and egg processing plants to realise higher economic return and control pollution hazard.

Poultry keeping in India was greatly handicapped in the past because of serious debility and mortality inflicted by various diseases, more particularly Ranikhet disease. The virus which causes this disease was first identified by Dr. Doyle in 1927 and vaccine against this disease was developed by Dr. G.S. Iyer and Dr. N. Dobson in 1940. This breakthrough made it possible to establish poultry farming on commercial scale not only in India but around the world. The expertise necessary for the diagnosis, prevention and treatment for some of these diseases was developed. With increase in poultry production in the country, more so new-high yielding varieties from west, several new diseases such as infectious bronchitis, infectious laryngotracheitis, Marek's disease, avian encephalomyelitis, infectious bursal diseases, egg drop syndrome, reo-viral arthritis and adeno virus infection have been introduced. Very recently diseases such as chicken infections anaemia and hydropericarditis syndrome (leechi disease) were reported in the country.

Research efforts undertaken at the IVRI during recent years has helped in making significant breakthrough in controlling many of the diseases. Vaccines were developed against several killer diseases. The Indian Veterinary Research Institute has played a dominant role in the development of poultry vaccines and immunodiagnostic methods for several poultry diseases; Marek's disease (Capt. S.B.V. Rao), IBD and EDS-76 (B.S. Rajya, G.S. Mohanty, K.C. Verma, CRD/mycoplasmosis, ILT (S. Kumar and M.P. Yadav), avian influenza (H.K. Pradhan et al 1985). The ICAR established a coordinated project on surveillance and monitoring of poultry diseases. All the Biological Product Institutes established in different states are manufacturing poultry vaccines against RD and fowl pox. The country has developed adequate expertise for containing of most of these diseases.

The growth and development of Indian Poultry Industry during recent years has been a painstaking combined effort of both Government agencies (DAH, ICAR, DARE, SAUs, SVUs, SDAH, Vet. Bio-institutes and private sector). While various programmes undertaken by the Government were instrumental in creating awareness and confidence among people about usefulness of poultry farming, the private sector and the financial institutions have helped to make available the necessary inputs that made commercial poultry farming a feasible and viable proposition. Providing employment to numerous people, through poultry farming is the hallmark of government programmes for inclusive growth. Efforts by the private sector on the other hand were directed towards large establishment for mass production. It may be worthwhile to mention here that the first commercial hybrid chick was hatched in November 1962 at Rani Shaver Farm, Delhi, a private enterprise. Private entrepreneurs have made it possible to supply the best layer and broiler stocks available in the world through import of the GP stocks. The entire process of compound feed manufacturing, processing of egg and meat, manufacturing of equipment, machinery, pharmaceuticals and biologicals, and marketing of egg and meat were also made possible by the private sector.

In recent years, a dozen of pureline breeding programmes were established in private sector for developing elite purelines layer and broiler stocks and to make the country self-sufficient in the availability of improved germplasm for commercial use. Most of the requirements of day-old chicks are met by private sector hatcheries which have either pureline operations or franchise arrangements with poultry breeding companies from abroad. Public sector breeding farms also produce hybrid chicks but they have limited replacements due to limited facilities for multiplication and supply and lack of specialised services such as marketing and sales promotion. In view of substantial participation of the private sector in pureline poultry breeding there is a need to reorient public sector research efforts towards those areas that are less attractive financially to the private sector at present but remain important for the overall growth and development of the industry. These include improved cost effective feed formulation, nutrient evaluation, identification of alternate energy and protein feeds, detoxification of toxin-infected feed, interaction among toxins and their effect on growth and production, improved management practices specially designed for small farm/free range production, development of methods for prevention and control of diseases suitable for free range/backyard chicken production, diagnostic technology which can be used under village conditions and development of value-added and designer poultry products, and production of anti-snake venom and other biological for human and animal use to meet the requirement of national and international market. There is also a need to intensify research in biotechnology, particularly in identifying genes for disease resistance, heat tolerance, adaptability etc. which call investments in genome analysis, molecular genetics, immunogenetics, genetic resistance to diseases, production of transgenics, evaluation and improvement of native germplasm and development of cost effective method for disposal of wastes.

Key weakness of public sector poultry research is the lack of client orientation, especially for small holder production including shortage of trained manpower and poor staff/operating cost ratios. Poultry extension services have also been completely neglected. Within the integrated public sector agricultural extension service the dissemination of poultry technology has not received adequate emphasis. The ability of the Indian Poultry Sector to fulfil its targeted growth in productivity and output will depend upon the quality, availability and assessability of these services. With increasing demand and declining government resources efficient allocation of resources to different areas is critical. To achieve these objectives, there is a need for re-defining of public and private sector roles in the poultry sector, creation of level playing field and establishment of appropriate incentives by the government.

The liberalised economic policy of the government has added a new dimension to our research and developmental effects. The protective shield of government regulations and stiff tariff barrier for import of goods is no longer valid. On the contrary, the market is very competitive and customer driven. Traditional quality means confirming the specifications and standards but in the changed scenario it also means productivity, competitive costs, timely delivery and total customer satisfaction. Development of human resources and upgrading the skill and capability of existing human resources will be necessary.

In order to achieve these objectives, the research institutes have to play a pivotal role for development of human resources and skills and for upgrading its research capability in various disciplines of poultry science to provide the necessary research, training and extension services comparable to international standard to the growing Indian Poultry Industry, which presently stands at number three and fifth place in the world in egg and broiler production, respectively, to meet the new challenges. This calls for total reorientation of research priorities and infrastructure.

2.12 DUCK

Asia is considered the homeland of domesticated ducks, but the best breeds and strains currently available for excellent egg/meat production were developed in Europe and America through systematic research in breeding, feeding, management, use of Technology and disease control. In the U.K., ducks have been kept as domestic birds for centuries, but were mainly kept as table birds until about a century ago. In South-East Asia, particularly Java, Malaysia and the swampy parts of Myanmar, egg-type ducks have long been kept. More than 100 years ago, proven layers from indigenous breeds were taken to European countries and America for developing high-yielding strains. Thus, modern strains of ducks have been developed from the ones originally available in South-East Asia. As compared to chicken, there are very few breeds of ducks of commercial value. Among layers, the two dominating duck breeds are Khaki Campbell and the Indian Runner. Among Western type Pekin, Aylosbury and Muscovy are popular. Egg production potential of modern-day ducks is 300 or more per year. This is significantly higher than that is possible by the modern hybrid genetically manipulated commercial layers (chicken). Meat-type ducks are as efficient as broiler- type chicken. In the West and other affluent countries, duck meat is a delicacy. In some of these countries, eggs are not popular because of the fear of possible *Salmonella* infection.

Number of ducks in India are estimated to be over 23.5 million (2012 Census) showing a decline of 14.85% compared to the numbers in 2007. Duck rearing in the USA is mostly confined to the Eastern Zone, around New York City. Compared with chicken, ducks are of minor importance in most countries of the world. In England, West Germany, Italy and Denmark, ducks represent no more than 2%–3% of the total poultry business. One exception is Poland where meat-type ducks represent approximately 10% of poultry. In the Far East, the duck production is important in many countries. In Taiwan, ducks account for nearly one-third of total poultry.

In India, ducks enjoy the most important position after chicken. According to the 1992 livestock census, duck population was approximately 22 million, which is 8% of the total poultry population. Egg production was approximately 4.4 million in 1991–92—5% of the total egg output in the country. West Bengal ranks first in terms of duck population, followed by Assam, Tamil Nadu, Andhra Pradesh, Bihar, Odisha and Kerala.

Although duck-keeping is confined to a limited area, it still contributes significantly to the total GNP at INR40 million/year. With appropriate scientific management, its potential can be raised by 60–250 million. Duck keeping provides occupation to 0.2–0.3 million rural farmers. Any scientific proposition to improve the production potential of indigenous ducks will have a tremendous impact on the lives of these farmers and will also create ancillary job opportunities to many others. Duck eggs are important source of food in India. They are very cheap to produce and can play an important role in balancing the diet. Consumption as well as production of duck eggs in India is mostly by the socially weaker section of the community. Furthermore, many people in India who do not consume hen eggs for religious reasons, can use duck eggs.

Although ducks, like chicken, are known in our country since time immemorial but remained neglected by research workers, developmental agencies and private sector. In recent years, however ducks have attracted attention of research workers throughout the world. In India as well, lead has been taken by a few research organisations. National plans are being formulated to ensure that ducks play their due role in the socio-economic development of rural masses. Tremendous scope of duck development in the country has caught the attention of many international agencies.

During the Fourth Five-Year Plan the Government of India decided to:

(i) establish a Central Duck Breeding Farm at Hessaraghatta in Karnataka where duck is of no significance. This farm has excellent flock of high egg layers of Khaki Campbel breed but has no clients to purchase its day-old chicks; and

(ii) establish State Duck Farms in states where duck farming has a major economic base.

Thereby, superior stocks from the Central Farm could be multiplied by the State Farms for upgrading the performance of existing farm stock. This unfortunately has not happened. Several strains and breeds were imported, including White pekines, White Minikos, New Hampshire and Khaki Campbell.

The management regimes, housing and animal health care system were standardised for layer as well as broiler types. No work is in progress on Indian strains or varieties. The bird has proven potential and is at the take off stage. Research and development departments need to interact to use this economically important bird in select areas in the country, particularly the states endowed with Lakes, ponds and other water bodies.

2.13 QUAIL

Coturnix Japanese quail belongs to the class Aves, family Phasianidae and genus *Coturnix* and is named *Coturnix coturnix japonica*. The Bobwhite quails belong to a different family. The male is lighter than the female and is identified by the cinnamon-coloured feathers on the upper throat and lower breast regions. The female is similar to male in coloration except that the feathers on the throat and upper breast are long /pointed and much lighter in cinnamon. Furthermore, the tan breast feathers are characteristically black-stippled.

Domesticated Japanese quails were first introduced in India by the Poultry Research Division of the IVRI, Izatnagar during 1974 as a pilot animal for avian research. After creation of the Central Avian Research Institute (CARI), Izatnagar in 1979, research efforts to increase the productivity under our agro-climatic conditions through multi-disciplinary approach of all the different research faculties were intensified.

Low volume, low weight, rapid regeneration rate, high rate of egg production and high acceptability as a source of egg and meat provided lot of promise for commercialisation of this species, Dr. B. Panda and his associates were pioneer workers to provide workable package of practices in breeding, reproduction, management, feeding and post-harvest technology. These workers developed equipment for raising quails under Indian conditions such as battery brooders, rearing cages, colony laying cages, portable individual cages, feeders, waterers, etc. Quail production technology developed served as the guideline for other tropical and sub-tropical countries.

Quail broiler and layer lines were developed and are being maintained at CARI. A line of laying white shelled eggs was also developed to meet the demand of scientists working in Avian diseases for production of vaccines.

Quails mature at 6–7 weeks of age and are in full production by 9–10 weeks of age. With proper care and management, hens can lay from 280–300 eggs/year. Eggs can be successfully incubated in incubators used for chicken by adding 1.2 × 2.5 cm strips of welded wire in the setting trays. Quail rearing is simple and is almost similar to that of chicken. The most critical period is the first 2–3 weeks of age when optimum brooding temperature must be maintained as the hatch weight of this bird is very low (approximately 67 g) and chicks need careful attention. Soft corrugated paper above the floor of the battery brooder for first week of rearing is best to provide foot-hold to the chicks. Young birds are very sensitive to temperature variations. Mortality is high during first two weeks. Quails establish well after two weeks. Their growth is also faster. Litter for battery brooding for rearing can be employed successfully. Colony cage of 60×30cm and 60×120×30cm can be used to accommodate 25 and 50 birds, respectively.

These birds need slightly higher levels of protein, energy and vitamins than chicken. A level of 27% protein and 2,800 KCal ME/kg is recommended for optimum growth up to three weeks of age and 24% protein and 2,800 KCal ME/kg during 4–5 weeks of growing period. Adult quails should be fed ration containing 22% protein and 2900 KCal ME/kg.

Considering the production potential and greater acceptability, quail production has increased significantly all over the country. NABARD has also formulated an innovative scheme on quails for refinance to the bank to meet the growing demand of quails around major cities as it has replaced Bater as diticacy and to assist farmers to expand their business activity of quail production. Wild life Act had created problems for rearing quails for commercial purpose as it could not distinguish between quails raised as domestic poultry and those captured in wild as the wild quails are classified as endangered species. This problem has now been sorted out.

The 19th Livestock Census 2012 does not separately provide the number of quails. The numbers for Turkeys and other birds other than fowls and ducks were successfully 13 million, growing by 277.32% over the number in 2007. In this phenomenon, quails have played major role besides Turkeys and pheasants.

2.14. ANIMAL HEALTH

The importance of protection of livestock against diseases and pests was realised by the British Government of India as early as 1868 when the Indian Cattle Plague Commission (ICPC) was appointed to report on the prevalent cattle diseases and suggest measures on their prevention and control. The ICPC recommended the establishment of Imperial Bacteriological Laboratory (IBL), forerunner of Imperial (now Indian) Veterinary Research Institute (IVRI). The IBL was established in 1889 at Pune and the Civil Veterinary Department came into being in 1891. At the time the IBL was started, rinderpest (cattle plague) was widespread among cattle and buffaloes in the country and thus early researches were concentrated on rinderpest. By 1926, an efficacious vaccine against rinderpest virus was produced. This helped to launch a countrywide programme to control losses due to this deadly disease. The IVRI undertook studies on other viral, bacterial and parasitic diseases of livestock, and developed effective methods of control and biological products for combating them. These researches made easier the task of the progressively expanding civil and military services, not only in combating several animal diseases but also in taking up programmes of livestock development. The Royal Commission on Agriculture (RCA), in its report submitted in 1928, while appreciating the work done remarked that 'in no sphere has scientific research conferred greater benefits on agriculture than by the provision of means of controlling livestock diseases'.

In the early period, the main concern of research was to develop suitable therapeutic and control measures to check the colossal losses resulting from high mortality and morbidity inflicted by the diseases such as surra, anthrax, rinderpest, tetanus, black quarter, haemorrhagic septicaemia and glanders in various species of livestock.

The outstanding research accomplishments were the development of a live attenuated goat-adapted vaccine for rinderpest, R_2B vaccine for Ranikhet disease and mouse brain-adopted vaccine for African horse sickness. Large-scale production of the first two vaccines helped in the planning and execution of countrywide immunisation programmes against rinderpest and Ranikhet diseases. This resulted in considerable reduction of economic losses due to these diseases. The vaccine developed against African horse sickness (AHS) played a vital role in the eradication of this disease from India.

With the planning of massive research programmes of cross-breeding with exotic animals during the Fourth Five-Years plan, it was realised that without adequate health cover all attempts to improve the production capacity were likely to result in failure, as cross-bred animals were susceptible to the diseases and pests prevalent in India. Thus, there was necessity for an efficient and well-knit animal health organisation. There was also a need to expand the present facilities for clinical and preventive medicine, and for diagnosis and investigation of diseases. Strengthening of veterinary biological production stations and animal diseases research centres thus became imperative. Further, it was found necessary to strengthen the quarantine organisation to prevent the ingress of exotic diseases. To safe-guard the health of human population from zoonoses and food-borne diseases, particularly those communicable through milk, meat, eggs, fish and shell fish, setting up of public health service was proposed. To prevent malnutrition and deficiency diseases in high producing animals, adequate supplies of feed and fodder of good nutritive quality would have to be made available.

The ICAR after its reorganisation in 1966, identified the IVRI as the nodal organisation for undertaking research in animal health. The major objectives of the Institute were to carry out research, education and transfer of technology programmes in animal health, animal production and animal products technology under one roof. The Institute has six Campuses/regional stations (Izatnagar, Mukteswar, Palampur, Kolkata, Bengaluru and Pune) and several divisions and sections in various disciplines of animal health and allied areas and other species. The Institute was accorded the status of a deemed-to-be university on 16 November 1983 under the UGC Act. A high security animal disease laboratory has been established with the assistance of UNDP and World Bank at Bhopal. Among the erstwhile campuses of the institute, Bhopal has been made a separate institute recently in 2014, while Srinagar Centre is non-functional and Port Blair Centre has been merged with CARI (Central Agricultural Research Institute, Port Blair). The Institute had the good fortune of being headed by eminent scientists who gave new direction to its development.

Dr. S. Datta (1947–54): A renowned pathologist took over as first Indian Director of erstwhile Imperial Veterinary Research Institute (IVRI) soon after the independence of India. Dr. Datta was instrumental in shifting of Headquarter of IVRI from Mukteswar to Izatnagar and renaming it as Indian Veterinary

Research Institute (IVRI) under Government of India (1947). During his tenure, Mukteswar Campus of IVRI celebrated Diamond Jubilee on completion of 60 years (1950). Under his leadership, an ambitious Rinderpest (RP) eradication programme formulated by Govt. of India was launched in October 1954. In this programme, 80% animals were vaccinated and Biological Products (BP) Division of IVRI made significant contribution in terms of supply of RP vaccine (1954). Dr. Laxmi Sahay (1954-57) and Dr. P.G. Pandey (1958-61) expanded the production of vaccines and diagnostics.

Dr. C.M. Singh (1966-82):A renowned pathologist and animal health scientist laid the foundation for infrastructure in all the three areas, namely Animal Production; Animal Health and Animal Products Technology. He was also responsible for the Institute being recognised as Deemed University.

Dr. P.N. Bhat (1982–92) was the first Indian to get the Ph.D in the subject of population genetics from Purdue University, USA. His major contribution was to consolidate the futuristic sciences like population genetics, biotechnology spread invarious health divisions and new major project on DNA Recombinent technology as a common platform for all biotechnologies including reproductive biotechnologies like biochemistry of semen, multiple ovalution and embryo transfer. All this was brought under one roof under new National Biotechnology Centre. This protocol was signed with several institutions in USA and UK. The largest grant was from UK government through University of Edinburgh and Institute of Animal Health at Pirbright and through this arrangement several countries were contacted to us by offering all the facilities under able guidance of Dr. Brian W J Mahy, who was at the then Director of Institute of Animal Health, Pirbright thus creating a model for similar other programmes including a platform for developing a common platform for vaccine research.

Dr. Kiran Singh (1995-1997): During his tenure, the first IVRI technology – 'Olinall', an effective herbal skin ointment, was commercialised (1997). Moreover, Centre for Advance Studies in Animal Physiology was upgraded, which later developed into Centre for Advance Faculty Training (CAFT) in Animal Physiology.

Dr. M.P. Yadav (2000–06): During histenure, the Institute witnessed all round progress; some of the landmarks included launch of ELISA based Rinderpest diagnostic kit validated by WRL, Pirbright, UK; dedication of HSADL, Bhopal–sixth best BSL–out of seven labs in the world, P3 facility for FMD vaccine quality control at Bengaluru campus; establishment of the Division of Temperate Animal Husbandry at Mukteswar Campus, and Kisan Call Centre at Izatnagar; commercialisation and patenting of the technologies developed, namely crysto-scope for determining optimum time for insemination in cows, area specific mineral mixture (ASMM) and PPR vaccine; creation of ATIC at Izatnagar; recognition for Best Institution Award of ICAR (2002), first ever OIE meritorious award to IVRI scientist (2000), best institute in seven categories by SCOPUS , ISO 9001-2000 Certification of CADARD etc.

Dr. Shiv Pal Singh Ahlawat (2006–09) was accredited with major transformation of the ambience and facelift of the institute with a popular slogan 'Clean IVRI, Green IVRI'. During his tenure, high milk yielding cross-breed cattle strain 'Vrindavani' was launched (2007). A video conferencing was established at Izatnagar Campus as well as a Kisan Helpline Centre was started at Mukteswar Campus (2007). The country got OIE certification for freedom from rinderpest (2006) and contagious bovine pleuro pneumonia (CPBB) infection (2007). The IVRI was reckoned as the best institute among the Veterinary Science Education institutes in the country and the 31st Most Productive Institution of the country by Scopus International Database 2008-09 (Elsevier Group) (2008). The Institute also developed a low volume saponified HS Vaccine, which was found safe and effective in farm cattle.

Dr. M.C. Sharma (2009–13), created history in the ICAR by winning a record eight national awards in 2009–10 (including the Prestigious Sardar Patel Outstanding ICAR Institution Award of 2009); 10 national awards in 2010–11 (including OIE recognition for Freedom from Rinderpest (Cattle plague) infection in India (2010) and nine national awards in 2011–12, including Best National Agribusiness Incubator Award-2011 and Best National Agribusiness Incubatee Award-2011. During his period ATIC was adjudged as the Best among all the 45 ATICs of the ICAR and SAUs. He was instrumental in establishment of ZTM-BPD Unit (Business Planning and Development Unit) (2009) at Izatnagar Campus and SPF Animal Facility and TEM Units at HSADL, Bhopal Campus (2009), Construction of Indoor ward for Pets at Polyclinic of Izatnagar campus (2011), National Referral Lab on Drug and Chemical Residues testing (ICAR, APEDA and MFPI) Govt. of India (2011) BSL-3 Facility Zonal Technology Management Centre (ZTMC) (2011) and On-line Exam Facility for NET and ARS Exams at Izatnagar (2012). He took several initiatives including the organisation of the First Institute-NGO Interface at IVRI (2010); and development of 'Single Window Delivery System' at ATIC, with a view to facilitate the sale of all the produce and services of IVRI to the livestock owners and farmers. During his period 77 patents, 4 extension-oriented software and 5 copyrights were granted/registered to IVRI credit, and got 13 technologies commercialised to 30 commercial houses. During his tenure IVRI remained top-most revenue earner among all the ICAR institutes with a steady increase in revenue from 9.7% to 41.5% over the target fixed by the ICAR.

With the reorientation of some of the priorities, the ICAR Research Complex for North Eastern Hills Region has been located at Shillong, Meghalaya and has been given the mandate of research on animal sciences including animal health, in addition to the other areas of agriculture and fisheries. Similarly, the National Research Centre on Equines (NRCE) was established at Hissar (Haryana) in 1985 to conduct research on production and health aspects of horses, donkeys and mules. An AICRP on Epidemiological Studies on FMD was initiated in 1968 at Mukteswar Campus with 18 regional centres to conduct research on this major disease. It was upgraded to Project Directorate of FMD

(PD-FMD) in 2001 as an independent institute in 2001 with 23 regional centres. The Bengaluru campus of IVRI was established in 1972 with the mandate for R&D on FMD vaccine.

2.14.1 Immunobiologicals

Research for developing new immuno biologicals and simultaneously improving the quality of existing ones continued. This led to the widespread application of several vaccines, *viz.* a tissue culture alternated vaccine for rinderpest, a polyvalent inactivated tissue culture vaccine for foot and mouth disease (FMD), an egg-adapted flurry vaccine for rabies, and attenuated cell culture vaccine for sheep pox, vaccines against HS, blackquarter, an irradiated vaccine for lungworm disease of sheep, a multicomponent clostridial vaccine against Black disease; vaccines for Marek's disease and canine distemper, a tissue culture vaccine for Gumboro disease and a cell culture vaccine for theilariasis, besides fowl pox, goat pox, blue tongue, PPR, classical swine fever, avian influenza H5 N1 (HPAI) and other diseases.

The major achievements made in the field of animal health comprise large-scale production of monovalent and polyvalent foot-and-mouth-disease virus vaccines by fermentor technology using BHK-21 cell suspension system of clone 13. The calf-hood vaccination methodology using split dose of the vaccine, developed by P.K. Uppal, P.N. Bhat and P.P. Bhat greatly helped the success of FMD vaccination to protect the highly susceptible young cross-bred calves, as it overcome the problem of interference in immune response to vaccine due to maternal anti-bodies. Equine influenza (A/Equi-2), equine infectious anaemia (EIA) equine coital exanthema (ECE), equine viral arteritis and equine rota virus were detected and identified for the first time in the country. Equine influenza outbreak due to A/Equi-2 serotype in 1987 was successfully contained through surveillance and controlling movement of horses. The virus was isolated and characterised. A cost-effective experimental split vaccine prepared with this isolate proved effective under field conditions.

A clear shift in the prevalence from FMD virus type 'A' to type 'O' was observed throughout the southern states. Asia 1 and A 22 are the types showing increased activity in the western and northern states. Subtype A5 was recognised after a lapse of 4–5 years. Its identity was duly confirmed by the World Reference Laboratory, Pirbright, UK.

A duck plague virus strain isolated from outbreaks that occurred in the North-Eastern region was attenuated by passaging in developing chick embryos. Further studies on developing an effective vaccine against duck plague are in progress.

2.14.2 Epidemiological Studies

Considerable work has been done in understanding the epidemiology of animal and poultry diseases such as FMD, rinderpest, Ranikhet disease, Marek's disease, bovine leucosis, EIA, equine influenza, blue tongue, haemorrhagic septicaemia, black quarter, anthrax, tetanus, salmonellosis, tuberculosis, Johnes' disease, rabies, Japanese encephalitis, leptospirosis, listeriosis, clostridiosis haemo-protozoan diseases, fasciolosis and mycotic infections. Much information is also available on major clinical entities such as mastitis, horn cancer, bovine haematurea, papillomatosis in cattle, pneumonia and respiratory diseases, gastro-enteitis, metabolic diseases, abortions, urolithiasis, anaemia, ruminal disorders, several surgical affections, egg drop syndrome and mycotoxicosis. These have paved the way for developing control measures for their management.

Studies on sero-prevalence and subsequent isolation of causative agents have conclusively proved the etiological role played by organisms such as mycoplasma, chicken infections anaemia, chlamydia, influenza, rota, Corona, BVD and certain enteroviruses. Investigations on buffalo-pox led to the isolation and identification of the virus elucidating the epidemiology of this disease. The presence of newer diseases such as blue tongue, infectious bovine rhino-tracheitis, malignant catarrhal fever, rhino-sporiodiosis, equine herpesvirus infection was also brought to light.

A DIVA test developed, based on FMD NSP 3ABC antibody detection has immensely helped to study the epidemiology of this devastating disease, as the test can discriminate between the immune response elicited by natural infection from the one produced in response to the vaccine. Foot-and-mouth disease has been recorded as a serious problem in mithun herds of Nagaland.

From 51 poultry farms in Uttar Pradesh, Rajasthan, Delhi and Chandigarh 456 birds were examined. Marek's disease, infectious bronchitis (IB), infectious bursal disease (IBD), infectious laryngo-tracheitis and Ranikhet disease were found to be prevalent. The incidence of salmonellosis, chronic respiratory disease (CRD) and *Escherichia coli* infections also showed an increasing trend. Morbid materials (176) received from 10 states showed almost a similar disease pattern. Serological studies revealed infectious bronchitis, infectious laryngotracheitis, chick embryo lethal orphan (CELO)—an adeno virus and antibodies to be quite prevalent in poultry flocks in different parts of the country. *Salmonella gallinarum* infection showed increasing incidence in different parts of the country.

In an organised equine breeding farm in northern India where 54 out of 360 (15%) pregnant mares aborted in the last trimester of gestation, equine herpes virus-1 (EHV-1) was established as the agent. EHV-1 was isolated in rabbit kidney (RK 13) cell-culture from all the 15 aborted foetuses examined. Rising antibody titres in paired sera samples against EHV-1 virus were also recorded in most mares. Screening of 100 equine serum samples, obtained from West Bengal and Tamil Nadu for Japanese encephalitis, employing indirect haemagglutination inhibition test indicated seven positive reactors.

Paratuberculosis or Johne's disease, a chronic infection, causes severe losses to goat and sheep flocks throughout the country. The disease also causes diarrhoea, poor condition and production losses in bovines. At the CIRG, 159 isolates were successfully isolated from faecal samples. This was the first documentary Mycobacterial isolates available in this country from goat and sheep origin. Four typical isolates from goat and four from sheep have been selected and adopted on to the artificial medium for characterisation by biochemical as well as molecular marker methods. Biochemical characterisation revealed that most of the isolates have similar characteristics as of *M. avium* complex. *M. paratuberculosis* also falls in this complex taxonomically.

Disease survey at various yak pockets in North Eastern States revealed the presence of pneumonic lungs in 29%, of which only 14% showed nasal discharge. Trombiculid mite infection resulting in alopecia in neck region was found in 19% animals in the pocket. Every year FMD outbreak in yaks is common in the locality. Other diseases were diarrhoea (10%) and hoof defect (7%).

2.14.3 Diagnostic Methods

To confirm the clinical diagnosis and to demonstrate the presence of antibodies in infected or vaccinated animals, attempts were made from time to time to adopt, standardise or develop suitable techniques based on detection of antigen and/or antibodies. Extensive use of agglutination, precipitation, complement fixation, conglutination complement absorption, gel diffusion, haemagglutination - inhibition and counter immuno-electrophoresis (CIE) have given very useful results of practical significance. Modification of the above tests and the use of micro-methods, development of radio-immuno-assays, ELISA (enzyme-linked immunosorbent assay) and FAT (fluorescent antibody technique), immuno peroxidate test (IPT) have added to the much-needed precision to the results. IPT can also be performed on formaline fixed tissues. These tests have led to the recognition of sub-clinical and latent infections and are now marketed as diagnostic kits.

Biotechnology has brought into veterinary practice new diagnostic methods for disease diagnosis precision and new treatments as well as vaccines. The basis of the new diagnostic methods lies in the discovery of ways to produce monoclonal antibodies (MABs) on a large scale which are relatively cheaper using tissue culture techniques. What is important is that the tests are usually employed outside the body (*in vitro*) using small samples of urine, milk or blood for the analysis. So far, approximately 100 tests based on MABs have been established.

Apart from *in vitro* diagnosis, MABs have an even wide application in *in vivo* diagnosis as well as in therapeutics (treatment). Through the linking of antibodies with specific therapeutic substances (toxins) and drugs, it will be possible to link diagnosis with therapeutics. Studies in this direction are extremely active, especially in the multinational pharmaceutical companies, due to the huge market expected.

Application of biotechnological methods in vaccine production represents another important field of new technology in animal health. Genetically engineered vaccines promise a higher grade of innocuity, and efficiency as well as a lower cost of production. Promising results in the preparation of sub-unit vaccines and vaccines based on peptide chemistry for veterinary use have been obtained. DNA recombinant technique allow expression of antigens, synthetic peptides, edible vaccines, monoclonal antibodies, RNAi and some other biotechnology products to be cloned and used. All this is opening up a new future land vista for the animal health immunisation programme. These developments are only beginning and their importance cannot be over-estimated. Latest technologies including therapeutic application of stem cells in canines, development of reverse genetic based vaccine against Avian Influenza, gene-deleted marker vaccines against *Bruccella* and IBR, DIVA based diagnostics for FMD, IBR; high-throughput SPR-bio-sensor chip based diagnostics, pen-side diagnostics and LAMP test in field conditions with high accuracy; peptide diagnostics, nano-technology based vaccine delivery system and mineral nutrition are being developed with satisfaction to face future challenges in animal health and production due to climate change and other factors.

The application of recombinant DNA technique led to development of new types of vaccines for many infectious diseases including virus diseases. The rinderpest virus (RV) is a member of morbillivirus sub-group of family Paramyxoviridae which includes measles virus (MV), canine distemper virus (CDV) and peste-des-petits-ruminants virus (PPRV). Except PPRV, these viruses are well known to show similar biological, pathological and serological characteristics. Recently their similarity in gene structure was also revealed. The interest in studying the pathogenicity of MV, CDV and RV from the comparative aspect by using animal models infected with these viruses is to produce a common vaccine. Among these models, rabbits infected with Nakamura III (L) strain of RV which had been adapted to rabbit at IVRI has especially attracted the attention globally because of its high virulence in rabbits. In this model, severe clinical signs such as high fever and diarrhoea are experimentally induced. Histologically this clinical course was shown to be associated with characteristic lesions, mainly consisting of giant cell formation and necrosis in the lymphoid tissues. Immune capacity such as the mitogen-induced proliferative response of the peripheral blood lymphocytes, delayed type hypersensitivity to tuberculin and antibody response to sheep red blood cells was also found to be markedly suppressed in RV-infected rabbits. Moreover, virus infection induces development of two types of auto antibodies, i.e. antinuclear antibody reacting with DNA and histone, and hemagglutinating antibody reacting with rabbit red blood cells. Thus, RV-infected rabbits provide an excellent model for the study on pathogenesis of morbillivirus and may lead to a common vaccine for humans and animals in near future.

In recent years, much effort has been made to apply modern methods for improving the quality of immuno-prophylactics. The introduction of tissue culture and cell-lines in vaccine production has revolutionised the production of vaccines. These vaccines are very potent, safe and cheap. Rinderpest tissue culture vaccine has all these three advantages over the original goat tissue vaccine. The large-scale production of this vaccine resulted in considerable reduction in the incidence and consequent economic losses due to these diseases. The tissue-culture rinderpest vaccine (TCRP) was instrumental in achieving global eradication of this malady-rinderpest or Cattle plague leading to historical declaration by FAO in June, 2011. India achieved the status of freedom from Rinderpest virus infection in 2006 by OIE. The use of cell-lines for production of FMD vaccine in large batches has also made it possible to flood the market with supply of excess amount of vaccine for cross-bred cattle. In this case, the vaccine is still not very potent due to the presence of many types and subtypes of virus as well as weak antigenicity of FMD virus for inducing immune response, more so in case of Type 'O' serotype; and the non-availability of live vaccine against FMD. The vaccine must be repeated every five to six months, and sometimes a monovalent vaccine against a particular type must be given. This needs immediate typing of virus. An effective control programme needs a good laboratory support for diagnosis of the disease.

New adjuvants were developed for improving effectiveness of vaccines. Potash-alum-precipitated and aluminium-hydroxide-gel-adsorbed vaccine against enterotoxaemia due to *Clostridium perfringens* type 'D' was developed. This is superior to the existing aluminium-chloride vaccine. In the same way alum precipitated HS vaccine is better than old HS broth vaccine. Development of double emulsion oil adjuvant vaccine against HS is easy to administer and provide lasting immunity.

The veterinary immunobiologicals are produced under 17 public centres and four private sector units. Vaccine production level has increased from 191 million doses in 1974–75 to 390 million doses in 1984–85. Still, the country is in short supply of HS and BQ vaccines.

For the control of FMD alone, adopting better technologies to produce the vaccine, *viz.* continuous cell culture using fermenter technique, concentration of the antigens and use of better inactivants and adjuvants, has raised hopes of launching a limited control programme in certain parts of the country. Moreover, any control campaign in this country will only be feasible when the cost of production of vaccine is substantially brought down to approximately 50 paise per dose. Keeping in view the 240 million susceptible bovine population (besides other susceptible species) in the country and short-lived immunity of vaccine, its high cost and present production technology available, a national control programme would cost approximately INR 10,00 million in the first year alone, with a recurring cost of INR 5,00 million/annum besides other overhead charges.

FMD occurs only in a mild form in indigenous cattle and buffaloes. It, however, assumes virulent proportions in exotic and cross-bred cattle. Exception to this was, however, noticed during 2013 when wide spread FMD outbreaks involved Karnataka, Tamil Nadu and Andhra Pradesh resulting in high mortality in indigenous cattle as well. Although annually approximately 0.2 million animals suffer from FMD in India, the mortality from this disease is only approximately 1,000. The economic loss due to morbidity, loss of milk production and reproduction capability and draft power depredation is estimated as INR 20,000 crores a year. FMD control is affected in two ways: (i) subsidy on FMD vaccine and (ii) mass vaccination. The subsidy programme envisaging 50% cost of the vaccine by the Government is meant for use by small and marginal farmers and agricultural labourers for their exotic, cross-bred and highly valuable indigenous cattle and buffaloes. Approximately six million vaccinations are arranged annually under this programme. The mass vaccination programme is implemented as a part of disease-free zone scheme in the four southern-most districts of the country. This programme is also carried out in some other districts of Kerala, Karnataka and Tamil Nadu with the Nilgiris district as the nucleus. Approximately ten million vaccinations are arranged under this programme.

The mass vaccination of bovines initiated since 2003 by the GOI with 100% assistance, initially in 54 districts of seven milk-shed states was being implemented in 313 districts in February 2014. By the end of the 12th Plan, it is envisaged to include all the districts (641) of the country under FMD—CP (Control Programme) following progressive control pathway (PCP-FMD) of OIE/FAO having seven phases. Initial 11 rounds of vaccine given at six-month intervals showed highly encouraging results in the control of the disease. Pigs are also to be included in the FMD-CP.

Besides the eradication of rinderpest, India got the status of freedom from CBPP in May 2007 and African Horse sickness in 2010s with the efforts of IVRI, DAD&F, GOI and other organisations. Dr. P.K. Uppal was instrumental in preparing the dossier on AHS for submission to OIE.

2.14.4 Sheep and Goats

The important point in disease management is that it not only rectifies reduction in output but improves efficacy of both reproduction and production. The magnitude of disease problems among sheep in India is enormous. Besides prevalence of many fatal acute contagious diseases, such as sheep-pox, enterotoxaemia, anthrax, pasteurellosis, PPR, foot-and-mouth disease, immature amphistomiasis, acute fascioliasis and black disease, several chronic diseases such as Johne's disease, brucellosis, gastrointestinal helminthiasis and other parasitic infestations create severe impediment for augmenting sheep production.

Moreover, emerging diseases such as bluetongue, viral pneurmonia, chlamydial and mycoplasmal infections are hindering sheep and goat development programmes in India. To combat the varied and complicated disease problems, efficient disease management system is necessary. The high susceptibility of imported and cross-bred animals to various diseases, in general, necessitates increased attention and investment in the health care/disease management programmes.

For efficient disease management, a link should be established between the incidence and control of diseases. This might be best accomplished if the diseases are forecast based on environmental parameters, quickly diagnosed, differentiated from other diseases, recognised in respect of their nature and behaviour and finally the etiopathologic pattern is determined for adopting effective control measures. To achieve these, efficient disease surveillance, monitoring and reporting systems are essential, besides quick diagnostic and disease investigation services. The popular proverb 'prevention is better than cure' is most pertinent to sheep disease management. Treatment of an individual case, except valuable exotic sheep (ram), is not economical and disease management in sheep is based on health cover measures for the entire flock by resorting to dips, bath, drenching and pasture management for the control of parasitic diseases and infestations, and vaccination against common infectious diseases. Several immunobiologicals and chemical agents are available for immunisation and protection against several sheep diseases. Application of these immunogens and chemicals are not pertinent and economical to all sheep-rearing zones. Although numerous diseases exist in India as potential hazards, yet distribution of some diseases is restricted to particular agro-/geo-climatic zones (for example, prevalence of lungworm disease in the hilly regions of Jammu and Kashmir, Himachal Pradesh and Sikkim). As such, priority for providing health cover would depend very much upon agro-/geo-climatic conditions and the type of animals maintained in that area. However, some diseases are prevalent in all sheep-rearing zones (sheep-pox, contagious ecthyma, fibrinous and suppurative pneumonia, enterotoxaemia and Johne's disease) for which health cover measures remain the same in all sheep-rearing zones. Again, some important disease problems crop up due to faulty and improper managemental practices which can be rectified by adopting proper scientific and hygienic disease management measures, *viz.* occurrence of pneumonia and coccidiosis can be prevented or minimised by avoiding over-crowding and providing hygienic and disinfected shed facilities for lambs). A comprehensive account of all sheep diseases is very vast and varied. As such, a detailed discussion on individual diseases is beyond the scope of this discourse. However, a brief and pertinent account of important infectious sheep diseases, related to rational disease management system is included.

Besides the infectious diseases, there are several other diseases which occur due to metabolic and nutritional disorders or toxicological factors. Management of deficiency diseases must be done with supplementation of required vitamins, minerals and trace elements in the form of licks/feeds or water mix after proper diagnosis. The management of toxicological disorders is based on detection of the toxic factor and its elimination (following detection) from its source. In addition to existing diseases, there always remains a risk for introduction of new diseases in a particular area or farm within the country or from outside the country through animals, animal products or other contaminants.

To prevent the entry of new diseases of infectious nature, animals should not be procured from a farm/country having a history of disease already under control in the locality or altogether new to that locality. All possible testing should be done on new animals to detect in-current infection before procurement. Before introducing newly procured animals to a locality or farm from within the country or from other countries, strict quarantine measures must be adopted and during the period of quarantine all relevant testing for seroconversion, clinical symptoms and prophylactic vaccinations and deworming operations are to be performed. If during the period of quarantine any infection is detected, the affected animal should be segregated, destroyed or treated properly for complete recovery depending on the disease condition and its status in the country, i.e. exotic, existing, re-emerging and emerging. In case there is a chance that the affected animal may act as a carrier even after recovery, it should be eliminated and if there is no such possibility, then recovered and healthy animals should be introduced in the farm or locality. Development of potent live PPR vaccine and adoption of mass vaccination programme on national basis by GOI in last few years has resulted in 75% reduction in the incidence of PPR in India.

2.14.5 Disease Management

For appropriate disease management, it has been repeatedly felt that an appropriate prophylactic calendar must be prepared with yearly schedules of health care practices. Preparation of such a calendar must be based on all available information regarding total disease picture of a farm or locality. For this morbidity and mortality data are to be recorded systematically.

Disease management is both science as well as art. It will vary from disease to disease, region to region, country to country, species involved, zoonotic/non-zoonotic nature of the disease, incidence of the disease; morbidity, mortality, production and economic losses, religious and other factors involved. Risk assessment, communication and management; disease forecasting, disease certification, traceability, quarantine, immune-suppressive agents in the environment, pasture management and deworming practices, balanced nutrition, productivity and adaptability of the animals, all influence the occurrence and extent of diseases. For forecasting of the diseases, study of epidemiology, developing state of the art of diagnostics, the ICAR established PD-ADMAS (renamed NIVEDI in 2014) during 1987 at Bengaluru. This institute is working on some of these aspects leading to disease forecasting, surveillance, molecular epidemiology and capacity building.

2.15 MEAT TECHNOLOGY

Meat technology is an essential component of animal production which contributes substantially to the economy of our country and deals with the production methodologies and post-harvest technological aspects of processing, preservation, utilisation, marketing and quality control of terminal products derived from slaughtered animals such as meat, meat products, offals, bones, blood, horns, gelatine, casings and other by-products.

Research and Development efforts were initiated in India in early 60s. Major efforts have been put in at the Central Food Technological Research Institute (CFTRI) and Defence Food Research Laboratory (DFRL) at Mysore, Indian Veterinary Research Institute (IVRI) at Izatnagar and Central Sheep and Wool Research Institute (CSWRI) at Avikanagar. Significant contributions have also been made by the Veterinary Colleges under various SAUs which have established departments for Animal Products Technology, Meat Science Technology Food Science Technology, etc.

The All-India co-ordinated Research Projects (AICRP) on poultry, sheep, goat and pig for meat, initiated by the ICAR at the various SAUs and R and D institutions have conducted research and generated data especially on feed lot performance and carcass quality evaluation of selected Indian breeds and their crosses with exotic animals.

This Central Drug Research Institute (CDRI), Lucknow, Central Leather Research Institute (CLRI), Chennai and Vallabbhai Patel Chest Institute, Delhi have also contributed through their work on utilisation of by-products of meat industry. There are many other government and private concerns which have also been involved, in one or other way, with the R&D of meat, meat products and by-products in the country.

The Union Ministry of Agriculture and Ministry of Food Processing Industries have contributed to the development of meat industry by extending support for modernisation of slaughterhouses and establishment of meat processing plants and allied industries throughout the length and breadth of the country. Five meat corporations have been set up by the State Governments of Andhra Pradesh, Goa, Karnataka, Tamil Nadu and West Bengal. A Sheep and Sheep Products Development Board has been established in Jammu and Kashmir. Now a few modern slaughter houses in public and private sector are operating under the Meat Food and Meat Products Order (1973) of the Government of India. Eight State Bacon Factories-cum-Composite Meat Plants are also operating under this order to ensure the production of quality meat and meat products. The Bureau of Indian Standards formulate standards for meat products and byproducts to rationalise the meat trade. Among the large abattoirs, the first one was established in Deonar, Mumbai in 1971. It has the capacity to slaughter 500 large and 7,000–8,000 sheep and goats/day. In 1978, a slaughter house with a capacity of 250 large animals per day was established at Aurangabad. This produces buffalo meat for export purpose only. In 1982, Goa Meat Complex was established. It has a capacity of dressing of 150 large animals/day. After this, several projects either for modernisation or establishment of new slaughterhouses had come up in the public sector. These include those at Durgapur, Mumbai, Delhi, Bengaluru, Srinagar and Hyderabad. A modern abattoir constructed with Danish assistance at the CFTRI, Mysore is being used for R&D work and training of personnel associated with handling of meat.

Poultry dressing lines with 1000 birds/day capacity have also been established at the CFTRI, Mysore and some of the SAUs, SVUs including HAU, Hissar. The CSWRI, Avikanagar has designed and constructed a low cost (less than 0.1 million) abattoir for hygienic meat production from small animals (50 sheep and goats/day).

Some business concerns in private sector have ventured into the area of establishment of modern slaughterhouse with processing facilities of meat and its by-products. Worth mentioning ones are AI-Kabeer group at Hyderabad, and buffalo slaughter houses at Derabashi (Punjab) and Aligarh. To promote the export of agricultural products and processed foods including meat and meat products, the Agricultural and Processed Food Products Export Development Authority (APEDA) was created by GOI in 1986.

For processing of carcass and meat by-products in the country, there are more than 250 bone crushing mills, 8 ossein manufacturing plants, 2 edible and pharmaceutical plants 1 photographic gelatin manufacturing plant, 3 surgical cat-gut manufacturing plants, 60 technical gelatin manufacturing units and 14 units manufacturing leather boards from leather shavings and trimmings besides a well-developed leather manufacturing industry. Manufacture of casings is on a cottage scale with very poor infrastructure. In metropolitan cities blood is partly being collected by pharmaceutical companies for the preparation of haematonics.

There are 27 agricultural universities with 30 veterinary colleges and 2 national institutes where Masters and Doctoral Degrees are awarded in Meat Technology and Hygiene, Meat Science and Technology/Livestock Products Technology. It is needless to mention that trained and qualified personnel are required to man various positions to effectively implement R&D programmes. Keeping this in view and the multifacial problems which the Indian meat industry faced, the ICAR established a National Research Centre (NRC) on Meat at Izatnagar. This NRC entrusted with the task of conducting basic and applied research on meat and to act as an apex centre for teaching, training, consultancy and technology transfer was subsequently shifted to Hyderabad. This centre is supported by a pilot plant abattoir and other meat processing and by-products utilisation facilities. The Centre supports the R&D programmes related to all aspects of meat and meat products to create sustainable technology for the growing meat industry in India.

Our animal resources for meat production are yet to be fully exploited. Except in the case of poultry, efforts towards evolving and developing special breeds of animals for meat production have not been made at national level as in case of milk production. However, due to increasing demand for quality meat, some studies have been initiated on production of farm animals for meat.

2.15.1 Meat Production

To improve meat production and quality, studies were conducted to evaluate meat yielding capacity and meat quality of selected breeds of animals available in

India and/or their crosses with exotic breeds. Cross-breeding done under the AICRP on Sheep for Mutton involving different sheep breeds of the region with exotic mutton breeds, *viz.* Sufflok and Dorset, proved that the half-breeds were superior to other grades with 3/4th and 5/8th inheritance in respect of performance as well as survivability. The half-breeds attain 30 kg live weight at 6 months of age when fed under individual feed lot trials.

Based on the performance of cross-breds at 50% and 75% levels of exotic inheritance and combining ability with two exotics, *viz.* Suffolk and Dorset, utilised in the cross-breeding programme, three mutton synthetics (strains), one at CSWRI, Avikanagar and two at AICRP, Palamner (AP) were evolved. The synthetics/strains show superiority in feedlot gain, efficiency of feed conversion and to a small extent in dressing over the indigenous breeds. They attained live weight of 30 kg at 6 month of age under intensive feeding conditions and are 60% more economical in term of meat production under intensive feeding conditions. Keeping in view the research results so far, it has been recommended that to augment body weights and the overall meat productivity, the native breeds with no wool or with very coarse wool may be crossed with Dorset and the exotic level of inheritance may be limited to 50% only.

The information available on the cross-breeding of Alpine, Saanen and Anglo-Nubian goats has shown that Anglo-Nubian crosses attained a boy weight of 23.98 kg at 5 month of age under intensive feeding. The large size indigenous goat breeds could serve as improver breeds for the small size breeds and non-descript goats. Although cross-breds are superior, the amount of gain was almost negligible.

Information on certain aspects in meat production in buffaloes such as growth rate, feed efficiency and carcass characteristics is available. Quality-wise buffalo meat contains higher protein and relatively less hydroxy proline content than beef. Usually meat from ruminant species is rich in saturated fatty acids. However, buffalo meat contains high proportion of polyunsaturated fatty acid. The need for salvaging male buffalo calves for meat production and studies regarding meat production potential of these animals in Indian conditions was felt. Studies indicated that male buffalo calves could attain weight gain of over 570 g/day on high protein and high energy diet.

A breakthrough was achieved in the production of broiler rabbits which can attain body weight of 2 kg in 12 weeks. Four exotic breeds, *viz.* Soviet Chinchilla, Gray Gaint, New Zealand White and New Zealand White Gaint, have been studied in this regard.

2.15.2 Meat Evaluation

A scientific score card system for evaluating the conformation of live buffaloes and goats, and their carcasses available in the market has been developed. Norms for slaughter, carcass cutability and separation parameters of male buffalo calves raised for meat production have been established at the LPT Division of IVRI.

Guidelines for rabbit and pig carcasses and meat evaluation were framed. Survey of indigenous cutting practices of buffalo/cattle carcasses at slaughterhouses of metropolitan cities in India revealed the prevalence of systematic splitting of carcasses into various commercial and retail cuts.

2.15.3 Improvement of Meat Quality, Meat Preservation and Shelf Life

Meat composition and lipid analysis were studied to determine basic meat characteristics such as water holding capacity, emulsifying capacity, extract release volume, pH, protein fractionation and texture, and their changes during post-mortem handling and storage. Improvement in slaughtering techniques specially scalding in poultry received much attention. Temperature between 12.5–12.8°F has been recommended for scalding of young birds and broilers for improvement of meat quality and shelf life.

Several efforts were made to improve the tenderness of meat from spent animals using electrical stimulation, enzyme treatment, pressure application, $CaCl_2$ phosphate etc. The work on electrical stimulation was carried out mainly at the Madras Veterinary College (MVC). The time, duration and voltage of electric current were standardised for chicken and mutton. Some work on post ex-sanguination E.S. was also carried out at the HAU, Hissar. Use of enzymes for improvement of tenderness was studied extensively at CFTRI, IVRI, PAU, HAU and MVC. Beneficial effect of papain in chicken meat reported higher drip and cooking loss in enzyme-treated samples. Use of papain for tenderisation of mutton; synergetic effect of papain enzyme and pressure treatment for tenderisation of spent hen meat were studied. Calcium chloride injection has beneficial effect on tenderness of goat meat.

Several workers have demonstrated the tenderizing effect of sodium tripolyphosphate (STPP). Increase in tenderness and shelf life in broilers soaked in polyphosphate solution. Histological changes caused by STPP with tenderness and beneficial effect of polyphosphate on drip los in quail have been reported. A conditioning technique of pelvic suspension of carcass instead of conventional way of hanging by achillies tendon has been derived at CFTRI, Mysore for sheep and goats. A minimum 4 h post-mortem ageing prior to hot or cold deboning of quail is recommended.

Studies were also conducted on thermal processing and cooking of meat. Stewing suited better for mutton while pressure cooker proved more beneficial for goat meat. Riboflavin was retained better than thiamine in cooked meat samples and per cent retention was maximum when meat was cooked from frozen state.

To increase the shelf life of meat, different sprays were tried. Buffalo meat sprayed with lactic acid and ginger extract in combination with sodium chloride has a shelf life of 48 and 43 h, respectively, at ambient temperature. Improvement in shelf life by lactic acid and/or acetic acid in buffalo and goat meat was also demonstrated. These technologies however, could not be adopted by meat processors due to problems in colour and texture. Dressed chicken chilled in 4% solution of polyphasphate, 5% brine and slush ice has a subsequent shelf life of not more than 6 days at 5°C. Shelf life of dressed quail is 8 days while 2% lactic acid treatment extended shelf life up to 10 days at 4°C.

The most common methods of meat preservation, *viz.* refrigeration storage and freezing, received maximum attention. Quality changes by adopting these methods were investigated. Freezing does not alter the textural profile of cooked mutton. Dressed broilers could be stored for 6 months at −18°C. Optimum time of freezing/method of freezing was also standardised at various laboratories.

Most of the studies on freeze drying were conducted in DFRL, Mysore. The feasibility of enzyme tenderised accelerated freeze dried goat meat was studied. Out of different enzyme tried, ficin was found best. Raw freeze dried goat meat is superior in quality than pre-cooked freeze dried meat. Freeze drying of even large pieces of poultry can be done.

Work on dehydrated chunks of mutton was carried out by Kumar et al (1976) and dehydrated rabbit mince were processed (Anthony and Ramanathan 1990). A dehydrated meat mince named 'Chevonette' was developed by Agnihotri and Prasad (1992), which remained stable and acceptable up to 45 days at room temp. Irradiation is also one of the methods for meat preservation. Work on irradiation was carried out mainly at Bhabha Atomic Research Centre, Bombay. Pushpa Paul et al (1990) reported enhancement in shelf life of lamb meat by using gamma irradiation. They have shown that pre packed fresh meat of buffalo, lamb and chicken, subjected to gamma irradiation at doses of 2.5–3 kg, could be stored up to 4 weeks at 0–3°C with remarkable improvement in microbiological qualities. Alur et al (1992) developed a microbiological method to identify irradiated meat. Kamat and Nair (1995) reported that irradiation not only eliminate common microorganism but also pathogens such as *Listeria monocytogenes* from frozen chicken meat.

Use of gases for preservation of meat has shown another promising area for extending shelf life. Reddy and Mondokhot (1987) at HAU, Hissar used SO_2 treatment (425–450 ppm) for increasing shelf life of minced goat meat. Abraham et al (1990) at MVC found liquid nitrogen useful for short term preservation of carabeef.

Principal of reduction in moisture content for preservation of meat was used by some workers. In CFTRI, Kadkol et al (1980) developed concentrated meat gravies on the same principal. A new product from pork named 'Meat preserve' with 38%–40% moisture, 6%–8% salt concentration and low pH (3.8–4.0) was developed by Puttarajappa et al (1990), which can be stored at ambient temperature (25–30°C) up to six months. At MVC, Prabhakar and Ramamurthi (1990) standardised a procedure by infusing/soaking buffalo meat sample in a solution of humectants glycerol (2%), sodium chloride (10%) and chemical preservatives. They reported acceptability of this product up to 2 months at room temperature.

Usefulness of curing and smoking as a possible method of storage at room temperature was tried at IVRI and CFTRI. It was found that by combination of curing and smoking, the shelf life of product could be enhanced to 14 days at room temperature and 30 days at refrigerated storage (Chatterjee 1969, Sharma et al 1973, Anjaneyulu et al 1995).

Studies on canning of meat and meat products were carried out at Madras Veterinary College and DFRL, Mysore. Various factors affecting pH, moisture, microbial load etc. were studied. Canning of some new products was also standardised. Harikrishnan et al (1983) prepared canned corned carabeef. Naresh et al (1986) reported that tin free steel and tin plate containers could be used for packing meat and poultry products. The quality of meat products packed in tin free steel cans was satisfactory and acceptable up to nine months at 37°C. On the contrary, Selvaraj et al (1989) reported that curried and canned sheep and goat meat remain acceptable up to 90 days at 28–30°C.

2.15.4 Identification of Meat

Identification of meat species has remained a subject of interest for many workers. This become more important under Indian conditions due to religious sentiments for consuming meat of certain species and ban on slaughter in some States and export of beef. Several serological and immunological methods were developed by Sherikar et al (1988, 1993). They have succeeded in developing reliable and sensitive technique for identification of meat of buffalo, cattle, sheep, goat and pigs employing species specific antisera to adrenal glands. The sera can detect adulteration to the minimum level of 5–10% of addition of fresh, frozen and partially heated meat and their organs. Use of polyacrylamide gel isoelectric focussing, agar gel diffusion and counter immune-electrophoretic techniques have been employed by same group of workers to identify adulteration in cooked meat. Shanmugam and Ranganathan (1972) at M.V.C. (Madras Veterinary College) evaluated precipitation test for differentiation of mutton and beef. At HAU, Hissar, Bansal and Mandokhot (1988) studied comparative efficacy of different serological methods. The differentiation between fresh and thawed meat is another problem of trade and quality assurance programme. Detailed work on this aspect was carried out by Sharma et al (1994) in LPT division of IVRI. Sen and Sharma (1996) developed a reliable test for differentiation of fresh, frozen and thawed and refrozen meats. Scientists at the Veterinary Collect, HAU, Hissar and NRC on meat have developed PCR test for differentiation of cattle and buffalo meat.

Microbiological analysis of meat has been the concern of many researchers and exhaustive data were generated. There is a need to compile these for framing standards and guidelines for improvement of hygienic status of meat and meat products. In their quest for rapid determination of microbial status and quality of meat, the scientists of DFRL, Mysore have developed an endotoxin assay for the first time using carcinoscopins amoebocyte lysate (CAL). The CAL test can be carried out in a moderately equipped laboratory and completed in just about 1.5

h. The test in the form of a portable kit has already undergone field trials. Thermo-nuclease test is an indication of growth of *S. aureus*. The occurrence of common pathogens in different meat and meat products was extensively studied. The organisms received much attention includes *Staphylococcus*, *Salmonella*, *E. coli* and *Campylobactors* specially, in poultry. Status of different market samples of meat and meat products has been studied at BAU, Ranchi; HAU, Hissar; CFTRI; IVRI etc.

2.15.5 Meat Processing

Several research efforts had been made to correlate/modify the emulsifying capacity and stability parameters in different laboratories in the country. Work conducted at LPT division of IVRI and CFTRI is worth mentioning. Sharma et al (1982, 1986, 1988) studied detailed functional characteristics of goat and buffalo meat. Kondaiah et al (1985, 1987) established relationship of different parameters in poultry meat.

Use of phosphates to improve the functional characteristics of meat products were tried by different workers. Kondaiah et al (1985) reported significant increase in pH, WHC, EC and decreased cooking loss in buffalo and goat meat by use of 1% sodium phosphate. Anjaneyulu and Sharma (1989) reported significant improvement in buffalo meat products by adding 0.5% polyphosphate which will also reduce the required salt level to 1.5%. They also reported that blends of phosphates were more beneficial than single phosphate. Madhwaraj et al (1983) reported that addition of wheat samoline (3%) and sodium pyrophosphate (1%) prevented water separation in the buffalo meat products. Anjaneyulu et al (1989) reported that addition of salt and phosphate during tumbling significantly improved extractability of salt soluble protein thus enhancing the cohesiveness, yield and sensory properties.

Use of hot processed meat for product preparation was demonstrated owing to saving of energy, labour, space, product yield and quality (Sharma et al 1988). Kondaiah et al (1985) found that hot minced and hot minced chilled goat meat had no significant difference in functional properties. Padda et al (1988a) reported that processing of hot and chilled goat meat into patties was equally good regarding most technological, organoleptic and microbiological properties. Padda et al (1988) also found that hot goat meat processed within 1 h PM had significant lower cooking loss, higher WHC, pH, water soluble protein and EC as compared to goat meat chilled for 24 h at 4°C.

Fat, which is one of the important constituents of meat emulsion, received much attention. It was established that not only level but type of fat is also important in determining the quality of product. Chatterjee et al (1979) found that addition of melted goat or sheep fat formed very unstable emulsion and they attributed this to the higher saturation of these fats as compared to pork fat. It was found that 50% peanut oil mixed with goat fat gave excellent emulsion. Padda et al (1986) reported two major problems in buffalo product manufacture, namely colour and mouth coating, which can be partially solved by replacing buffalo fat with vegetable oil or fat from other meat species. Several other studies conducted on type and level of fat will be discussed product wise.

In meat processing, major research is being conducted on emulsion based products. Among these types of products, meat patties received much attention. Major work on these has been conducted at LPT division of IVRI. Patties made from combination of chicken and mutton had shown significant better flavour and higher overall acceptability than mutton patties. It was also found that chicken fat produced significantly better flavour products than mutton fat (Kondaiah et al 1987). Similarly, for buffalo meat patties, combination of hydrogenated vegetable oil (15%–20%) and buffalo fat (0%–5%) were reported better than buffalo fat only. Incorporation of 20% whole egg liquid in mutton patties (Kulkarni et al 1988) or 15% soybean oil in buffalo patties (Anjaneyulu 1988) was also recommended. Sharma et al (1988) reported higher overall acceptability of goat meat patties when coarse mince (9 mm) plus one fine mince (4 mm) was followed by 2 min chopping in bowl chopper. Use of STPP was also recommended for overall improvement in the quality of patties (Anjaneyulu and Sharma 1989). Beside these, several other studies were conducted on the use of different binders/extenders on development of indigenous or indigenous flavour profile by use of onion, garlic, ginger and their combination for patties. Recently the use of textured soy protein (10%) in mutton patties was also recommended. Attempts were also initiated to develop patties from rabbit meat alone or in combination with chicken meat.

Sausages are most popular meat product in India. Sizeable amounts of these are also exported from India. Appropriate lean: fat ratio's and spice mix for Indian and export market have been standardised in different laboratories. It was demonstrated that during the processing of buffalo meat sausages, addition of 100–200 ppm sodium nitrite, prevent the development of dark brown colour (Sharma et al 1982). Highly acceptable sausages from buffalo meat by incorporation of offal meat were prepared by Krishnan and Sharma (1993). Use of different extenders/filters, were also tried in different laboratories to reduce the cost. Incorporation of potato was reported to improve acceptability of pork sausages as compared to wheat flour and soy. Incorporation of 20% egg liquid into mutton, mutton and chicken sausages have been reported to improve the quality attributes of the sausages (Gupta et al 1993). Anjaneyulu et al (1996) recommended use of skim milk co-precipitates in chicken sausages to increase the yield and acceptability.

With the increasing cost of energy, meat fermentation as a means of preservation appears promising. Among these types of products, dry and semi dry sausages having very long shelf life at room temperature were developed at LPT division of IVRI. Quickly ripened fermented sausage using lactic acid bacteria (LAB) having low salt and low nitrite was prepared by Salahuddin et al (1993). Use of dairy starter cultures in the processing of fermented sausages was found effective by Sharma and Mukhopadhyay (1995). They concluded that *L.*

plantaraum alone or mixed with *S. lactic* and *L. acidophilus* can be used successfully for processing of fermented sausages. To achieve better results and quality of sausages it would be desirable to adopt these dairy cultures on meat.

Processing technologies and recipes were also standardised for meat nuggets. It was demonstrated that in buffalo meat nuggets, emulsion technology was preferred over the restructuring process (Anjaneyulu et al 1993). Research was conducted to replace lean meat with extender/filters, e.g. pea flour, rice flour and milk proteins. Incorporation of rice flour up to 10% level was reported beneficial without any adverse effect on the quality of nuggets. Anjanyulu et al (1993) used raw pork rind without significant effect on nuggets quality. In another study, Anjaneyulu et al (1995) reported better acceptability of buffalo meat nuggets by replacing 50% of added fat with whole egg liquids.

A novel non-emulsion type product named meat ball was developed in LPT division of IVRI. The main advantage of this product lies in the non-requirement of bowl chopper. A simple hand mixing of minced meat can easily be done even at house hold level. Fat dispersion problem in meat ball can easily be solved by mincing the goat fat twice through 4 mm plate of meat grinder/mincer rather than once (Sharma et al 1988). Use of *Maida, Suji* and roasted *basen* at 5% level was recommended to enhance the cooking yield (Padda et al 1989). Attempts were initiated to develop low fat meat balls for health-conscious customers by use of carragenon.

Recipes and technology for chevon samosas were developed by Sharma et al (1988). The samosas with 10% inclusion of binders (Cooked mashed potatoes, black gram or black gram paste) in the stuff had better binding and three times more protein as compared to all vegetable samosas. A unique meat product 'meat tikka' of very good textural and organoleptic properties was developed by Sharma et al (1989) through the incorporation of various level of shredded potatoes, wheat flour, rice flour and bread crumps. Further attempts to improve the quality of meat tikka were made by Sekhon and Bawa (1991). They used *maida* and semolina for preparation of tikka from spent hen meat. The recipe for preparation of meat Kofta was also standardised in LPT division of IVRI. Padda et al (1988b) recommended use of 2.5% *maida* and 5% whole egg liquid for desirable binding in pork Kofta.

Tandoori chicken and kabab are popular traditional poultry products. Attempts have been made to standardise the processing method for these products to ensure quality and safety to the consumers. Salahuddin (1988) prepared better quality buffalo meat kabab from hot, chilled and frozen conditions of buffalo meat. The recipe for goat meat kabab was also standardised by Salahuddin (1988) and use of combination of goat and chicken fats was recommended. Recently in CFTRI, the processing of tandoori chicken was standardised. Baking of chicken for 90 min at 28°C after marination for 120 min was found optimum.

Recipes for pickle preparation from meat of different species and from poultry gizzard were standardised at CFTRI and IVRI. The peculiarity of this product lies in its shelf stability convenience and good acceptability by the consumers. Padda and Sharma (1982) developed a shelf stable 'bone-in' and 'bone-out' goat meat pickles. Low pH (approximately 5.0), low moisture content (30%–40%) and high salt content (4%–6%) were indicated to contribute to the long shelf life of such products. A very good quality chevon pickle from old buck meat with shelf life of 60 days at ambient temperature was developed at CIRG by Pal and Agnihotri (1993). Scientists at CARI have demonstrated that chicken gizzard pickle have shelf life of 45–75 days at ambient temperature during summer, monsoon and winter seasons (Sachdev et al 1994). Pickle from light fried chicken nuggets with better flavour, texture and acceptability was prepared in LPT division of IVRI. This product did not show visible fungal growth up to six months of storage.

Studies were conducted on several Kashmiri Wazavan products such as Gostaba, Rista and Nate-yakhani. The Indigenous methods are laborious, time consuming, limited to hot meat and are less relevant to suit large hygienic production. Effect of mutton fat level, hot vs. cold boned meat and manual vs machine mincing was studied on processing of Gosthaba and Rista by Samoon and Sharma (1994). It was concluded that machine minced, chilled mutton and fat up to 20% can be used after addition of 0.5% STPP for the processing of these products and cooked Goshtaba can be stored for seven days in LDPE bags (0.25 mm) at 4±1°C. Anotified recipe for Nate-yakhani was developed by Samoon and Sharma (1988).

Several turkey products were prepared at HAU Hissar by Panda (1985), comparative studies on processing and cut up yield of chicken, duck and turkey were conducted by Narahari et al (1988). They also prepared different products from Japanese quail meat using STPP.

2.15.6 Utilisation of Slaughter House By-Products

Bones constitute approximately 15% of the body weight of animal. In India, out of 0.4 million tonnes of available bones, only 0.1 million tonnes are collected and utilised (this data also includes bones from dead and fallen animals). The major portion of the bones is utilised by the bone-crushing mills for producing crushed bones, bone-girts and bone meal. In India, there are approximately 98 crushing mills with an installed capacity of approximately 0.5 million tonnes per annum. These mills crush bones primarily for export of crushed bones and bone grits. Bone meal is prepared by grinding crushed bones and then used as fertiliser. Steamed bone meal is used in poultry ration.

Another major by-product from slaughter house is hide and skin. More than 2,000 tanneries process approximately 5 lakh tonnes of hides and skins annually. Indian leather industry is growing fast to meet the demand of domestic and export market. Export from leather industry increased from INR 180 crores for the year 1973–74 to 3,760 crores by 1992–93. During 2014–15, leather and leather products exports were to the tune of USD 6.5 billion (approximately INR 40,000 crores). Presently, total value of Indian leather industry stands USD 12 billion. Increase returns from hides and skins compensate for meat production costs. The major work on improvement of quality and processing of leather was carried out at CLRI, Madras.

Blood is a major slaughterhouse by-product. Approximately 30,000 thousand litres of edible blood could be made available from slaughtered animals in certified slaughter houses in India, which amounts to approximately 5,800 thousand kg of high-quality protein annually. Several were made to utilise blood for preparation of blood meal and its utilisation in poultry feed. Efforts were also initiated to use blood protein in meat products. Hazarika and Biro (1993) concluded that addition of 5%–10% plasma, 1.5% blood, 1.5%–2.5% PPI or GPI in meat products would not affect the sensory qualities of sausages. Studies were also conducted in IVRI and the levels of different blood proteins in meat products were defined.

Besides blood, efforts were also made to utilise edible by-products into meat products. Krishnan and Sharma (1990) produced good quality emulsion type ready to eat beef sausages with 70 parts of skeletal meat and 30 parts of offal meat (heart meat and rumen meat 1:1). Anjaneyulu and Kondaiah (1990) reported that 15%–20% edible by-products (rumen and heart in 3:1) can be incorporated in meat products without affecting quality. Kondaiah et al (1988) successfully incorporated chicken by-products, *i.e.* skin, gizzard and heart in chicken sausages. The incorporation of these by-products resulted in better acceptability. Kowale and Rao (1995) isolated protein from ovine lung and rumen and incorporated up to 20% level in mutton patties without affecting their quality characteristics. Koil-Raja et al (1995) prepared snack type food from rumen musculature of goat. Agnihotri et al (1995) also attempted to develop a value-added product from goat stomach in the form of ready to eat crisp snack. The product packed in poly propylene bag was stable up to eight months at room temperature. Excessive fat in meat is also considered undesirable. Efforts have been made to utilise fat of animals, especially, pork into products prepared from other meat. A better technology for processing of edible lard has been evolved at the IVRI.

This resulted into increased export of casings from India. The Central Leather Research Institute (CLRI) developed a new method for processing of animal casings known as 'dry ready to wet' method. This organisation also developed an edible glue for joining the small pieces of casings together to bring the desired constant calibre. The processing of intestines into wet casing within 3–4 hours of slaughter was found superior in yield than processing of intestine following overnight ageing. Radhakrishnan and Ramamurthi (1987b) tried different chemicals to improve the quality of hog casings. Recently, Jaman Lal (1989) and Lakshmanan and Mendiratta (1996) studied processing and marketing aspects of buffalo and buffalo calf casings. They reported that the casing industry has great potential. However, immediate attention is required to improve quality. Specially, a suitable method is needed for removal and recovery of fat preparation of catgut from goat casings was studied by Singh et al (1995).

Research was carried out to prepare meat protein concentrate (MPC) through solvent extraction process from beef scraps and trimmings. The relatively stable nature of MPC and the absence of any objectionable taste and odour rendered it to be conveniently mixed with popular food without altering their organoleptic characteristics (Kowale and Rao, 1995). Better techniques for manufacture of carcass meal and hydrolysed feather meal were evolved for livestock and Poultry feed at IVRI and CARI.

Technology for extraction of neets foot oil from buffalo hooves was developed at BVC and IVRI, Izatnagar. Fractional crystalisation of the oil was performed by Mittle et al (1990) and reported to contain three fractions out of which one was non-solidifying at 0–10°C and could be used as a lubricant in fine machines operating under low temperature.

Technical know-how for isolation and purification of heparin from lungs and intestinal mucosa and oxytocin and vasopressin hormones from posterior pituitary glands of slaughtered animals were developed at the Division of LPT, IVRI, Izatnagar. Work on other by-products of the meat industry was carried out at CDRI, Lucknow (insulin, pepsin, pancreation, cholic acids and bile salts and albumin) and Vallabhai Patel Chest Institute, Delhi (hyaluronidase and heparin).

CLRI, Madras has developed a concept of 'two tier technology' for animal wastes utilisation and found suitable under our prevailing conditions (Muralidhara Rao, 1995). Firstly, the soft tissues are converted into meat meal. In the second phase, all kinds of offals are ensiled under anaerobic conditions. The fermented product develops an agreeable flavour to livestock and is fit to be used as a feed material. Both the approaches are claimed to be simple with minimal requirements of energy. Moreover, all the available raw materials including the ruminal contents could be made use of.

2.16. TRADITIONAL (HERBAL) VETERINARY MEDICINE: A PREVIEW OF THE LAST TWO DECADES

Ayurveda or traditional Indian system of healthcare having diversity, flexibility, accessibility and affordability, has been dealing with animal health care also have the potential to meet with the new challenges in health care due to changing life styles of humans and animals. Herbal medicine is today used with allopathic system of medicine for the benefit of human and animal health.

Based on the Ayurveda (life-knowledge), a medical art which had its roots in vedic literature (1500–1000 B.C.), Indian veterinary medicine is known for its specialised literature, which provides information on ancient methods of preventing and treating diseases of animals before the advent of modern medicine (Mazars 1994). A tradition of veterinary therapy developed very early in India and this has survived to the present time. The veterinary science was possibly more evolved and specialised as is indicated by various divisions of veterinary Ayurveda. There had been evidence on existence of literature on veterinary science in *Rigveda* (4000–2000 B.C.). In India, several Sanskrit treasuries on veterinary medicine such as *Asvâyurvedasiddhânta* (Complete Ayurvedic system for horses), *Hastyayurveda* (The Ayurveda of elephants), *Carakasamhit* and *Rajamartanda* (Nakula 1952, Mukhopadhyaya 1926, Sharma 1983, Mishra 1966) respectively. Pandav brothers Nakula and Sahadeva were experts in horses and cow (cattle) science respectively including description and treatment of disease ailments. Shalihotra and Nakul Sanhita have detailed literature on horse science.

Sustainability in Traditional Medicine. Where is the key?

India's ancient scriptures, teachings of sages as well as innumerable sayings and proverbs contain profound store-house of ideas, concepts and practices that are designed to address the process of building harmonious relationship among man, animal and nature. Indian medical treasuries such as *Charaka, Sushruta* and *Harita Samhita* contain content about animal care. India's traditional systems of medicine are part of a time-tested culture that intrigues people even today. The end of 20th century was earmarked with renewed interest and popularity of traditional system of medicine. This recent resurgence of natural therapies has been mainly due to the realisation of limitation of modern drugs in many conditions, increasing side effects, multiple drug resistance and presence of antibiotic residues in food chain and/or associated allergies and ever escalating cost of the modern therapeutics (Dhamaet al 2005). According to the FAO, lack of drugs to treat diseases and infections, results into losses of 30%–35% in the breeding of livestock in many developing countries, where poor animal health remains the major constraint to breeding (Confessor et al 2009). Traditional medical practices have sensed this gap and presented themselves as a dependable alternative. In recent years, uses of alternative medicines have increased significantly around the world. Herbs are more compatible with body because of their nature and having medicine homologues components together; which usually lack unwanted side effects; therefore, they are most suitable, especially in cases of long consumption as well as in chronic diseases (Borimnejad et al 2008, Sharma et al 2007).

Traditional medicine in principle is the medical practice, meant for resource poor people, utilizing local resources and offering an opportunity to get healed without access to many detailed investigations. The importance of Traditional Veterinary Medicine as an adjunctive form of care cannot be underestimated. Ethno-veterinary system is based on folk beliefs, traditional knowledge, skills, methods, and practices used for curing diseases and maintaining health of animals (Ahmed et al 2005). The information on treatment of ailments such as infection of horns, ears, tooth, throat, heart, navel, rheumatism, haemorrhagic enteritis, dysentery, digestive ailments, cold, parasitic/verminous diseases, stomach worms, rabies, abscess, anaemia, wounds, epistasis, retention of urine, urinary colic, constipation, lacrimation, arthritis, rhinitis, sprain, haematuria, skin infection and increasing milk production has been given in detail by (Sharma et al 2007, Somvanshi et al 1993). The use of this ancient practice, which is thousands of years old, often improves recoveries in comparison to conventional therapies alone. In short, traditional medicine is sustainable medicine in its true sense as it takes care of the environment from where this is originating.

Role of Ayurveda in Livestock Health Management

To be sustainable, livestock sector growth needs to simultaneously address key environmental, social, and economic challenges; such as growing natural resources scarcity, climate change, widespread poverty, food insecurity and global threats to animal and human health. In next 40 years, the demand for livestock products will continue to grow, and it will become increasingly challenging to meet it. At some point, perhaps as soon as 2050, it is estimated that there will be 9.15 billion people to feed, 1.3 times as many as in 2010 (UN Population Division 2009). As per FAO estimates in 2050, 2.3 times as much poultry meat and between 1.4 and 1.8 times as much of other livestock products will be consumed in 2050 as compared to that in 2010 (FAO 2006). The world economy has also been growing dramatically over recent decades, with a twenty-fold increase in global gross domestic product (GDP) between 1970 and 2012 (World Bank 2012). The world economy is projected to nearly quadruple by 2050, leading to a very significant increase in the demand for energy and natural resources (OECD 2012). Agriculture, particularly livestock production, accounts for a major share of these resource-use dynamics, as rising income levels tend to shift dietary patterns towards increased milk, meat, egg and fish consumption (FAO 2013). The World Bank World Development in its Report on Agriculture for Development (WDR) assesses the potential contribution of livestock sector to agricultural/economic growth and poverty alleviation, with particular focus on the policy implications (Pica et al 2008). An estimated 70% of the world's rural poor, whose livelihoods depend on livestock (FAO) must be taken care of.

The use of *homeopathic* and *phytotherapeutic* remedies in veterinary medicine has gained interest, among other reasons, due to increasing demands on the quality of meat and milk products such as the requirements for producing organic food goods (Pica et al 2008). The use of medicinal plants in the treatment of diseases as well as for increasing the productivity of high yielding animals such as cattle and buffalo has generated renewed interest in recent times, as herbal preparations are increasingly being used in cattle healthcare practices (Pieroni et al 2004,Wanzala et al 2005). In rural India and tribal societies, use of plants as veterinary medicines is very common. The ethno-veterinary and medical knowledge offers a range of herbs to be evaluated for treatment of livestock ailments. Application of turmeric on udder for mastitis treatment and linseed oil and turmeric externally for treatment of foot and mouth ulcers are commonly used. For Tympany treatment, drenching linseed oil along with a mixture of ginger, turmeric and asafoetida are used (Sharma et al 2008). In case of retention of placenta, bamboo leaves are fed (Gonzalez et al 2011). Different herbs, *viz. Caesalpinia crista* (karanjwa), *Melia azedarach* (bakain), *Saussurea lappa*, *Morringa oleifera* (sohanjna), *Trachelospermum jasminoides* (zard chambeli), *Butea frondosa* (dhak), *Fumaria parviora* (shahterah), *Nigella sativa* (kalonji) etc. can be used as anthelmintics. Several plant species *viz. Capsicum annuum*, *Lepidium sativum*, *Allium sativum*, *Sesamum indicum*, *Citrus limon*, *Zingiber officinale*, *Citrullus colocynthis*, *Curcuma longa*, *Amomum subulatum*, *Sesamum indicum*, *Cuminum cyminum*, *Rosa indica*, *Centratherum anthelmisticum*, *Triticum aestivum*, *Nigella sativa* and *Peganum harmala* have been documented and tested for the treatment and prophylaxis of mastitis in cattle (Dilshad et al 2009). Furthermore, these plants have been used for their immuno modulatory, appetite stimulatory, liver tonic, antiseptic and galactogogue properties. In rural Tami Nadu *Carissa spinarum*, *Arisaema tortuosum*, *Careya arborea*, *Bonzbax ceiba*, *Trichodesma indicurn*, *Baulzinia variegate*, *E. geniculata*, *Dalbergia latifolia*, *Azanza lampas*, *Syzygium heyneanum*, *Plumbago zeylanica*, *Wendlandia exserfa*, *Colebrookea oppositifolia* are used for different cattle ailments (Balakrishnan et al 2009). Of the plants recorded, *Abrus precatorius, Andrographis paniculata, Cassia fistula, Dendrocalamus strictus, Gymnema sylvestre, Strychnos nux-vomica* and *Wattakaka volubilis* are recognised as very commonly used ethnoveterinary medicinal plants (Torri 2013). Similarly in Rajasthan, Bihar, Maharashtra and Madhya Pradesh the herbs of ethnoveterinary importance *viz. Aegle marmelos, Annona reticulate, Zingiber officinale, Curcuma longa, Allium sativum, Cissus quandrangularis Miliusa tomentosa, Lagerstroemia microcarpa, Holoptelia integrifolia, Soymidia febrifuga, Butea superba, Merremia gangetica, Limonia acidissima, Santalum album, Datura metel,* are used in the treatment of various complication in cattle (Selvaraju et al 2011, Galav et al 2013, Mishra et al 2011).

Multilateral Policies and Promotional Initiatives

Traditional knowledge on health can range from home level understanding of nutrition, management of simple ailments and reproductive health practices to treatment of serious chronic illnesses or addressing animal health requirements. Interlinks to geography, community, world views, biodiversity and ecosystems make traditional health practices diverse and unique. Traditional medicine forms a central health seeking arena for a large part of the world population in developing countries (WHO 2002). The interdependence of humans, animals, and their environment has never been more important than now. The most prominent issues putting pressure on global health today include the **dramatic emergence and spread of zoonotic diseases, contamination of food, water and soil, bioterrorist events, and degradation of resources and habitats.** Current global health challenges have prompted a call for more holistic, collaborative and action-oriented approaches towards the goal of logical and practical solutions. Veterinarians have pivotal obligations, opportunities and contributions to make in enhancing public health, recognizing and responding to zoonotic disease transmission, maintaining food and water quality, and promoting wildlife and ecosystem health (Osburn et al 2009). Over the past three decades, WHO has produced several policy documents under different names for traditional medicine relating to their promotion and development, regulation, guidelines and standards for safety, efficacy, good manufacturing, research, assessment, clinical trials, rational use, training of health practitioners and standardisation of terminologies (WHO 2002). A single medicinal plant may be defined as a food, a functional food, a dietary supplement or a herbal medicine in different countries, depending on the regulations applying to foods and medicines in each country. This makes it difficult to define the concept of herbal medicines for the purposes of national drug regulation and confuses patients and consumers. To meet these challenges, the WHO Traditional Medicine Strategy was developed, with its four primary objectives: framing of policy enhancing safety, efficacy and quality; ensuring access and promoting its rational use. The World Health Assembly (WHA) has also adopted several resolutions relating to traditional medicine as below:

(i) In 1988, referring to the Chiang Mai Declaration, the WHA resolution on medicinal plants, placed the medicinal plants, their rational and sustainable use, and their conservation, firmly in the arena of public health policy and concern (WHA 41.19).

(ii) In 2003, a WHA resolution on traditional medicine, referring to the WHO traditional medicine strategy, requested the WHO to collaborate with other organisations in the UN system and NGOs in various areas related to traditional medicine, including research, for protection of traditional medical knowledge and conservation of medicinal plants resources (WHA 56.31)

(iii) Regarding the GSPA-PHI, the WHO identified traditional medicine as one of the areas to be addressed in its Quick Start Programme. The programme aims to support research and development and to promote standard-setting for traditional medicine products in developing countries (WHA 61.21).

(iv) In 2009, the Beijing Declaration, a solution on traditional medicine, urged national governments to respect, preserve and widely communicate traditional medicine knowledge while formulating national policies and regulations; promote appropriate, safe and effective use; further develop traditional medicine based on research and innovation; and consider the inclusion of traditional medicine in their national health systems (WHA 62.13).

According to WHO guidelines for herbal medicine, the countries should establish a process for manufacturers to acquire GMP status within an established time frame. At all levels of handling of raw plant materials, clear and accurate identification and labelling is important. Countries should also consider mechanisms for controlling contamination of raw plant materials with pests, microorganisms, aflatoxins and other mycotoxins, pesticides, heavy metals and other foreign matters. A special unit for herbal medicines should be created under the responsible agency for the regulation of medicines (WHO, 1998). An innovative research effort to define the advantage of traditional system of medicine with respect to their safety and efficacy could result in a better utilisation of these complementary systems of medicine. Internationally several pharmacopoeias have provided monographs stating parameter and standard of many herbs and some products made from these herbs. Some pharmacopoeias are given below:

(i) United States Herbal Pharmacopoeia

(ii) British Herbal Pharmacopoeia

(iii) British Herbal Compendium

(iv) Chinese Herbal Pharmacopoeia

(v) The Ayurvedic Pharmacopoeia of India (API)

(vi) Japanese Standards for Herbal Medicine

These Pharmacopeias lay down monographs for herbs and herbal products to maintain their quality in their respective nations (Verma and Singh 2008). The World Health Assembly - in resolutions WHA31.33 (1978), WHA40.33 (1987) and WHA42.43 (1989) - has emphasised the need to ensure the quality of medicinal plant products by using modern control techniques and applying suitable standards. Standardisation and quality control parameters for herbal formulations are based on following fundamental parameters: 1. Quality control of crude drugs material, plant preparations and finished products; 2. Stability assessment and shelf life; 3. Safety assessment; documentation of safety based on experience or toxicological studies; 4. Assessment of efficacy by ethnomedical information and biological activity evaluations (Shinde et al 2009). Chemical fingerprint is used to analyse the chemical constituents, aflatoxins, heavy metals, pesticides, foreign organic matter, in herbal medicines, consisting chromatographic fingerprint, such as thin layer chromatography (TLC), high performance liquid chromatography (HPLC), gas chromatography (GC), capillary electrophoresis (CE) and spectral fingerprint, for instance, UV, IR, MS, X-ray and their hyphenated techniques (Yongyu et al 2011 and USP Monograph 2013). The quantitative tests, e.g. total ash, acid-insoluble ash, water-soluble ash, alcohol-soluble extractive, water-soluble extractive, ether-soluble extractive, moisture content, volatile oil content and assays are the methods upon which the standards of Pharmacopoeia depend (The Ayurvedic Pharmacopoeia of India, GOI).

Several research institutions of the ICAR and Agricultural Universities have conducted various investigations on plant types; its botanical assessment, toxic-pharmaceutical assessment and experimental trials on animal models and concerned species. This has also been done by various Veterinary drug producing companies such as Ayurvet Limited, one of the leading organisations in this area.

Contribution of private sector R&D for Animal Health Care Management

Herbal product development: Private sector has taken a key initiative in this arena of combining empirical knowledge about the popular use of medicinal plants with establishing scientific basis of quality, safety and efficacy of products. By its combined approach of traditional knowledge and modern research, it has modernised the pharmacy into branded herbal ready-to-use formulations for livestock.

Quality standards: Phytopharmaceuticals could only be considered as rational drugs if these are standardised and their pharmaceutical quality is approved. At standard private sector firms, quantitative and qualitative multi-component analysis by advanced analytical techniques is done which serves as a specific tool to set quality standards and specifications to ensure therapeutic efficacy, safety and shelf life of herbal drugs. The application of high-technology oriented advanced hyphenated techniques in its R&D serves as a rapid tool in unravelling the hidden medicinal benefits from complex herbal formulations.

Every product formulated, undergoes stringent quality-control tests. A major emphasis is given to the right selection, scientific processing and efficacy evaluation of different herbs. Toxicological and pharmacological profiles of each herb are thoroughly documented before inclusion in product formulations. Each product undergoes testing for dose standardisation, palatability, pre-clinical and clinical efficacy evaluation and scientific validation.

Quality control of herbal formulations is done according to standard WHO guidelines. Thorough macro- and microscopic examination is done for the raw material to be used for manufacturing products. The development of authentic analytical methods which can reliably profile the phytochemical composition, including quantitative analyses of marker/bioactive compounds and other major constituents, is a major task performed by R&D scientists. Therapeutic activity of a herbal formulation depends on its phytochemical constituents.

Standardisation of Herbals: Standardisation is an important step at R&D for the establishment of a consistent biological activity, a consistent chemical profile, or simply a quality assurance program for production and manufacturing of an herbal product. Standardisation based on a single or small number of chemical markers or classes of compounds serve mainly to promote quality control and batch-to-batch consistency in terms of efficacy. Methods of standardisation are considered keeping in view all aspects that contribute to the quality and pharmacological efficacy of the herbal preparations.

Herbal formulations of private sector have reached extensive acceptability as therapeutic agents for several diseases. Therapeutic activity of herbal formulation depends on its phytochemical constituents.

In view of the above, standardisation is an important step for the establishment of a consistent biological activity, a consistent chemical profile, or simply a quality assurance program for production and manufacturing of herbal drug.

The development of authentic analytical methods which can reliably profile the phytochemical composition, including quantitative analyses of marker/ bioactive compounds and other major constituents, is a major challenge. Standardisation is an important step for the establishment of a consistent biological activity, a consistent chemical profile, or simply a quality assurance program for production and manufacturing of herbal drugs. Standardisation is an important aspect for maintaining and assessing quality and safety of polyherbal formulation as these are combinations of more than one herb to attain the desired therapeutic effect. Standardisation minimises batch-to-batch variation; assures safety, efficacy, quality and acceptability of the polyherbal formulations.

Clinical Validation: Clinical trials are an integral part of a new product discovery and development. Clinical trials are initiated on new formulations and existing products to demonstrate product benefits and safety in targeted species with a cost-effective study design. With the holistic approach to rigorously pursue the product scientific validations and quality control, R&D collaborates with the premier government agencies such as the ICAR and its institutes, educational institutes, universities, veterinary colleges, and the scientific research and clinical efficacy studies on the formulations are carried out by the competent scientists and investigators at various research centres in India and abroad. The objectives of these clinical studies are largely three-fold: (First, the verification, demonstration and substantiation of clinical benefits. Second, the products performance across diverse geographical locations. Third, to understand the mode and possible mechanism of the product action, including the aspect of pharmacological evaluation). The clinical studies continue even after the product has been introduced. To fulfil these objectives, clinical trials are executed to assess product tolerance, determine maximum tolerable dose, estimate product efficacy for targeted indication, standardise dose rate, exclude the possibility of any serious toxicities and thereby to establish product safety and efficacy profile. The data generated and the results of these investigations are shared with the veterinary fraternity for the benefit of the community. The trial reports and the results of clinical research studies are also essential for gaining approval for registration of dossiers and for marketing new formulations or new indications of existing products from regulatory agencies throughout the world.

The private sector has pioneered the launch of many scientifically validated herbal formulations and trusted brands for poultry, swine, aqua and livestock in the last one decade.

Role of phytogenics and herbal feed additives for animal health care: Feed additives depending on the specificity of function of additive improve the health, digestibility, productivity, product quality and immunity. It is scientifically established that inclusion of herbal feed additives to ration improves overall growth, performance and productivity of animals, enhances nutrient utilisation and feed efficiency, possess immune-potentiating and anti-oxidant effect, improves gut microflora and have anti-microbial activity.

Improves gut function: Improvement in gut function is mainly attributed to the possible stimulatory effect of phytogenic substances on digestive secretions, such as digestive enzymes, bile, and mucus. Phytogenic substances from certain herbs, *viz. Aegle marmelos, Plantago ovata, Acacia catechu, Corriandrum sativum,* many more other herbs and their extracts improve gut microflora and exert pharmacologic actions within the digestive tract, as evidenced by their gut function modulating efficacy (Satyavati and Gupta 1987). Pathogenic bacteria are always present in the gut, but the balance between pathogenic and beneficial bacteria determines whether a disease will occur. Maintaining a healthy balance between all microflora within the gut is known as eubiosis and can be influenced by bacteria endemic to the microflora (Kolte et al 2009). In the intestine, bacteria considered beneficial to the gut, including lactic acid forming bacteria such as *Lactobacillus* spp, prevent proliferation of pathogens, such as *Salmonella* spp., through competitive exclusion for nutrients and for receptor sites on the gut wall. Beneficial bacteria can also produce an adverse environment for pathogenic bacteria to colonise and grow, for example, by the production of short-chain fatty acids which lower the pH and prevent growth of pH sensitive pathogenic bacteria (Kolte et al 2009). The microflora also have functions in the development of the digestive and immune tissue in the host animal as these can produce nutrients that can be used by the host and can neutralise some feed toxins and promote an environment in the gut where anti-nutritional factors and toxins are minimised. In the past antibiotic feed additives were used for manipulation of the microflora to create eubiosis. However, with the severe restriction of antibiotic feed additive use in the EU and increasing consumer concern, alternatives to antibiotic feed additives have been investigated and found to significantly influence this balance also.

Increases feed intake: The stimulatory effect of herbal feed additives on feed intake is due to the claimed improvement in palatability of the diet resulting from the enhanced flavour and odour. The addition of *Woodfordia fruticosa, Zingiber officinale, Allium sativum, Trigonella foenum graecum* etc. to poultry and pig ration improve feed efficiency (Satyavati and Gupta 1987). Increased palatability of the diets associated with the addition of phytogenics also may be due to their anti-oxidative effects, which might contribute to preserving the desired organoleptic qualities of the diet.

Digestive tonic and growth promoter: Role of ruminal microflora in digestion of nutrients is vital, so interactions of the normal microbial flora with the host can be manipulated to improve the efficiency of nutrient utilisation (Waghmare et al 2009). Supplementation of herbs that can modulate the rumen function may be siren to animals for efficient cellulose breakdown and digestion, maintenance of normal rumino reticular function, intestinal movement, optimum utilisation and absorption of nutrients, improving feed conversion ratio, productivity and body weight gain in animals. A potent herbal formulation containing some minerals and 28 different herbs such as *Allium sativum, Azadirachta indica, Calotrophis orocera, Centratherumanthelmenticum, Commiphora mukul, Eclipta* *elba, Embelica ribes, Picorrhiza kurora, Zinziber officinale* and *Piper longum* is used as an appetiser, restorative, stomachic, digestive tonic and growth-promoter (Bhatt et al 2009, Kolte et al 2009).

Hepato efficiency enhancers: Supplementation of certain liver tonic preparations also increase secretion and flow of bile for better digestion and to treat anorexia by maintaining the liver parenchyma in healthy state and regulating liver functions such as detoxification of metabolic products, toxic drugs, chemicals and treatment of hepatic dysfunction. The herbal ingredients such as *Andrographis paniculata*, *Eclipta alba*, *Picrorhiza kurroa*, *Phyllanthus niruri*, *Tephrosia purpurea*, *Tinospora cardifolia* and *Boerhavia diffusa* improve feed conversion efficiency, body weight gain and reduce mortality in poultry and swine; owing to their hepato-protective, hepato-stimulants and growth promoting properties (Dwivedi et al 1986).

Another polyherbal liver tonic and growth promoter product for cattle comprises herbs namely, *Andrographis paniculata, Eclipta alba, Picrorhiza kurroa, Phyllanthus niruri, Tephrosia purpurea, Tinospora cordifolia* and *Boerhaavia diffusa*, each with documented hepato-protective and hepato-stimulant properties. It counter act hepatopathy and restore liver functions in bovines (Pradhan and Dey 1996). Polyherbal liver tonic formulations also protect the hepatic parenchyma from various parasitic and liver fluke infestations such as visceral larvae migrants.

Anti-oxidant, immunomodulator and anti-stress effect: Various stressors such as high ambient temperature and relative humidity influence the performance of animals and birds by reducing feed intake, feed efficiency, nutrient utilisation and feed conversion ratio (Sahin et al 2003).

Ayurvedic formulae that treat stress contain herbs with adaptogenic (antistress) effects. Herbs with high anti-stress and anti-oxidant activity such as *amla* (*Emblica officinalis*), *ashwagandha* (*Withania somnifera)*, *tulsi* (*Ocimum sanctum*), *Shilajit* and many more have proved to be potent oxygen free radical scavenger *in vitro* and *in vivo* models. Medicinal plants or herbs owing immuno-potentiating properties can provide an alternative to conventional therapy for a variety of diseases, especially when the host's defence mechanism must be activated under conditions of impaired immune response.

Herbs such as *Mangifera indica, Echinacea purpurea, Phyllanthus emblica, Uncariatomentosa, Withania somnifera, Ocimum sanctum, Tinospora cordifolia,* and*Asparagus racemosus*, among many others, are scientifically validated to possess anti-stressor, immuno-stimulant or immuno-modulant and adaptogenic properties. Mangiferin, which is the active constituent of herb *Mangifera indica*, is a strong inducer of *invivo* and *invitro* activation of peritoneal macrophages. Induction of interferon release from macrophages by *Mangifera* has a potent lymphoproliferative effect on macrophage activation, thus establishing the therapeutic potential of *Mangifera* as a immunomodulator. Another herb *Echinacea purpurea* stimulates macrophages but does not increase resistance to a wide variety of stressors such as physiological stress. However, *Uncaria tomentosa* stimulates interleukin-1 and-6 in macrophages, stimulates endothelial cells to produce a lymphocyte proliferating regulating factor and enhances recovery of leukopenia induced by doxorubicin. Similarly, *Withania somnifera* is well established to augment endogenous antioxidants, maintenance of myocardial antioxidant status, significant restoration of most haemato biochemical and oxidative stress marker parameters by its free radical scavenging activity in addition to potentiating agglutinin antibody titres and complement fixing antibodies (Ziauddin et al 1996). The antioxidant potential of methanolic extracts of *Ocimum sanctum* indicated their prime role in free radical elimination, thus combating immunosupression. It can be interpreted that the characteristic dysregulation of stress hormones and neurotransmitters; thereby, the immuno-supression can be successfully ameliorated by the botanical adaptogenic remedies including either single herb or combination of different herbs (polyherbal formulation) (Manoharan et al 2004). Another such a polyherbal immunomodulator, adaptogenic and anti-stressor formulation 'Stresroak' was scientifically validated in amelioration of various stressors, e.g. production, vaccination, overcrowding, water deprivation, lead toxicity, cadmium toxicity and to potentiate immune response during diseases conditions, e.g. infectious bursal disease and New castle disease in poultry.

Antimicrobial effect: The medicinal or antimicrobial properties of plant-derived substances have been well known for centuries. This property is mainly attributed to the essential oils of these plants, namely, *Trichyspermum ammi, Cinamonum camphora, Mentha piperita* and many more. *Oregano* and T*hyme* are among those which have received a great deal of interest. These plants contain monoterpenes carvacrol and thymol, respectively, and have demonstrated high efficacy *in vitro* against several pathogens found in the intestinal tract, suggesting that phytogenic feed additives may be suitable replacements for in-feed antibiotics to improve pig health and growth performance, particularly during the first few weeks post weaning.

Herbal formulation for cure of skin diseases

A herbal formulation which contains Onion extract, Garlic extract, Karanj oil, Camphor powder, Lemon extract, Turmeric powder and Sesame oil is very effective against mange, ring worm, pyoderma, eczema, FMD lesions, foot rot lesions, surgical wounds, burns, cuts, abrasions, etc. It was developed by M.C. Sharma and S.K. Dwivedi during 1985–92 at Indian Veterinary Research Institute, Izatnagar. This herbal formulation 'all-in-all' has been commercialised in the year 1997 and available in market for treatment of skin ailments. It is the first technology of IVRI which has been commercialised to private sector. IVRI had developed many other herbal formulations, which are very effective against hepatitis, diarrhoea, dysentery, renal disorders, gastric ulcer, diabetes, etc.

Toxin binder: Aflatoxins reduce growth and feed efficiency, suppress immune system, reduce antibody titre and cause mortality and morbidity leading to severe economic losses to poultry farmers (Sakhare et al 2007). Synergistic action of herbal toxin binders along with liver tonic formulations is scientifically well proven to improve growth, reduce mortality in broiler chickens by better protein and energy utilisation by limiting the adverse effects of aflatoxins. Additionally, other polyherbal toxin binders namely Vilocym and Vilocym Z are popular in poultry industry, among farmers and veterinarians for control of mycotoxicosis.

Natural methionine: Methionine is required in free as well as conjugated form to perform various metabolic functions. It assists in the breakdown of fats and thereby prevents the build-up of fat in liver and arteries, as well as assisting the digestive system and removing heavy metals from the body since it can be converted to cysteine, a precursor to glutathione, which is of prime importance in detoxifying the liver. Its deficiency may manifest in symptoms such as fatty liver, retarded growth, reduced feed efficiency, weakness, edema, skin lesions, poor feather condition, decreased egg production, small egg size, immuno-suppression, low carcass yield and disturbances in various metabolic pathways (Ramarao et al 2003). Improper conversion of methionine can lead to atherosclerosis. Though most plants contain very little amounts of methionine; some have significant amounts. High levels of methionine are found in fish, meats, and some plant seeds. Almost all feed ingredients of plant origin used for compounding poultry ration are deficient in methionine used and hence DL-methionine is commonly added as supplement in poultry feeds. In the past, the requirement of amino acids of chickens used to be met from more than one sources of protein from plant and animal origin.

However, due to increase in prices of quality fish meal and availability of comparatively cheaper soybean meal, use of all soya feed started with addition of methionine as soybean meal is deficient in methionine which again is a costly supplement. In poultry ration, just like protein, the vitamins and minerals have equally important role in development of musculature. In the list of vitamins, choline (growth promoter) is playing vital role by acting as lipotropic agent, thereby preventing abnormal fatty infiltration in liver, thus ensuring proper metabolism of the body and effective utilisation of nutrients. Moreover, it helps in formation of an excitatory neurotransmitter acetylcholine, which is responsible for proper functioning of nervous system and maintains harmony. There is no consistency in the choline content in the natural feed stuffs and their bioavailability is not predictable. Therefore, in broilers as well layers, choline in combination with chloride (CC) is added as an important feed ingredient.

However, herbal or natural products, which have proven results in various areas of poultry production, can be combined as a mixture and premix can be prepared, which may be used in feed formulations. Supplementation of polyherbal formulations containing ingredient herbs that mimic methionine, choline and biotin such as activity prevent fatty liver syndrome and improve overall carcass quality, yield and egg production in broilers. These herbal amino acid supplements have an added advantage over harmful effects of synthetics, use of which is totally prohibited in organic poultry production.

Conclusion

Traditional systems of medicine have stood the test of time for over 4000 years but should not be considered as an alternate to the modern veterinary medicine; rather it complements and enhances the production of livestock. Since ancient times, herbs and their products have been constantly used for curing illness. However, herbal additives have aroused much scientific interest over the past few years to explore their role as performance enhancers in livestock production. The cited instances in this text are just glimpses from the vast and virgin world of Ayurveda through the clear eye of modern science. It is up to our scientists and the industry how best they use this precious gift from Mother Nature for health and welfare of livestock and in turn mankind.

2.17. QUALITY CONTROL OF HERBAL PRODUCTS

Nature always stands as a golden mark to exemplify the outstanding phenomena of symbiosis. In the western world, as the people are becoming aware of the side effects of synthetic drugs, there is increasing interest in the natural product remedies. With the advancement in science and development of new quality control methods, a global resurgence and renewed interest of scientific community in herbal products has occurred (Barnes 2003). With an increasing global demand for herbal formulations, it is necessary that the quality and the consistency of these drugs are maintained for their maximal efficacy. Quality is of paramount importance when it is specifically related with drugs (De Smet 1999).

Throughout the history of mankind, many infectious diseases and non-infectious conditions have been treated with herbals. Scientific investigations have highlighted the importance and contribution of many plant families, i.e. Asteraceae, Liliaceae, Apocynaceae, Solanaceae, Caesalpinaceae, Rutaceae, Piperaceae and Sapotaceae, which contribute medicinal plants. Medicinal plants play a vital role for the development of new herbal products/formulations. The bioactive extract should be standardised on the basis of active compound and should undergo safety studies (Wani 2007). Herbal products/medicines originated from Ayurveda. Ayurveda a centuries old traditional Indian system of health care, is an evidence based science and one of the oldest medical disciplines. Ayurveda is a complete system which emphasises living in harmony with the environment. Ayurveda is the synthesis of *Ayur*—meaning life—and *Veda*—meaning knowledge. Ayurveda, based on the strong pillar of positive health, is a holistic approach to total health care by means of preventive and curative medicine to maintain complete internal-milieu (*dhatusamaya*), 'Homeo-stasis' or equilibrium of the various *dhatus* (De Smet 1999). It insists on the pro-host approach: 'Strengthening the body defence system to fight infection'. Ayurveda emphasises mainly on prevention of illnesses and maintenance of health. Relevance of science of Ayurveda, as applied to animal health, can be traced back to Mahabharata period. Herbal medicine is today a valid and proven science, with great potential for integration with allopathic medicine to safeguard human and animal health (Bodeker 2005).

Recently, Ayurveda has been documented and practiced on animals as an alternative to synthetic medicinal supplements. Herbal veterinary products are available for prophylaxis, therapeutics and as nutraceuticals and feed supplements for prevention and treatment of various diseases and for maximizing productivity from livestock.

Quality Control of Herbals

Quality control refers to processes involved in maintaining the quality and validity of a manufactured product. The term 'Herbal Drugs' denotes plants or plant parts that have been converted into phyto-pharmaceuticals (Arun et al 2012). Phyto-pharmaceuticals could only be considered as a rational drug if these are standardised and their pharmaceuticals could only be considered as a rational drug if these are standardised and their pharmaceutical quality is approved. For the transformation of 'Herbalism' into a 'Dynamic, scientifically validated and evidence based science', a major emphasis is required to be given on the right selection, scientific processing and efficacy evaluation of different herbs (Schulz et al 2001).

The methods of testing for quality control involve macro and microscopic examination as the first step towards establishing the identity and degree of purity of raw material/herbs. Macroscopic inspection is based on odour, taste, size, shape and surface of characteristics. The microscopic examination is more important for the identification of the herbs for quality control (ICMR 2012). Another important aspect on quality control of herbal formulations is determination of foreign matter, including tests for hazardous contaminants such as pesticide residue, arsenic and heavy metals, micro-organisms, including aflatoxins and sulphur dioxide residues.

Quantitative and qualitative multi-component analysis by advanced analytical techniques such as HPTLC, HPLC, GC/MS, LCMS etc., is done which serves as a specific tool to set quality standards and specifications to ensure therapeutic efficacy, safety and shelf life of herbal drugs. This provides simplest and quickest mean to establish identity and purity of herbal drugs. Further standardisation techniques *viz.* characterizing constituents/chemical entities/ markers/TLC/DNA fingerprints of herbal products are done for quality precision checks. A variety of molecular techniques including restriction fragment length polymorphism (RFLP), random amplification of polymorphic DNA and DNA sequencing are used to authenticate plant material and detect adulterated plant species. Each technique has advantages and disadvantages in terms of cost, accuracy, reproducibility, time and taxonomic level of identification (Wankar et al 2009).

Standardisation of herbal drugs

With the realisation of the growing popularity and demand for phyto-pharmaceuticals, standardisation is becoming a mandatory part for the regulation of herbal drug industry (Gaedcke 2003). At present, owing to advancement in the chemical knowledge of crude drugs various methods such as botanical, chemical, spectroscopic and biological methods are used for various estimating active constituents present in the crude drugs in addition to its physical constants (Wani 2007). Researchers today are emphasizing on evaluation and characterisation of various plants and plant constituents against several claims given in Ayurveda. These unique plant phyto-constituents are used to standardise botanical preparations during all aspects of manufacturing processes, and to obtain stability profiles of the formulations being developed (Kunle 2012). Standardisation is an important aspect for the establishment of a consistent biological activity, a consistent chemical profile, or simply a quality assurance programme for production and manufacturing of an herbal product. Standardisation based on a single or small number of chemical markers or classes of compounds serves mainly to promote quality control and batch-to-batch consistency in terms of efficacy (Eskinazi 1999). For standardisation of natural product drugs, single chemical entities, 'marker compounds', may be used as potency standards in high performance thin layer chromatography (HPTLC) or HPLC analysis. These marker compounds may be used to help identify herbal materials, set specifications for raw materials, standardise botanical preparations during all aspects of manufacturing processes and obtain stability profiles.

Salient examples of standardisation of herbals

A proprietary poly-herbal formulation, Toxiroak premix, a mycotoxin inhibitor, acts as a mould inhibitor for poultry. It inhibits the biosynthesis of mycrotoxins, effectively binds the toxins in the GI tract, and helps in bio-neutralisation of the toxins in liver. The mould inhibition and bio-neutralisation activity of the premix is due to the presence of certain herbs such as *Azadirachta indica, Curcuma longa, Allium sativum* that are known to inhibit the growth of fungus and block the biosynthesis of mycrotoxins. *Andrographis paniculata* or Kalmegh also known as 'king of bitter' is one prioritised medicinal plants in India and this herb is reported to possess anti-hepatotoxic, antifungal activities besides its general use as an immune-stimulant agent. The key chemical constitutes of the plant are andrographolide, bisandrographolide, andropanolide, 14-deoxy-11-oxo-andrographolide, 14-deoxyandrographolide, andropanolide, 14-deoxy-11, 12-didehydro-andrographolide, etc. To ensure the batch to batch consistency in terms of biological efficacy and chemical profiling, a new simple, rapid, precise, accurate and specific RP-HPLC method was developed to carry out the quantification of andropapholide in the formulation vis-à-vis its standardisation to ensure the batch to batch consistency in mycotoxin inhibition activity of the poly-herbal formulation premix.

Another proprietary polyherbal formulation, Methiorep premix is available for poultry, a phyto additive containing herbal ingredients, mimics methionine-like activity. It is natural analogous to the synthetic counterpart and is first limiting amino acid in poultry diets which is very essential for all vital biological functions (Vijay et al 2014). Its constituent herbs namely *Cicer arientinum, Phaseolus mungo, Mucuna pruriens, Triticum sativum*, and many more are scientifically well established for having highly bioavailable methionine and for having potential role in promoting growth (Rajmane and Sonawane 1997). New high performance thin-layer chromatography (HPTLC) methods were developed for the quantification of two main ingredients of the formulation, i.e. *M. pruriens*, a herb having rich source of protein with neurotransmitter precursor L-3, 4-dihydroxyphenylalanine (L-DOPA) as one of its major constituents, and the plant *Azadirachta indica*, a hepatoprotective, aflatoxin synthesis inhibitor (Reddy et al 2011); bird's growth, and feed efficiency promoter (Wankar et al 2009) having rutin as one of its bioactive phyto-constituents. The methods were validated for linearity, accuracy, and precision in accordance with the statistical method of validation given in ICH Q2R1.

A proprietary poly-herbal formulation, stresroak liquid is available for combating stress in poultry. It is a blend of extracts of medicinal plants, *viz., Phyllanthus emblica, Withania somnifera, Mangifera indica, Ocimum sanctum* and *shilajit*. This combination has excellent anti-stress properties and immuno-modulatory activity (Oyagbemi et al 2008), resulting in increased adaptive immune response in birds and non-specific modulation of the immune system. The major active components of the formulation for example, gallic acid, hydrolysable tannins (Ghosal 1996), mangiferin (Andreu et al 2008), with anolide-A (Mishra et al 2008), eugenol, ursolic acid (Anandjiwala et al 2006), dibenzo a-pyrones and fulvic acid (Islam et al 2008), contribute to the anti-mutagenic, anti-cancer, anti-oxidant, immune-modulatory activity of the product (Bauer et al 1983). These compounds and their quantification have been reported in various parts of the plants (Sharma et al 2007). Standardisation of the product with respect to the bioactive phyto constituents was taken up to ensure the batch to batch consistency in efficacy. New HPTLC and HPLC methods were developed for the quantification of two main ingredients of the formulation, i.e. *Phyllanthus emblica* with gallic acid and *Mangifera indica* with mangiferin having anti-oxidant activity (Charoenteeraboon et al 2010).

Bio-standardisation

In one of the studies, an anti-stress product, a blend of extracts of medicinal plants *viz., Phyllanthus emblica, Withania somnifera, Mangifera indica, Ocimum sanctum* and *shilajit*, having excellent anti-stress and immune-modulatory properties was bio-standardised with respect to the free radical scavenging anti-oxidant potential which in turn substantiates its anti-stress property and ensure the batch to batch consistency in terms of biological activity (Ravikanth et al 2013).

Conclusion

Determination of pharmacological properties of an herbal formulation, which depend on phytochemical constituents present therein, development of authentic analytical methods which can reliably profile the phytochemical composition, including quantitative analysis of marker/bioactive compounds and other major constituents, is a major challenge to scientists. Without consistent quality of a phytochemical mixture, a consistent pharmacological effect is not expected. Standardisation is an important aspect for maintaining and assessing the efficacy, quality and safety of the poly-herbal formulation as these are combinations of more than one herb. The concept takes into consideration, all aspects that contribute to the quality and pharmacological efficacy of the herbal drugs.

2.18 SIGNIFICANT ACHIEVEMENTS IN ANIMAL PRODUCTION AND HEALTH IN INDIA DURING THE PAST 25-30 YEARS

The advances made in different spheres of livestock and poultry development sector during the past two and half decades in the country are described in brief here. Special emphasis was laid on the impact of major achievements observed in the researches carried out under the umbrella of the ICAR system in addition to various SAUs' and SVUs' in improving the livestock productivity and disease control in different species of livestock and poultry; along with different approaches made in achieving these major objectives. The role of biotechnology that promoted growth in the livestock sector in the recent past received special attention. The impact of feed processing for strengthening the feed resource base in the livestock and poultry feeding received maximum attention to improve their overall performance. Success stories have been mentioned wherever the impact was remarkable and more on job creation and financially remunerative.

2.18.1 Animal Nutrition

Livestock enterprise plays a multi-dimensional role in the agricultural development in India and has contributed substantially to the national gross domestic product' (GDP). During 2010–11, India produced an estimated total milk production of 121.8 million tonnes, total meat production of approximately 4.83 million tonnes, 61.45 billion eggs and 43.13 million kilograms of wool. Productivity and income growth in this sector witnessed a strong income multiplier. Income through livestock farming accounts for 15%–40% of the farm households' income in different states of the country.

The challenges currently posing the dairy production in India include lower productivity from indigenous stock, small herd size, insufficient quality of male germplasm coupled with poor artificial insemination (AI) network, inadequate quality health coverage, fast-declining grazing lands superimposed with poor quality feed and fodder resources and a strong drift in the rural youth force to be away from livestock farming. However, India has rapidly positioned itself as the world's largest producer of milk. The per capita availability substantially increased from 226 g/day during the 90s to approximately 280 g/day by 2010–11 against the world average of 285 g/day. In 2013–14, per capital/day availability of milk was approximately 307 g against the world average of approximately 294 g. The growth in meat production was faster as compared to milk production. Total meat production has increased from 0.9 million tonnes of the 80s to the current level of 6.2 million tonnes. The meat production structure has changed dramatically during the 90s with monogastrics, more precisely poultry, which has emerged as the major contributor. Cattle, buffalo, sheep and goats and pigs and poultry contributed to 55%, 17%, 11% and 16%, respectively, to the total meat production. The egg production quadrupled from 10 billion in the 80s to 62 billion by the 2010s, leading to per capita availability of 55 eggs with total egg production at 74.752 billion in 2013–14, per head/year egg availability has become 61. During recent years, wool production is showing a declining trend since a sizable number of sheep are slaughtered at an early age due to more remunerative price obtained through sale of mutton.

Livestock research receives approximately 20% of the total agricultural research resources. Despite the low intensity of investment in research, animal science research, over the last few decades, has generated several technologies in animal genetics and breeding, feed and nutrition, health and management. The technical feasibility of many of these was proven under experimental conditions. Examples include cross-breeding and replacement in cattle, pigs and poultry; chemical and biological treatment of cereal straws through fermentation technology; and vaccines against rinderpest, influenza and foot and mouth disease, improved livestock management practices and post-harvest management. Additionally, processing technologies were developed to strengthen the vertical linkages between the farm and dairy industry. The adoption pattern of these technologies varies widely across species, farm typologies and regions.

In compiling the information furnished in the following section under various sub-heads, salient features from relevant and selective research publications, annual reports of species specific national institutes, vision documents, final reports of completed research projects and 2013 edition of handbook of Animal Husbandry of the ICAR have been consulted.

Ration balancing

Livestock feed is a scientifically balanced blend and nutritionally superior to homemade feeds. Of late, a concept of 'Total Mixed Ration' (TMR) has become popular consequent to enormous research efforts made by researchers in the country as an improvement over traditional feeding system of providing concentrates and roughages as separate entities. Research results support the view that the TMR feeding improves the milk yield by 10%–15% and the yield remains uniform over a long period with a longer persistent peak yield.

Food-feed system (FFS) is also a recent concept developed due to inter-collaborative research with national and international research institutes such as

International Livestock Research Institute (ILRI) and ICRISAT to produce foods for human and feeds for animal production and this FFS has been receiving more attention under smallholder farming systems in the pursuit of providing feeds on a year-round basis. Good example is the recent development and release of improved varieties of groundnut by the ICRISAT at Hyderabad.

In the organised sector, balanced compound feeds are blended and fed in such a manner that 3.5–4.0 kg of it may support 10 kg of milk production when fed over and above the maintenance level. The National Dairy Development Board (NDDB) has recently developed a user-friendly software for advising small holders at their doorsteps to balance the ration for lactating cattle with locally available feed resources including area-specific mineral mixtures (ASMM). This model has been accepted by the Food and Agriculture Organization (FAO) for implementation in other developing countries. A training program was organised in Columbia in association with NDDB, Anand to popularise and extend the technique to the needy stake holders. The outcome of data generated under this project in India involving approximately 12,000 dairy animals in seven locations indicated that feeding a balanced ration can increase net dairy income by 10%–15% for those having 1–2 cows and buffaloes. This has been possible by an increase in milk production and a decrease in feed cost besides reduction of enteric methane emissions by 10%–20% per kg of milk produced. Large-scale implementation of such models can further enable enhanced productivity and net income of smallholder dairy farmers.

The Indian feed industry has grown remarkably during the past three decades reaching approximately 7 million tonnes by the Compound Livestock Feed Manufacturers Association (CLFMA) members alone, which is 5% of the total cattle feed requirement for the year 2012. The remaining 95% of the compound cattle feed is still manufactured under unorganised sector spread over the country. There will be a large cattle feed demand in the future and the agro-industrial and forest by-products could play even a bigger role as an unconventional source of energy and protein. Equally important is the need for the effective participation of private sector in adopting and spreading the modern technologies developed in India such as crop residue based feed blocks and area-specific mineral mixtures.

The technology of straw-based densified total mixed ration (DTMR), also called as densified complete feed block (DCFB), offers a good opportunity to remove regional imbalances in feed availability, because of easier transportation. This is a novel approach of supplying balanced feed to ruminants at a lower cost. Results of on-farm feeding trails conducted recently on buffaloes around Karnal, showed a significant increase in milk production by 10%–15% compared to conventional feeding by the farmers. Although, feed cost increased by 30% with block feeding, the profit earned was approximately 20% higher. The growth rate of calves was increased by 25%–30% resulting in early maturity of growing stock thus improving overall life time productivity. The optimum supply of nutrients, including micro-nutrients has a positive effect on health status of animals due to immuno-protection against many infectious diseases resulting in significant savings in the cost of medicating animals.

Of late, the term precision animal nutrition (PAN) is gaining important role under Indian context to produce quality animal products to the consumers as a profitable enterprise with least disturbances to climate change. This new concept leads to effective utilisation of available feed resources more judiciously with the major objective of maximizing animal response to nutrients. The tools to achieve PAN include improved feed processing techniques, precise ration formulation, implementing phase feeding and the optimum use of feed additives. It also attempts to balance the end products of rumen fermentation in lowering enteric methane production from ruminants. Efficiency of ruminal fermentation and digestibility of nutrients are found to be the key factors in improving the efficiency of feed use.

Biotechnological approaches

Shortage of feed and fodder resources has been the major constraint in improving livestock productivity in India. Productivity of Milk, egg and meat has consistently increased over the past 25 years. Improvements in overall feed availability have no doubt been the major contributor to achieve the improved productivity. It is against this background that the National Institute of Animal Nutrition and Physiology (NIANP), Bengaluru undertook a systematic study to assess the availabilities and requirements for various feed resources in different regions of the country. Increasing livestock production by improving the germplasm of livestock and strengthening health care is possible only when there are adequate feed resources. The precision of feed resource assessment can be judged by relating the total livestock numbers and livestock produce with the feed resource availability. Addition of new feed resources, changes in extraction rates of any of the by-products due to fluctuations in processing and changes in export/imports must also be accounted for to obtain precise data.

The ICAR, New Delhi led Indo-Dutch project on bio-conversion of crop residues was implemented in the 90s through a network of centres spread over the country. This project demonstrated the use of systems and participatory approaches to understand the farming systems in general, with in-depth understanding of livestock subsystems in various agro-eco-zones (AEZ) of India. Studies on the involvement of women in livestock production showed that women have good knowledge of feeding and management aspects of livestock, optimum use of local resources, and that women take decisions related to these aspects. However, the current research system rarely involves women or considers their knowledge and use in planning studies, conducting trials or assessing the research outputs. This needs to be changed.

Substantial research studies have been conducted in improving the feeding value of cereal straws through different chemical treatments resulting in standardisation of urea-ammonisation technique being the most practical processing method to be popularised under field conditions. The major constraints in the field scale application of this developed technology were reported to be the cumbersome and labour-demanding aspects of treatment, marginal returns, requirement of large quantities of clean water and covering material, spoilage of straw when not properly stored besides poor extension linkages existing with the farming community. If the spraying of the urea solution could be mechanised, the adoption of the technology appears to have a bright future.

Urea-molasses-mineral block (UMMB), a strategic feed supplement for ruminants, provides a constant source of fermentable nitrogen throughout the day to promote growth of rumen microbes. Feeding of UMMB increased feed intake, milk yield and growth rate and as such is a cost-effective approach to maximise the utilisation of locally available feed resources for improved productivity. The NDDB has successfully come out with a technology using cold process of manufacturing these blocks and has designed a plant with automatic pressing device for commercial manufacturing of block licks. The technology is currently being made available to dairy co-operatives, private sector organisations and international agencies for commercial production of blocks. Currently, approximately 3 lakh blocks of 3 kg are being used in India annually.

Feeding different crop-residue-based complete feeds in mash or pellet form, on average increased milk production by 15% and supported daily weight gain up to approximately 110 g in kids and lambs. Feed wastage in lactating and meat yielding stock were reduced by approximately 23% and that of cost per kg milk yield and weight gain by 17%–20% under on-farm conditions. A complete feed-processing unit suitable to rural India was developed by ANGR Agricultural University, Hyderabad and distributed to several organisations and farmers as a component of transfer of technology to improve nutrient use efficiency in the animal food chain. This technology, if transferred for use on a community or on co-operative basis, could play a key role in enhancing livestock-based rural economy in the country.

More than two decades of research involving several studies conducted for ruminant production has established the relevance of by-pass protein technology in the country. The NDDB has standardised the process of producing rumen by-pass protein supplement by treating the protein rich meals through formaldehyde treatment to enable rumen by-pass of 75%. This technique has emerged as one of the cheapest feeding technologies meant for high yielding ruminants. This proved to be a viable commercial feed-technology in India and has increased milk yield by 10%–15%. Experimental proof clearly indicated that farmers feeding by-pass protein can earn 10% more income in case of cows and 15% in buffaloes.

Adding protected fat (by-pass fat) to the rations affected efficiency of dairy cows through a combination of caloric and non-caloric functions. Feeding calcium soaps of vegetable oil to high yielding dairy cows resulted in higher milk yield and fat. Inclusion in the rations of high yielding dairy cows during yearly lactation enhanced tissue mobilisation of nutrients before peak production. The work carried out at NDRI, Karnal revealed that supplementation of protected fat at 2.5% of dry matter significantly increased the proportion of unsaturated fatty acids in addition to increased level of long chain fatty acids in the milk fats. Protected fat supplementation at 10g/litre of milk in cows maintained on existing feeding practices under on-farm situations improved milk yield and reproductive efficiency with a net profit of INR 12/cow/day.

The emerging molecular biology techniques such as sequence-based metagenomic approaches have been used to isolate novel microbes including methanogens and GMOs with high fibre degrading capacity in ruminants. However, it is a long way to go for their field application.

Azolla, which is a free-floating water fern that grows naturally in stagnant water with the ability to double its weight in a week's time, can produce 9 tonnes of protein/ha of pond per year. The potential use of azolla, as a bio-fertiliser and animal feed both for livestock and poultry, makes it an attractive and effective input for both the vital components of integrated farming (crops and livestock). When used in feed for lactating cattle, azolla was reported to increase the milk yield by 15%–20%. Its use in India is in its infancy, nevertheless, it is being promoted by non-government organisations, co-operatives and government agencies and interest in its use is gradually catching up, although sporadic.

Attempts were made to reduce the adverse effects of high tannin fraction in leaves by adding polyethylene glycol at feeding. Some larger farmers in Punjab treat wheat or paddy straw with urea at the time of harvesting itself with good success.

Emerging Feed Resources

The bio-fuels industry has evolved rapidly during the recent past with developments in improved processing techniques and the expansion of the range of plants and other natural energy sources being considered as feed stocks. Majority of research studies are concerned with the feeding value of co-products; the need for standardisation of products identified; and the search for new feed resources, particularly those indigenous to an area but underutilised and their safety standards. Two major co-products from the biodiesel process are protein-rich cake/meals and glycerol. Adult cattle can consume 1 kg of glycerol/day, as a source of rapidly fermentable carbohydrate. The two major by-products from palm oil processing are palm kernel cake (PKC) and crude palm oil. The PKC could be successfully incorporated in the ruminant diets up to 30% level. Micro-algae are rich in protein and vitamins, therefore, are useful as food and feed. Micro-algae containing astaxanthins were also used as potential feed resource in aquaculture production, where they can be fed with, or replace fish meal, acting as colouring agents in such species in addition to improving yolk colour, growth and survival when fed to poultry.

Distillers grains (DG), a by-product of the alcoholic drinks and beverage production, have been fed to livestock in the recent past. This by-product is regarded as a cost-effective energy feed that also contains substantial amounts of protein with useful levels of amino acids and is also rich in digestible phosphorus compared to other feed resources.

The potential uses of sweet sorghum co-products for livestock feeding are unequivocally established. These new feed resources in ensiled form offer unforeseen opportunity not only for meeting livestock feed demand for poor farmers but also for offering an environmentally sound agro-enterprise that has tremendous implications for organic recycling related to carbon sequestration, green house gas (GHG) emissions and ecological balance.

The ILRI and ICRISAT centres have carried out extensive research on the feeding value of sweet sorghum by-products meant for bio-ethanol production. Sweet sorghum bagasse and stripped leaves, the residues after juice extraction, have fodder qualities that allow incorporation into feed blocks without prior chemical or physical upgrading.

In a classical participatory research study conducted in Vaishali area of Bihar, it is evident that more than 4,000 farmers adopted urea-ammonisation of paddy straw during 2010. The study also indicated that such technologies fitted well with the straw handling system and benefited well with the straw handling system and benefitted the entire farm operation of the farmers (not milk production alone). The large network of Krishi Vigvan Kendras, Universities, Research Institutes and ICAR centres should enable adoption of such approaches without much difficulty. However, there would be a need to orient research scientists to develop close linkages/interaction with the extension system and farmers.

Nutrient requirements and Indian feeding standards

Extensive research contributions made by distinguished animal nutritionists representing several national institutes of the ICAR and State Agricultural/ Veterinary Universities spread over in India worked out the nutrient requirements of different livestock species in India during the past two decades which led to the publication of 10 bulletins that dealt on the 'Nutrient Requirements' of cattle and buffalo; sheep, goat and rabbits; poultry, pig; finfish and shellfish; camel; equine; yak and mithun; companion, laboratory and captive animals besides nutrient composition of Indian feeds and fodder by the ICAR during 2013. This mammoth task was the result of contributions made by a high level technical committee appointed by the ICAR, New Delhi under the Chairmanship of Dr. K Pradhan with NIANP, Bengaluru as the nodal agency. These publications took into consideration the decadal changes in the national feed resource base, dynamism in livestock population trends and their product specific utility. Computer models, considered to be the most effective means of taking animal variations with changing needs of different nutrients have been used wherever needed. The contributions made through these publications will serve as a preamble and a navigating guideline for the livestock and poultry keepers and future researchers involved in formulating scientifically blended balanced rations.

Animal Physiology

Several reports indicated low reproductive efficiency in cattle and buffaloes owing to area-specific mineral deficiencies, as defined by analysis of soil, water, forage and blood of cross-bred dairy cows. Information from 10 locations in Karnataka state was used to formulate supplements containing common salt and sources of calcium, phosphorus, copper and zinc. As subsequent on-farm testing proved to have better conception rates and improved profitability, the mineral supplement was made available to the farmers through the Karnataka Milk Federation (KMF). The success story led to implementation of are-specific mineral mixtures in eight states at present.

Agricultural biotechnologies are prominent among the suite of innovations available to livestock producers. In addition to involvement of public sector organisations, few private agencies are also involved in undertaking quality research to improve animal productivity. A good example is the classical research study initiated by Nimbkar Agricultural Research Institute (NARI) in Maharashtra in 1996. In this research rams and ewes of Deccani breed of sheep carrying the *Fec B* gene for prolificacy were introduced into 36 smallholder flocks. Several are now designated as multiplier flocks providing *Fec B*—carrier ewes to other farmers. The sheep produced 27%–46% more saleable lambs compared to sheep without this gene.

Scientific programs such as frozen semen artificial insemination, embryo transfer technology (ETT), multi-ovulation embryo transfer technology (MOET), marker assisted selection (MAS) and development of semen sexing technology leading to use of sexed semen are the technologies that were aimed at faster propagation of elite germplasm and for increasing livestock productivity. It is, therefore, necessary that research in newer biotechnologies is intensified, so that traditional genetic improvement programs and improved health management systems are enhanced further. At the same time, waste management is going to acquire important concerns with increase in the number of intensive farming units coming up. Hence, there is a need to consider making anaerobic waste management obligatory for larger dairy farms, poultry industry and dairy/meat processing plant. To take full advantage of the innovations and bio-nano-technologies throughout the length and breadth of the country, capacity building at state level field institutions with both public and private partnership mode is a must with needed investments.

Breeding technology

India has a broad spectrum of native breeds of cattle, buffalo, goats, sheep, swine, equine, camel and poultry with merits of adoptability to climate and nutrition and resistance to diseases and stress. Genetics and breeding researches have helped in evolving many new breeds, strains, synthetic population of cattle, pig, sheep and poultry using cross-breeding techniques. These breeds have better production coefficients compared to indigenous ones. Cross-breeding of low-yielding indigenous breeds with high-yielding exotic breeds has been widely acknowledged as a valuable strategy to improve animal productivity. Sporadic attempts to improve the genetic potential of different livestock species by selective breeding (poultry, sheep, and cattle) in India were initiated during the latter half of the 19th Century, but no significant achievements were reported. In the 50s, systematic cross-breeding research and development programs were initiated. Cross-breeding research has focused mainly on cattle because of their dual role of milk production and use as draught animals in the crop subsector of agriculture. Severalcross-breeds with improved production potentials were evolved. These, include Hariana x Friesian, Hariana x Jersey, Hariana x Brown Swiss, Rathi x Jersey, Gir x Jersey, Gir x Friesian and Sahiwal x Jersey. Adoption of cross-breeding technology has been slow. Only 7.5% of the cattle population consists of cross-breeds. However, their adoption in the field is limited and sporadic; only approximately 8% of cattle, 5% of sheep, 15% of pigs and 33% of poultry populations belong to the cross-bred/improved category. The adoption level is higher in peri urban areas compared to rural areas. In general, wide scale adoption of cross-breeds is restricted due to their non-acclimatisation to the tropical climates prevailing in most parts of the country. In the dairy sub-sector, the buffalo could emerge as a promising alternative to cross-bred cattle because of its adaptability to varied ecological conditions, higher milk yield and higher fat content, realising a premium price. Unlike cattle, buffalo slaughter is not banned and animals have a disposal value. These factors have favoured a higher growth in buffalo population even without improved breeding interventions. However, in certain regions, non-descript low-yielding buffaloes have been upgraded using high-yielding breeds such as the Murrah. Scientists of National Dairy Research Institute, Karnal, have developed the landmark technique, i.e. 'Handguided Cloning Technique' and produced the world's first cloned buffalo calf.

Dairy products

The dairy co-operative network in the country includes 177 milk unions covering 346 districts and over 1,55,634 village-level societies with a total membership of nearly 15.1 million farmers till March 2013. Fifteen major State Co-operative Federations have 173 dairy plants (of total 254 co-operative processing units) with a handling capacity of more than 335 lakh litres of milk/day in 2013–14. Besides liquid milk, these plants manufacture value-added products such as *paneer*, *dahi*, dairy whitener, cheese, ice cream, infant food, *khoa*, *peda*, *misti-doi*, sweet curd, milk-cake, *kalakand*, *lassi*, *shrikhand*, *gulabjamun* mix, *rasagolla*, condensed milk, *burfi*, buttermilk, *chhach*, sweets and UHT milk, besides *ghee*, milk powder and flavoured milk. The private-sector plants manufacture mainly dairy whitener, cheese, milk-food, condensed milk, UHT milk, and curd. Testing of milk for safety and quality parameters at the collection centres is almost non-existent. Lack of appropriate anaerobic waste treatment and dairy by-product utilisation are the other concerns. value addition and export potential has not been fully exploited due to the quality concerns of milk.

Post-harvest technologies help producers to realise better gains from technological changes in the primary production sector. However, the postharvest processing facilities are lacking in the country. Only approximately 20% of the total milk production is processed into value-added products. The bulk of it, however, is processed into *ghee* and curds by the producers, and a large proportion of this is consumed at source. Although considerable efforts have gone into developing infrastructure for milk processing in the co-operative sector, only approximately 5% of the total milk output is processed into table butter, cheese, milk powder and baby foods. Information on the proportion of other livestock products entering the value-added chain is not available.

Bulk of the investment for livestock development comes from the state governments. The central government contributes approximately 10% to the total investment. There is hardly any private-sector investment in animal husbandry. The dairy sector, however, has attracted considerable private investment in processing, value addition and marketing. Flow of institutional credit, mainly the commercial banks is approximately 10%. More than 70% of the refinance disbursement by NABARD goes for dairy development. Investment-linked tax incentives and attractive credit facility to private investors are missing.

Growth in small ruminant research

Goats and sheep constitute approximately 39% of total livestock population. In India, 40% of sheep slaughtered are of spent category with low body score, dressing yield and carcass fat. There is ample scope of bridging the gap between demand and supply by improving meat yield and quality of culled sheep by feed supplementation which has not been explored to the extent needed.

During the past three decades, commendable work on nutrition of goats and sheep on native ranges and pastures has been carried out in the species-specific national institutes in the country. More specifically, in rehabilitation of land by perennial grasses, silvi-pasture, legume pasture and its utilisation by rotational grazing, stocking rate, carrying capacity and grazing management were developed. The studies further supported the concept that range lands promote full spectrum of production systems from secondary to traditional nomadic and semi-nomadic system. Nutrient digestibility in general was found to be higher in goats relative to sheep in view of faster functional rate of passage of digest a through the rumen, leading to higher dry-matter intake.

It has been estimated that if a farmer spends approximately INR 300/animal for three months, one can earn a net profit of approximately INR 200/animal from the culled stock. At national Institutes such as CSWRI and CIRG, modern technologies such as defaunation, probiotic supplementation, protected fat supplementation complete feed blocks and limiting nutrient supplementation have been tried successfully in kids and lambs. A short duration feeding strategy of 45–90 days with minimum feed input and maximum output in terms of live weight gain may fetch a higher price. This will compensate for the feeding and rearing expenses besides minimising the recurring expenditure on these animals.

Meat production trends

During 2010, India produced 0.29 million tonnes of mutton by slaughtering 24.10 million sheep out of a total sheep population of 74 million. Contribution of small ruminant sector to the Indian economy is estimated to be approximately INR 3,000 crores/annum, contributing approximately 5.40% of Gross National Product (GNP) of agriculture sector. The NRC on Meat has established during 1999 at Hyderabad. Research and development efforts made at the Centre would contribute for developing an organised meat sector in India to benefit meat animal producer with sustained earnings and consumers with safe and quality meat products of choice which, in turn, contribute for increase in per capita animal protein consumption and employment generation opportunities among weaker sections of the society. Value addition to meat sector and direct as well as indirect contribution to related sectors invariably would support GDP growth positively. Another important area where a major breakthrough was achieved is salvaging male buffalo calves and spent animals to enable the domestic meat needs as well as increasing export potential to as many as 75 Middle-Eastern and other countries. A new concept of consortia model project on 'Value Chain for Clean Meat Production from Sheep' was carried out successfully during 2008–12 under the ICAR system with the World Bank's assistance by the NRC on Meat, Hyderabad as the Consortia Leader. Different value added products were developed from meat and by-products of sheep, which included a novel shelf stable rumen crackles and a good range of edible products from sheep blood having commercial viability. It was for the first time in the country that field studies involving approximately 200 farmers and 5,000 ram lambs, covering 94 villages spread over five districts of Andhra Pradesh State were undertaken to promote commercial sheep production for value addition, based on the technology developed from on-station studies. The program included development of a feeding system that included development and evaluation of creep feeds from weaner stage followed by testing the impact of economic complete feeds in combination with silage developed from locally grown cereal forages to attain early slaughter and quality improvement of meat and value addition through several by-products developed.

At several institutions emulsion based meat products were developed as prime choice and economy with variable quantities of lean, fat, by-products and extenders. Comminuted products also allow inclusion of non-meat extenders such as soy, cereal floors, and meat by-products which will help to reduce the product cost and improve the yield and quality of the finished products. In India, only 1%–2% of total meat is processed into products for trade as compared to more than 60% in developed countries.

Average carcass weight of most livestock species is still lower; 104.27 kg/cattle, 119.59 kg/buffalo, 13.36 kg/sheep, 11.01 kg/goat, 40.26 kg/pig and 1.3 kg/poultry. During 2000–01 to 2009–10, the meat yield of cattle, sheep, goat and poultry grew at an annual rate of 1.5%–2.0% and 0.8% for pig, while meat yield of buffalo remained constant despite best efforts put up in improved animal production practices and implementation of better health care programmes. Meat production from registered slaughter houses increased from 3.6 million tonnes during 1992–93 to 6.2 million tonnes by 2013–14 at an annual growth rate of approximately 1%. India exported 545,731 tonnes of bovine and 52,251 tonnes of sheep and goat meat by 2009–10, which amounted to INR 5,436 and 737 crores, respectively. In 2011, India edged out the US as the third largest exporter of beef. India is also the largest exporter of buffalo meat in recent years. This marks a significant growth since 2009 when 0.61 million tonnes were shipped from the country. The market is growing for ready-to-eat and semi-processed meat products because of a changing socio-economic scenario as seen from increase in exports to neighbouring countries, especially the Middle-East. The mechanised slaughter houses produce huge quantities of offal and digesta, which could be profitably utilised for producing value-added products such as meat-cum-bone meal, tallow, bone chips, pet foods and methane.

Poultry production improvement

The poultry sub-sector has responded well to technological changes and has grown faster than the dairy and ruminant meat sector. Enhancement of genetic potential has been the most important factor in the growth of this sector. However, this has been complemented by health and nutrition technologies. The growth trends are more prominent in specialised peri-urban/urban poultry systems, because of higher demand for poultry meat and eggs in urban areas. Here, the entry of the private sector has boosted the adoption of technology and growth in poultry outputs.

The poultry population in India is 729.21 million which is approximately 3% of the world's population; with the growth in annual egg production being 4% and that of chicken meat approximately 3% of the world's production. Genetic improvement has made a major contribution to the progress of poultry industry during the last few decades. The most important effect on commercial success is the improvement in feed conversion ratio (FCR). During the past 15 years, there has been a 35% improvement in FCR and 4.5% increase in meat yield. Currently, it is possible to produce a 3-kilograms male broiler at 40 days of age, 1.7 kg in females at 30 days while white egg layers capable of producing more than 300 eggs in 52 weeks of lay. In broilers, annualgains of 0.1%, i.e. 100g in body weight and a reduction of 2% units in FCR are obtained. A 300% increase achieved with commercial diets is quite extraordinary compared to other farm animals.

Contract farming was started in broilers in late 90s. Fluctuation in bird prices encouraged the individual broiler farmers to enter the contract farming business with broiler integrators. Expansion of poultry enterprise in India is being driven by rising incomes and a shift in industry structure towards integrated ownership in poultry production. In the past, poultry production focused on the nutritional requirements of the birds. In the recent past, poultry nutrition is closely linked with biotechnology, microbiology and computational technology to produce quality poultry products at the most economical cost. Least cost ration formulation using computer programming is an advancement making the poultry enterprise as a highly remunerable sector in the modern livestock farming. Computational knowledge includes data mining and data analysis, using techniques such as bioinformatics, meta and holo-analysis. Bio-informatics are being used in poultry nutrition through applications such as gene sequence analysis, measuring biodiversity and systems biology, genomics, proteomics and metabolomics.

Research studies are in progress to develop a chicken feed involving nanotechnology to remove *Campylobacter*, a bacterium that provokes cramps and diarrhoea in people ingesting the contaminated under-cooked meat. It is expected that in the near future the nano-biotics will substitute the antibiotics, in the poultry feed.

Under the All India Coordinated Research Project (AICRP), in the recent years, certain strains of layers and broilers were used in developing coloured birds with better performance over the native chicken, important ones being Giriraja, Vanaraja, Gramapriya etc. Giriraja and Vanaraja were extensively tried in several states, which proved their worth leading to their demand from all over the country. However, the newly developed birds are less resistant to diseases such as Ranikhet requiring more frequent vaccinations compared to native chicks.

Emu birds were recently introduced into India a decade back for commercial value of its meat, oil and skin. India is the second highest and next to China in global emu production. During the last decade, the farmers have learnt scientific methods of rearing emus particularly their housing, feeding, incubation and hatching. An average daily gain of 110–120 g/bird up to maturity has been recorded with feed conversion ratio of 2.53 at 14 weeks of age with 10 kg body weight; 3.85 at 35 weeks of age with 26 kg body weight and 6.00 FCR at 64 weeks with 44 kg body weight. Viable emu unit for better economic returns was worked out to be 50 birds. However, lately the emu-farmers are closing these units/farmers due to lack of adequate market for emu eggs and other products. Designer eggs, such as low cholesterol or cholesterol free eggs and similar products from poultry meat will be in demand in future requiring research. Anti-snake venom and other biologicals for human use can be produced through eggs of poultry.

Turkey production also picked up well in specific areas, especially where people prefer meat. The bird is quite suitable for upliftment of small and marginal farmers as they can be easily reared under free-range or semi-intensive system with little investment on housing, equipment and management practices.

Unlike chicken, turkeys are reared in India for meat purpose, as they have ability to grow fast. The potential for turkey meat is high in areas where people are health conscious about eating lean meat and in Christian dominating areas.

Meat and other products

Small ruminants provide much needed livelihood support to the landless and weaker sections and hold considerable potential for commercialisation. A stable sheep population in the last two decades produced approximately 40 million kg wool annually, of which only 4 million kg was of fine quality. Goat population grew faster than any other species of livestock and has been a major source of meat. There is a considerable demand for processed meat products; however, it remains constricted due to inadequate processing facilities. The same applies to export of these products. Furthermore, not enough attention has been paid towards sanitary and phytosanitary measures. Slaughterhouses are often ill equipped and unhygienic. Low-cost processing technologies have been developed for both cottage and large industries. Meat production from the recognised sources is estimated to be 6.2 MT in 2013–14 and has increased at 4.1% annually during the last five years. Buffalo meat has grown at approximately 8% annually. Cattle and buffalo, sheep and goat, pigs and poultry contribute 55.0%, 17.1%, 11.4% and 16.3%, respectively, to total meat production. Inedible offals and animal wastes from the meat plant have large potential to be used as valuable proteins/materials for export. There is also huge demand of Indian ethnic meat products in the international market. However, lack of international processing standards is the hindrance. Unfortunately, schemes on modernisation of slaughter houses and by-product utilisation have not been effectively implemented. As in the developed countries, the responsibility to inspect meat and milk should rest with veterinary public health specialists and not with medical professionals.

There are 30 export-oriented modern abattoirs and 77 meat processing plants registered with Agricultural and Processed Food Products Export Development Authority exporting raw meet (chilled and frozen) to approximately 56 countries. India exported 5,45,731 tonnes of bovine and 52,251 tonnes of sheep and goat meat (2009–10), amounting to INR 5,436.1 and 737.27 crores, respectively. The export of sheep and goat casings and processed meat (put together) has been approximately INR 100 crores during the last 10 years. It is mainly the buffalo meat that has been increasing at the rate of 8% per annum during the last five years. Constraints in the meat sector include poor infrastructure of slaughter houses both registered (3,894) and unregistered (25,754), lack of cold chains and animal traceability, appropriate meat inspection system and poor utilisation of by-products.

Health management strategies

Electron microscopy, immunobiology and molecular characterisation involving the use of protein and nucleic acid based studies, have helped in understanding viral pathogenesis and immunogenesis. The use of tissue culture, application of hybridoma technology for the development of monoclomal antibodies, the versatile ELISA and PCR and nucleotide sequence analysis have proved to be most important modern tools for developing precise and early diagnostics, field diagnostic kits, improved vaccines, strain differentiation, selection of candidate strains for vaccine production and understanding of mucosal immunity. The parallel development of bio-security, bio-informatics, computer applications and adoption of good manufacturing practices have placed the country well poised for undertaking effective disease control measures leading to the eradication of rinderpest, AHS and CBPP in the last decade.

The application of biotechnology in turn resulted in development of improved diagnostics and effective vaccines against economically important animal diseases including rinderpest, bluetongue, bovine herpes virus-1 infection, pox virus infection, infectious bursal disease, clostridiosis, salmonellosis, pasteurellosis, etc.

In recent years, R&D activities under aquatic animal health have been initiated for fish and shell fish pathology, disease diagnosis and control of these diseases. Not-withstanding the greater resistance of Indian livestock species to tropical diseases, the Indian livestock sector faces a major constraint relating to prevalence of infectious diseases such as foot and mouth disease, which restricts trade opportunities in livestock products. This disease is considered as the biggest impediment to growth of the livestock sector in India as it affects 1.37% of livestock units per annum. It adversely affects the animal performance, resulting into direct economic losses to the extent of INR 20,000 crore/annum. Currently, India produces approximately 300 million doses of trivalent vaccine per annum which can target only 54% of cattle and buffalo population against the target of 200% for a vaccination schedule of two times a year, excluding the requirement for pigs, sheep and goats.

Deadly diseases such as PPR, Avian influenza, foot and mouth disease, sheeppox, goatpox, haemorrhagic septicaemia and black quarter have been the major threats to profitable livestock production in the country. Livestock disease control has undergone a paradigm shift during the recent past and now India has become a rinderpest-free nation. Several biological products in the form of vaccines have been developed for preventive/curative disease control measure. The infrastructure for disease control has been expanded significantly and currently disease diagnostic kits, including ELISA based and PCR tools, have allowed rapid and precise diagnosis of diseases for timely effective control.

The main limitation to effective livestock health management under field conditions includes an inadequate focus on preventive measure, diagnostic facilities, surveillance and monitoring, lack of medicines and equipment in the veterinary clinics, and ignorance among the farming community about the diseases and their effective control measures. This is reflected in the frequent occurrence of many of these diseases in several regions across the country. In India vaccines, antisera, diagnostic allergens and antigens are produced in public sector entirely in the individual states and some private sector companies with IVRI, Izatnagar, is still being the centre for research for development and refinement of vaccines and diagnostics and their production. Currently, the biological products have entered the commercial level of production under PPP mode with poultry sector dominating in disease control management programs.

Impact of technological change

Technological change improves the production potential of livestock and is reflected in productivity growth. The improvements in production lead to increased welfare of the producers as well as consumers of livestock products. In India, although the intensity of adoption of different technologies is low, the technological changes as discussed above, together with improved management practices have contributed to the increased output of many livestock products.

Conclusion

The significant achievements made during the past two and a half decades in the livestock sector with special reference to improved animal productivity and effective disease control management are due to the tireless and dedicated efforts put up by the thousands of veterinarians and research scientists spread over the country. This is also partly attributed to the phenomenal growth achieved in the expansion of veterinary education and advanced research infrastructure base created by several species specific national institutes of the ICAR, State Agricultural/Veterinary Universities and participation of private sector research and feed manufacturing organisations. Currently, there are over 46 veterinary colleges functioning in the country both under public and private sector under UGC system of India.

2.18.2. Achievements in Animal Nutrition

India recorded peak milk production of 137.7 million tonnes (MT) in 2013–14 and continue to remain the top milk producer globally since more than a decade. Milk production in India has increased by approximately eight fold in 2013–14 since 1950–51 (17 MT) Milk production further enhanced to 163.74 MT during 2016-17. This phenomenal increase could happen due to several policy and technological initiatives and implementations such as breed improvement, better feeding, improved health care and management. There are several research initiatives and innovations which singly and in combination have contributed to this growth in livestock sector in India. During the last four decades, several animal-nutrition-based technologies and practices have been developed and applied both on station and farm with varying degrees of success. Some technologies have produced profound beneficial effects and have been widely used, while others have shown potential on research stations and require validation at field level.

In India, ruminant production is based on grazing on community pastures/ barren lands and crop residues. Small ruminants subsist entirely on open grazing/

browsing and large ruminants depend on grazing and on crop residues. This feeding system is likely to continue and hence it is essential to improve the forage production and better utilisation of crop residues for animal feeding.

Forage Production

Research efforts of fodder scientists in the country have resulted in the development of high-yielding fodder varieties of grasses, legumes and other fodder crops. The green fodder availability has increased due to evolving of newer fodder varieties with higher biomass, improved protein value, multi-cut cereal varieties (sorghum), less water requiring varieties, drought resistant varieties etc. Several high-yielding perennial tropical grasses such as hybrid Napier varieties CO_3 and CO_4 yield over 300–400 tonnes of greens per hectare annually under multiple cuts.

Technologies for Improving the Utilisation of Crop Residues

Deconstruction of ligno-cellulosic feeds: Due to land use conflict for production of food/fodder due to population pressure, ruminant production in tropical countries would continue to be based on crop residues. In this context, research should focus on enhancing the fibre digestibility in the rumen through lignin degradation *in-situ* or *ex-situ*, use of fibrolytic enzymes or fungal culture supplementation etc. Solid state fermentation technology to achieve breaking down of complex lingo-cellulosic bonds has met with varied degrees of success and has been in the process of upscaling. Recently, lignin degrading enzymes have been produced using immobilisation technology and found effective in improving the digestibility of lingo-cellulosic feed materials. Supplementing exogenous fibrolytic enzymes and rumen fungal culture, the utilisation of ligno-cellulosic feeds have been improved. Urea treatment of straw improved the nutritive value of cereal straws by increasing crude protein content, palatability and digestibility, and weakening the bonds between lignin and cellulose. This technology has the potential to be widely adopted at field level but due to some of the limitations of the technology such as being labour intensive and yielding marginal returns, the technology has not spread large and wide. Urea Molasses Mineral Block (UMMB) is a strategic feed supplement for ruminants. The use of UMMB for supplementing crop-residue-based diets for livestock has the potential to increase livestock production and net daily income. Supplementation with UMMB licks significantly increase feed intake, milk yield and growth rate and as such is a cost-effective approach to maximise the utilisation of locally-available feed resources for improved productivity. The UMMB technology is suitable for areas where dry fodder is the predominant source of roughage for animals and in places affected with natural calamities.

Technology of straw densification and total mixed ration (TMR): This is an innovative feed technology and a novel nutrient delivery system by which animal can get all the required nutrients in a balanced form. Preparation of TMR comprising of roughage/forage, concentrate and other supplementary nutrients in required proportion made into uniform mixture either in the form of mash or pellets/blocks (complete feed) helps in delivering all the nutrients to achieve synchrony in the rumen for better nutrient use. Crop residues, fibrous fodders and unpalatable but nutritious unconventional feed ingredients can be effectively used in total mixed ration. This enhances the digestibility of roughages, reduces the nutrient loss and results in increased rumen microbial protein synthesis and thereby improves the productivity and profitability, particularly under intensive system. However, preparing TMR or complete feed requires machinery (grinder and mixer), which involves investment and cannot be taken up by individual farmers. These are to be installed at community level involving co-operative milk societies and self-help groups who can play an important role in successful implementation of this technology. The complete feed blocks are now commercially available and more importantly, the machinery is also being manufactured in India and exported to other Asian and African countries. The response from farmers is quite encouraging. This technology has also a great potential to deliver feeds to areas affected by floods and draught.

Alternate feed resources to enhance the animal feed basket: Animal nutritionists in the country are continuously working in search of alternate feed resources not only to make the animal diets and thus animal production cost-effective but also to enhance the availability of feed resources to meet the urgent demands. Incriminating factors have been identified in unconventional feeds and methods for their detoxification have been evolved. Protein cakes after oil extraction from seeds of neem, castor, karanj, palm, mahua, cashew nut etc. have been evaluated and found suitable after detoxification to use for feeding as livestock feed.

Areca sheath as an alternate dry fodder: Over the years, in some regions of Karnataka, Kerala, Andaman and Nicobar Islands and Assam, the areca (*Areca catechu*) cultivation as a commercial crop has replaced the traditional paddy and other cereal crops to a large extent due to its higher economic returns. This has consequently resulted in deficit of dry fodder, especially in the costal zones and livestock farmers are procuring paddy straw from adjoining districts at much higher cost. Converting the areca sheath as a dry fodder source by shredding and chaffing in the form of total mixed ration along with suitable proportion of concentrate feed has been highly successful. The technology is economical and has been found to enhance milk production by 8%–10% at field level. Suitable machine to disintegrate and shred the dried areca sheath has been developed and is commercially available.

Hydroponic green feed: Growing the fodder through germination of grains such as maize, barley, oat or wheat produces consistent quantities of green feed throughout the year at the rate of 6–8 kg per kilogram of seed within a one-week period. Hydroponic fodder is nutritious, rich in micronutrients, palatable and has been found to substantially increase milk yield in dairy cows. Of late, much economical models of hydroponic units have been fabricated by few entrepreneurs in India and are becoming popular. This technology could be more useful for organised farms in peri-urban areas where the land for growing fodders is limited. More research to develop cost effective solar operated hydroponic machines may be required as it has great potential in the present-day scenario of reduced resources such as land and soil. Ayurvet India has taken a lead in producing hydroponics and developing cost effective hydroponic machines.

Pineapple fruit residue silage: In India, an estimated quantity of 1.3 million tons of pineapple fruit residue (PFR) is available annually and is mostly wasted. Due to high moisture and sugar content the keeping quality of PFR is low and is a disposal problem and an environment issue. Silage process was developed from PFR and nutritive value in terms of energy and minerals was superior to maize green fodder. In regions of pineapple cultivation and processing, pineapple fruit residue silage is a valuable alternative to conventional green fodder for feeding livestock.

Technology of Bypass Nutrients

Bypass Protein: Among the several technologies available today, the one having the greatest potential for commercial production is bypass protein technology. Bypass protein refers to dietary proteins escaping rumen degradation and it aims at decreasing the wasteful ammonia production in the rumen from highly degradable cakes and thereby increasing the total availability of amino acids in the host intestine to meet the increase demands of high milk yield. The highly rumen degradable cakes such as ground nut, mustard, rapeseed and sesame cakes are subjected to heat or formaldehyde treatment to protect the protein. This technology has successfully been adopted by the feed industry in India with NDDB taking the lead and setting up several plants to produce bypass protein feed. This technology is mostly useful for high yielding animals.

Bypass Fat: The bypass fat technology is quite effective for high and medium producing animals. Bypass fat is prepared by using some protective mechanism by converting the fatty acids into calcium salts of fatty acids by fusion method or by double decomposition process for preventing bio hydrogenation of fat in the rumen. As the dietary energy requirement of high yielders are not so easily met with cereals and oil cakes, particularly during post-parturition and peak lactation, it is necessary to feed animals with high energy dense diets. In this regard, bypass fat has the potential to meet the energy demand and thus improve the milk production and reproductive efficiency.

Area-specific mineral mixture technology: Majority of the reproduction related problems such as repeat breeding and anoestrus in dairy animals have been implicated due to the deficiency and/or imbalances of micronutrients, particularly trace minerals such as Cu, Zn, Mn and Co. The approach of free choice mineral supplementation could sometimes lead to deleterious effect, as some of the minerals may be available in excess than requirements affecting utilisation of other minerals due to antagonism and this approach of mineral supplementation is expensive for farmers. Moreover, it has been found that mineral deficiencies vary from region to region of the country. The method of supplementing only the most deficient minerals through area specific mineral mixture by assessing the mineral content in soil, feeds and fodders and in animals in different agro-climatic zones appears to be the best approach. It has been found that the most limiting micronutrients are calcium, phosphorus, copper, zinc and manganese. This approach has been found to improve the reproductive efficiency and health in cross-bred cattle under field conditions. This technology now has been well accepted by the livestock farmers and widely adopted in the country. Seven entrepreneurs are manufacturing area specific mineral mixture for use by farmers. This area specific mineral mixture was developed by Prof. M.C. Sharma under the ICAR National Fellow Project during the year 2000–06 at Division of Medicine, IVRI, Izatnagar, as such this study was conducted in three states, i.e. Uttarakhand, Uttar Pradesh and Delhi which increases milk production on an average of 25%–30% in lactating cattle and buffaloes and also increase 22%–34% body weight in calves (6–24 month). The daily requirement of this mineral mixture is 50 gm/day/adult animal and 25 g/day/calf. In fact, it is in great demand by farmers, state and central government organisations, military farms etc.

Chelated minerals and Nano-technology: The trace minerals are now available in chelated form. This is a newer and more efficient technology. In chelated form, mostly the trace minerals such as Cu, Zn, Fe, Mn and Co are bound to some inorganic or organic ligand, so that their absorbability in the intestine is improved and their bioavailability at the tissue level is enhanced. Because of this, these minerals can be used more effectively in smaller quantity without interaction with other minerals or dietary nutrients and reduce the environmental pollution. There is high commercial potential for this technology.

Research on Nanominerals indicated that nano-technology has the potential to improve the bioavailability of minerals (e.g. nano-Zn, nano-DCP, Nano-Se) as their size is very small and due to larger surface areas, the absorption of minerals is more and could be the right approach for targeted nutrient delivery.

Sheep and goat mineral mixture: Small ruminants such as sheep and goat have a different mineral requirement than dairy animals and hence need specific mineral supplementation. Requirement of zinc and sulphur is more in sheep, while that of copper is less as compared to cattle. Depending on the most limiting minerals in feeds and fodders and feeding practices of sheep/goat, specific mineral mixtures have been formulated and tested for their efficacy and have been found to be highly beneficial. This technology is now commercially available. Special mineral mixture pellet has been developed for goat.

Feed additives and nutraceuticals: Several feed supplements including probiotics have been developed and some of these are now commercially available in the market. Live yeast cultures of *Saccharomyces cerevisiae* and *Aspergillus oryzae* as feed additives are effective in enhancing fibre digestion in the rumen. However, this technology may be useful at large farms where animals are fed high concentrate diets and may be effective in urban dairies with high yielding animals requiring high concentrate feeding. Similarly, recently rumen anaerobic fungal cultures have been found effective in enhancing the fibre digestibility. Commercially produced fibrolytic enzymes such as cellulase and xylanase are now available in the market for use as feed additives.

For regulating the rumen fermentation and to increase propionate production ionophores such as monensin, lasolacid, salinomycin, nigercin and gramicidin are used. Prebiotics are the non-digestible food ingredients that beneficially influence the host by selectively promoting growth of limited bacteria in the gut. Prebiotics such as xylo oligosaccharides or fructo oligosaccharides are of plant origin in nature and known to protect gut integrity even at low concentrations. They are anti-carcinogenic, and do not have any residual effect besides having capability to stimulate immune system. Numerous microbes (*Lactobacillus, Saccharomyces, Enterococcus, Bifidobacteria*) are currently used as an alternative to chemical antibiotics, particularly in pigs and poultry. Use of probiotics has shown to reduce pathogenic microbes in gut and decrease scours in young stock. With ban on the use of most antibiotics in the diet of livestock coupled with growing concern for health and environmental protection, the role of feed additives such as probiotics, prebiotics and enzymes is very crucial. Other novel feed additives include anti-oxidants (curcuminoids, rosamarinic acid, resveratrol, carvacrol, catechins, flavonoids) from plant origin to overcome oxidative stress. Feed emulsifiers such as lecithin and lysolecithinmay be used to aid in better digestion and absorption of lipids and proteins. Similarly, toxin binders such as aluminosilicates, bentonite, and esterified glucomannans are also widely used to counteract mycotoxins in feeds.

Protocols for production of xylooligosaccharides (prebiotics) from agricultural wastes such as coconut pith, corn cobs, sugarcane baggasse and natural grasses have been standardised, thus bringing down the cost of nutraceuticals considerably. Feeding of these synthesised pure xylo oligosaccharides to poultry birds improved the gut health and function.

Technology Concerns on Environment and Global Warming

Since the methane contributes approximately 25% of the green house effect and ruminants produce approximately 25% of methane, a major issue is how to reduce methane emissions per unit of animal product. Mitigation of methane emission not only protects the environment from green house gas effect but also improve feed conversion efficiency in animals. The database on methane production potential of different feed ingredients used for animal feeding has been developed. Several plant secondary metabolites have been tested for their efficacy to reduce the methane production in animals. Ration balancing strategies and use of tree leaves in the TMR or Feed block have been adopted to reduce the methane emission from ruminant animals at farm-gate level.

Nutrigenomics: Over the last decade, advances in the biochemical technologies available for examining functional genomics have provided several new molecular tools for evaluating responses to nutritional strategies. These tools are largely based on an understanding of the expression and control of specific genes and gene products. A molecular marker for copper deficiency has been developed based on CCS (Copper chaperone for superoxide dismutase) gene expression in sheep which could help in developing field level diagnostic kits for assessing the copper deficiency.

Metagenomics of rumen microorganisms on supplementing plant bioactive compounds while reducing methanogenesis could help in mitigating methane production.

Gut geography of different animal species: The nutrient digestibility has been found altered in different species of animals. In this context, for identifying the novel microorganisms, gut geography is being studied in different species such as cattle, buffalo, sheep, goat, camel yak and mithun in different agro-eco zones of the country through metagenomic approaches to identify the microbial population having better fibre digestibility, low methane production, better detoxification etc.

Strategic feeding approaches: As crop residues continue to be the major source of roughage for Indian livestock, there is a need to improve their nutritive value both in terms of their intake and digestion. Enriching the crop residues with urea-ammonisation has shown promising benefits but has limitations for adoption at field level. Strategic supplementation of nitrogen/protein/select synthetic amino acids and minerals using locally available feed resources would help in improving the rumen fermentation and microbial activity, thereby improving feed utilisation. These strategies are helpful in small and medium holding production systems.

Supplementation with concentrates: Protein and energy supplementation is necessary to improve both intake and digestibility of dry fodder. For better utilisation of the basal roughage, the supplements must be provided in small quantities to get positive associative effect. Cereal grains (maize, ragi, jowar) and by-products (bran, rice polish) and oil cakes (groundnut cake, soybean meal) can be fed to improve dry fodder utilisation. Supplementation of ragi grain to cows suffering from anoestrus resumed estrus cycle by reducing the blood urea nitrogen (BUN).

Supplementation with forages: The leguminous trees such as glyricidia, erythrina and leucaena have very good potential as supplements because they are perennial and yield higher biomass. The legume will have a better protein content to supply both fermentable and bypass protein. There will be additional benefits as the legume contains other critical nutrients (minerals, vitamins and other plant compounds) which enhance the rumen ecosystem to increase microbial growth and fibre digestion. Supplementary feeding of azolla or hydroponic baby fodder as green feed along with concentrate mixture is a strategic approach to provide additional nutrients.

Compensatory growth: Growing calves and highland animals (Mithun, Yak) lose their body weight substantially during lean season due to shortage of feed and fodder or during natural calamities. Research has shown that these animals when given adequate nutrients as per the requirement subsequently gain weight and compensate for the earlier loss.

Residual feed intake as a measure of feed efficiency in animals: Typically feed makes up the larger portion of non-fixed costs for cow-calf producers and 65%–70% of total production costs. Therefore, seemingly small improvements in feed efficiency can result in dramatic cost reductions. This reality, along with new technology, has led many researchers to take a new look at feed efficiency. Residual feed intake (RFI) or net feed efficiency has been proposed as a measure of efficiency that is independent of growth and body weight. Residual feed intake is moderately heritable and, as such, is a good candidate for genetic improvement. RFI is the difference between an animal's actual feed intake and their predicted feed intake. An animal with a negative RFI value is more efficient and, conversely, an animal with a positive RFI value is less efficient. Residual feed intake values can prove to be very useful as a selection tool in a breeding programme.

2.19 GENETICS OF CROSS-BREEDING

Artificial selection to improve economic characteristics in domesticated livestock was practiced even before genetics as science came into being. A scientific approach to such efforts however, ushered in due to the advent of principles given by Gregor Mendel in 1865 and the reconciliation between biometry and genetics in the hands of Fisher in 1918 through his infinitesimal model based on a numerous genes with small effects and environmental modification. The heritability of a trait as a proportion of the phenotypic variation that is due to genetic causes became a prime indicator to make selection decisions for genetic improvement of economic characteristics. The prediction of response to selection in terms of intensity and accuracy of selection and genetic variability was found to be successful across several crop plants, livestock, poultry and fishery. The relationship between phenotype and genotype was, however, considered like a black box wherein an inferential approach based on statistical principles was the only way to look into it.

In developing high yielding plant varieties, the concept of inbred lines/isogenic lines and their interbreeding and utilisation became a new tool and was named as hybrid breeding. The principles of crossing of hybrid varieties to utilise heteroisis to produce higher yields and repeated back-crossing of isogenic lines/highly inbred lines to higher yielding phenotypes would get converted to the better parent in view of the facts that both the lines are isogenic/highly inbred in nature.

Applying similar principles with cattle and other livestock within the same or similar breed populations did not however respond similarly as in plants. It was, however noted that the Mather's principles applied to isogenic lines of plants where highly inbred plant lines (isogenic lines) did response, as lines were homozygous, where all the genes were fixed when back cross-bred with parents of superior type, genes of original parents became fixed in those of original parents. This does not happen in livestock particularly in cattle.

When these phenomena were compared based on the experiment done on livestock, it become obvious that in the case of livestock, when you cross-bred the two random breeding segregating populations the F_1 so produced had the 50% complement of genes from each population to produce a hybrid population (F_1) which were theoretically similar to an F_1 cross of two isogenic plant lines. Once these F_1 progeny are interbred, there is random segregation of genes on chromosomes and new genotype are formed, which on average at population level may carry half the genes from each line but at genotypic/phenotypic level are totally different from each other varying from genotype of 1st parent on one side to almost genotype of the 2nd parent on the other.

In cattle when two breed populations one inferior with respect to milk yield and other superior where crossed the resultant F_1 population had half the genes from each of the parental populations, when the F_1 generation was interbred the resultant population (F_2) was segregating as per Mendel's Principles of random segregation, i.e. on an average the F_2 population had half the genes of male parent across the population but individually genes from an inferior gene type to the individual genes (F_2) of superior parent quality and all the best would fit in between these two limits of the present population.

If F_1 offspring of this population was back-crossed continuously with superior genotype the position became more complex resulting in ¾, 7/8, 15/16, 31/64, etc. Theoretically the genes, under ¾ grade or 7/8 grade on an average have 7/8 genes limit on population level are due to mendelian segregation but the individual would have any gene combination as dictated by random segregation. The situation that environment has large component of the effect on product (output of milk) the position becomes more complex because F_2 was only genetic component which could be used for genetic improvement of economic traits in cattle given the breed type (new synthetic) is result of phenotype, i.e. $P=G+E+GE$.

The most cross breeding experiments in the topics are comprised of crossing indigenous cattle with the better developed dairy breeds of European origin. The F_1s thus generated would have half the number of genes coming from the better parent and the second half coming from indigenous parent. The fact of the case is that this cross involves two segregating populations where males from one population having higher gene frequency for production are being crossed with females having low gene frequency for that trait. A cross-bred population when further mated would then show segregation and linkage between blocks of genes affecting production traits which would then generate a series of genotypes with different proportions of inheritance from either parent population.

The situation becomes more complex when backcrossing with the better parent is continued generate genotypes with a higher proportion of exotic inheritance: namely, ¾, 7/8, 15/16, 31/32 and 63/64. In fact, the very validity of calling a certain grade — 3/8, ½, 5/8 or ¾ — is questioned because individual within that grade may not in reality have that level of exotic inheritance.

In a somewhat-planned cross breeding conducted with cattle (*Bos taurus* x *Bos indicus*) at Military dairy farms in India, it has been generally observed that with an increase in exotic inheritance, the advance made in production in later crosses has fallen short of theoretical expectations (Bhat and Khanna 1970, Taneja 1973). This led to backcrossing with the indigenous parent. Due to forward and backward crossing, a series of different genotypes having varying levels of exotic inheritance were generated. Therefore, the problem is to identify each grade with respect to another assuming that these two are from different populations, if any, in the phenotypic expression of economic traits and their genetic basis.

The data for the study were taken from breeding records of animals maintained at eight Military Dairy Farms: Ambala, Bareilly, Dehra Dun, Jabalpur, Jullunder, Lucknow, Meerut and Mhow. The breeding policy in operation prior to 1952 was that of criss-cross breeding. Since 1953, a policy of backcrossing with Sahiwal bulls was introduced with a view to reduce the exotic inheritance gradually.

A four-way fixed effects model with grades, farms, periods and seasons was used for analysis following the procedure given by Harvey (1966). Using least-square constants, the records were adjusted for significant effects due to farms, periods and seasons. The adjusted records were used to estimate genetic divergence among the grades.

The results of this investigation reveal, that in cattle when two breeds of diverse origin, different in their production levels, one of superior merit and the other relatively inferior, are crossed in a scheme of forward crossing, there is no linear increase in production level with the increase in the level of inheritance of the superior parent. The genetic reasons for this are not obvious. In view of this, it cannot be assumed that by grading up a total replacement of genes can be obtained, at least in cattle.

The result of the above study has given several leads about the behaviour of genes of livestock, in special reference to randomly segregating populations of several cattle breeds and their crossing behaviour. Based on these results and those of similar nature done under similar tropical environment, the several breeding plans can be proposed to stabilise various breed types.

Feed and fodder play a central role in providing proper nutrition to livestock. The feeding of a diet balanced in all nutrients and at a level that meets the production objective considering the animal's physiological state is imperative for achieving high and sustained livestock productivity. The success of animal production and health programmes rests on proper nutrition.

Deficient and improper feeding leads to productivity losses and increase in emission of pollutants in the form of methane (up to 12% of feed energy is lost in the form of methane) and nitrogen and phosphorus release (60%–70% of the nitrogen and phosphorus fed in intensive production systems is lost to the environment) in soil and water channels. If these pollutants are not managed properly could cause water pollution, resulting in erosion of biodiversity, deterioration of human health and decrease in agricultural productivity. Balanced and proper animal nutrition therefore plays an important role in addressing ongoing and emerging challenges imposed by increasing human and livestock population, global warming, land degradation, water shortage and pollution, biodiversity erosion and increasing energy prices.

During the last couple of decades both production and consumption of animal products have substantially increased and therefore an assessment of feed and fodder for the present and likely growth in the livestock and poultry population is most urgent.

Availability of Feed and Fodder

Availability of feed and fodder is a major constraint in promotion of animal husbandry in the country. The rise in demand for animal protein is difficult to meet with the present livestock farming system, as India is facing shortage of dry fodder, concentrate and green fodder to the tune of approximately 11%, 35% and 45%, respectively. It has been estimated that only 880 million tons of dry fodder was available including greens, which can meet only 35%–40% of the demand. On a dry-matter basis, deficiency of dry fodder, green fodder and concentrate was estimated to be 40%, 36% and 57%, respectively in 2007 (NABCONS, 2007). This clearly indicates that as most of the livestock are underfed, they are not able to perform optimally. Out of the available dry matter, most of it is available in the form of agricultural by-products and dried grass collected from community wastelands and forests which are of inferior quality. Similarly, the concentrates required for feeding the livestock are also in acute shortage. Thus, even the high yielding animals, which are presumably well-fed are suffering from nutritional imbalance and producing 26%–51% below the attainable yield. Regarding cultivation of forage crops, hardly 3%–4% of the area is being utilised in selected pockets where dairy husbandry is prospering as an important source of income. If we take feed conversion ratios of approximately 2, 4 and 9 for poultry, pigs and cattle respectively and consider carcass percentages, a high demand for feed will ensue by 2050. It is a challenge especially when we are faced with increase in population, decrease in arable land for crop production, water shortage, food-feed-fuel competition, frequent climate extremes, increasing animal and human health risks and economic instability.

From the Table below, one thing becomes obvious that the total available livestock and poultry population cannot be fed at the required level of feed due to non-availability of (shortage) concentrate and fodder. It must survive only on the crop residues (880 MMT) and surplus grains after meeting the human food requirements and spoiled grains, etc.

Feeding of humans is the priority, i.e. production of grains/fruits/vegetables and whatever is leftover along with crop and industrial residues is left for feeding to the animals.

The obvious conclusion is that except for crop and industrial by-products, the animals main net feed requirement must come from crop and industrial residues.

Unproductive livestock, feed burden and climate change

Owing to the numerous unproductive animals that there has been severe shortage of feed and fodder resources (Table 2.10). The population of cattle and buffalo has increased 29% and 143%, respectively, between 1951 and 2007. The continued increase in population has serious implications on the limited natural resources of the country. The crop yield/ha in India is less than 50% that of USA and China with continued problem of land fragmentation, degradation, nutrient depletion, salinisation and expanding urbanisation. The per capita crop land is only 0.128 ha in India against world average of 0.23 ha globally. Therefore, the production of animal feed can be considered as one of the major hotspots for any future attempt to increase efficiency in the existing environmental impact from livestock production.

Table 2.10: Showing Population of human and Livestock and their requirements for maintenance

1.	Human population	1.22 billion
2.	Total livestock population	512 million
3.	Total number of bovine population (cows and buffaloes)	299.9 million
4.	Total number of bovine population (cows and buffaloes) in milk	80.52 million
5.	Total number of bovine population (C.B.) in milk	7.01 million
6.	Total number of cow/buffaloes in milk	17.15 million
7.	Buffalo in milk	56.36 million
8.	Unproductive animals (cows and buffaloes)	181.31 million
9.	Human population to be fed by 2030	1.25 billion
10.	Total available land for Grains	142 million ha
11.	Total land area needed to produce feed for Animals	512 million ha

With regard to the demand for milk and meat, it is likely to increase significantly because of increase in income and consequently improvement in local consumption. This may enable the rural families to take active part in animal husbandry provided it is economically viable and the necessary infrastructure is available to support this programme. To transform the current status from negative contribution to profitability, particularly to benefit the weaker sections of the society, there is a need for a drastic change in the future livestock husbandry framework and management with a focus on mitigating the ill-effects of unproductive livestock and necessary policies and practices to control them.

It may be estimated from the table above that for milk production at the needed rates and standards (as per Indian Council of Medical Research (ICMR) standards of human requirement, approximately 35 million breedable cows and 35 million breedable buffaloes are required at the present level of production (add some take some) i.e. at the most 100 million cows and buffalo females to produce 146 MMT of milk as against the existing 199 million cattle and 108 million buffalo cows for producing this amount, having a net surplus of 200 million cattle/buffalo to the requirement. These 200 million cattle and buffalo are competing for survival and consuming an amount of feed and fodder which costs INR 4983873.5 million annually. Therefore, if the feeding cost of productive and un-productive animals added together, the estimated cost of production of 146 million tons of milk then works out roughly to INR 943717 million. If this burden is eliminated, the huge savings can be diverted to benefit families below the poverty line and R&D efforts for addressing problems of livestock sector particularly feeds and feeding. Further, the reduced burden of unproductive livestock will pave the way for closing the 'yield gap', i.e. the difference between realised productivity and the best that can be achieved using current genetic material and available technologies and management. It is estimated that the average productivity of farm animals can be increased by 30%–40% merely by balanced nutrition, good health and best practices generating surplus of livestock products for export.

It is now well known that livestock has an adverse impact on environment. Firstly, livestock demands huge fodder and feed. As there is a severe shortage of cultivated fodder and feed resources, farmers let their livestock for free grazing on community lands and forests. Such stray livestock not only denude vegetation but also accelerate soil erosion. Secondly, livestock, particularly ruminants, produce methane and carbon dioxide while digesting feed in their rumen. Livestock dung also releases methane and nitrous oxide during anaerobic decomposition. It is estimated that livestock contributes 18% of the CO_2 equivalent to green house gases, which is a serious concern. The GHG emission from animal feed production comes from both the primary stage of crop production and from use of fossil energy in the processing of the crop in to animal feed. With careful planning of livestock inventory and its balanced feeding, it is possible to reduce the ill-effects of global warming at an affordable cost.

It is strongly felt that proper nutrition of such a large animal population is not possible without destroying the natural resources. Policy makers, scientists and society at large has not looked at these issues and taken up the challenge to generally improve the Indian Livestock by selecting the best of the bovines by the best method of selection and mitigate the surplus population. Although the most effective and humane method of elimination of surplus livestock is through slaughter, there may be several other alternatives to control population including selection and culling, breed improvement, use of sexed semen and conservation and export of livestock. Taken together, these strategies could lead to more feeds available for consumption of productive livestock, and sustain the water, forests, and soil that food production depends on.

It is to be understood clearly that there is no silver bullet — we need to incorporate the best of what we know now into solving the food-feed problems and protecting our natural resources. Can we do it? We have to — it's absolutely necessary. It's up to us to decide what's scientifically appropriate and politically feasible. We can change how we govern, tax, ship, produce, etc. What we can't change are only the laws of physics.

The problem of feeding the world and not wrecking the planet is a huge challenge and it's going to shape a lot of the 21st century. Solving it will require huge co-operation, innovation, and hard work.

By close examination of the Table 2.10, one thing becomes obvious that the total livestock population cannot be fed at the required level of feed due to non-availability of (shortage) of concentrate and fodder. It has to survive only on the crop residues (880 MMT) and surplus grains after meeting the human food requirements and spoiled grains, etc.

In case of products of livestock origin, the reason they are produced are obvious: (i) since they exist with the farmers; (ii) some of them reproduce and produce products which meet human needs. If we prioritise their need for requirement they stand after food grains. Priority wise the first group is production of grains/fruits/vegetables for feeding of humans. From livestock products milk, meat and its by-products have first priority. The second group comprises of products associated with apparel coming from wool, fibre, hair, pelt, leather and other products. The 3rd group comprise of drugs, surgical thread and packaging material which cannot be produced either from plants or from chemicals, etc.

Conclusion

The estimate cost of production of 146 million tonnes of milk works out to INR 4,439,843.7 (P) + 4,983,873.52 (unproductive) which works roughly to INR 9,423,717 million.

Thus for producing 146 million tons of milk a year, we will spend INR 9,423,717 million, which is highly uneconomic.

If this burden of unproductive animals is eliminated, we can have this money to benefit all the below poverty line families and get them out of this mess in the next four years.

There are many possible solutions as to how to feed 100 million cows/ buffaloes. Convert the large amount of crop residues (around 880 mmt/year) into livestock feed, using solid state fermentation technology, to feed rest of livestock.

No government has looked at these issues and taken up the challenge to generally improve the Indian livestock by selecting the best of the bovine by the best method of selection and eliminate the surplus population. The most effective and humane method of elimination is through slaughter.

Chapter - 3

Services and Policy Support

3.1. CATTLE

Up to 1960, the policy of the Government of India pre and post-independence was to improve breeds of cattle and buffaloes better adapted and resistant to prevalent tropical diseases and their improvement for milk and draught through within breed selection and grading up of non-descript animals and castration of scrub bulls. After the 60s, for augmenting milk production, superior bulls, semen of exotic breeds *viz.,* Holstein-Friesian, Brown Swiss and Jersey, were imported on experimental basis and mass crosses was estimated. Facilities for undertaking large-scale artificial insemination, deep freezing of semen and semen banks were established and it was realised that the emphasis on draught was reduced. Similarly, infrastructure facilities for embryo transfer technology were made available to facilitate early selection of bulls. These input services have resulted in increase in milk production from 17.40 MT in 1959-51 to 163.74 MT in 2016-17. The number of cross-breds available in the country in 2012 was approximately 33.7 million.

3.2. SHEEP

Main policy frame for improvement in wool production was through within breed selection of sheep in plains and through cross-breeding in the temperate Himalayan region and Deccan Plateau. For this, rams were produced through the public sector and distributed to farmers. The Central Sheep and Wool Research Institute (CSWRI) was established in 1962. To develop appropriate breeding strategies for improving sheep for fine wool and mutton, research trials through systematic breeding were planned and initiated. All India Co-ordinated Research Projects on improving sheep for fine wool and mutton were launched in 1970. Infrastructure in the form of sheep breeding/production facilities such as sheep breeding research stations with adequate numbers of animals were established and research technologies generated. These technologies, however, could not be fully adopted by state Animal Husbandry Departments/Sheep Husbandry Departments in the field except in Jammu and Kashmir, Himachal Pradesh and parts of Rajasthan because of problems of artificial insemination under field conditions and due to illiteracy among sheep farmers. Further, there were problems relating to non-disposal of cross-bred apparel wool because of short staple length and less production due to inadequate nutrition both at the lamb and growing stage. There was also a mismatch between the wool produced and machines which were to use it for processing. The new research institutes and centres were established to sort out the problems.

3.3 GOATS

The policy in goats has been to introduce milk traits through cross-breeding with Saanen and British Alpine breeds, and improve local goats through selection for meat and hair. Despite creation of infrastructure facilities for undertaking organised research on goats in the country, desired progress could not be achieved as the infrastructure in the development departments was not upgraded. The Central Institute for Research on Goat came into existence in 1979.

3.4. POULTRY

In building a modern and organised poultry industry in India, the first major step was to create awareness that poultry farming could create jobs and wealth, and that a technology package was available which needed skills and a minimum level of financial input. This phase lasted for one decade *viz.* 1960–70. The second major policy push was to provide unrestricted import of Grand Parent (GP) stocks of layers and broilers for which custom duty exemption was granted (1970-1980) by the Government of India. By 1980, the import of G P stock was banned but pureline breeding was encouraged by offering concessions in tax rebate and loans for infrastructure. This was meant to accelerate the growth and development of commercial poultry farming in the country. The Government of India established four regional poultry farms, one each at Mumbai, Bengaluru, Bhubaneshwar and Chandigarh for multiplication and supply of the exotic stocks. The ICAR sanctioned several poultry improvement programs such as popularisation of baby chick scheme, random sample units, improvement of poultry through provision of soft loans, etc. The ICAR setup several All India Coordinated Research Projects (AICRP) of Poultry including animal health to become self-sufficient.

During the last three and a half decades, the entire scenario of poultry farming in the country has changed. It is now recognised as an organised and scientifically based industry with tremendous employment potential and a tool to fight poverty and malnutrition, livelihood and inclusive growth. Thousands of poultry farms, both small and large, have come up all over the country and indigenous *desi* birds have given way to highly specialised layers and broilers. A network of hatcheries was established both in public and private sectors to supply commercial hybrid chicks. An integrated industrial model was developed by the Industry, which used chick to egg system using small and marginal farms who are providers of chicks for rearing with inputs, raw material and markets to become self-reliant.

3.5. RINDERPEST ERADICATION

A country-wide programme to control losses due to rinderpest was launched in 1926. With the launching of massive research programmes on cross-breeding in livestock for improving productivity, efficient and well-knit animal health institutions were provided by expanding facilities on clinical and preventive medicine, diagnosis and investigation of diseases. Vaccines were developed against rinderpest, foot and mouth disease, sheep pox, New castle disease, fowl pox, AHS, rabies and other important diseases of livestock and poultry. The systematic approach is animal health management and application of planned animal health programme in livestock has resulted in significant reduction in wastage due to morbidity and mortality. India produces vaccines and diagnostic kits in 15 public sector and another four in private sector. The country is by and large self-sufficient in the preventive area in respect of vaccines and diagnostics. However, some poultry vaccines and vaccines for pet animals are also imported.

However, there is a need to integrate the health protection system at the village level and to provide the services at the farmer's door. Besides, if proper investments both by public and private sectors are made to keep pace with global trends, research would lead to satisfactory level of protection.

3.6. DEVELOPMENT OF FEEDS AND FODDERS

Policy on the development of feeds and fodders has not been satisfactory. The condition of pastures has been very deplorable, as there was no proper policy to develop and reseed alpine pastures. The condition of common (*gochar*) lands in arid and semi-arid areas has also not been satisfactory as there was no policy to develop them. There has also been no policy to develop wastelands available, through reseeding, and their area has been on the increase over time. It is pathetic to note that while milk production in the country has been rising fast (4%–5%) in the last decades and India holds the prestigious position of the largest producer of milk in the world for more than 10 years now, the feed and fodder situation in the country has been deteriorating very fast making a bad situation infinitely worse. The country is currently short of concentrates by 47% and in roughages by 50%. The current cropped area under fodder is a bare 3% and the 12 million ha of pasture lands are in a miserable condition with hardly any grass. The present available nutrients can at best sustain only 60% of the present large ruminants. With approximately 70% of all livestock in the country in the hands of 67% of small and marginal farmers, the future of the dairy industry can be better imagined than described. The population of large ruminants has also been steadily increasing due to the highly restrictive slaughter policies in most states. The large unproductive ruminant population monster comprising of 40% (200 million) of the total bovines including stray cattle is a serious deterrent to animal production, a serious threat to crop production, a hazard to human health and a severe drain on the national economy, and need to be dealt with a sense of urgency.

■■■■

Chapter - 4

The Path

The contribution of agriculture sector to the total GDP in 1950–51 was 50.5%, of which the share of animal husbandry was 15.5%. During 1980 to 2012–13, the contribution of livestock to agriculture and allied sectors rose from 18.6% to approximately 26% at current prices in contrast to a reduction of agricultural output from 41% to approximately 15%. During 2014–15, the value of output from livestock sector as per the estimates of the Central Statistics Office (CSO) at current prices was about INR 7,33,054 crores which is approximately 28.7% of the value of output from agricultural and allied sector (BAHFS 2016).

Table 4.1: Production of milk, eggs, meat and wool

Year	Milk (million tonnes)	Eggs (billion)	Wool (million kg)	Meat (million tonnes)
1990–91	53.9	21.1	41.2	NA
2000–01	80.6	36.6	48.4	1.9
2009–10	116.4	60.3	43.1	4.5
2010–11	121.8	63.0	43.0	4.9
2011–12	127.9	66.5	44.7	5.5
2012–13	132.43	69.7	46.1	5.95
2013–14	137.7	74.75	47.9	6.2
2014–15	146.3	78.48	48.1	6.7
2015–16	155.5	82.92	43.60	7.00

A close look at the Table 4.1 shows that between 1990–91 and 2012–13, the dairy sector had an annual growth rate of 6.5% with per capita availability of milk of 296.5 g per day which is more than the world average of 294 g/day. Per capita availability of eggs increased from 30 to 57. Availability of meat and fish in 2012–13 was 11.78 kg per capita. Meat production registered a growth rate of 7.87% in 2012–13 over 2011–12; during 2014–15 milk, eggs, wool and meat further increased to 146.3 MT, 78.48 billion, 48.1 mkg and 6.7 mt, showing a growth of 6.27%, 7.2%, 0.48% and 6.7% respectively over the previous year. Per capita availability of milk and eggs is 322 grams per day and 63 eggs per year, respectively in 2014–15.

The greatest contribution animal husbandry has made to the national economy is its support to the green revolution, particularly the motive power from bullocks valued around 37,000-31556 MW between 1961-1981 (Table 4.2). Such stable motive power was achieved through the reduction and control of rinderpest and other diseases. An All India Rinderpest Eradication Programme was launched in 1954. India achieved rinderpest eradication in the year 2006. Global eradication of this dreadful viral disease was declared by FAO in June 2011. Thus, uninterrupted bullock power was provided to crop agriculture. It accounted for 23.3% of all power by 1991.

Table 4.2: Power availability from energy sources on Indian farms

	1961			**1971**			**1981**			**1991**		
Source of Power	**Number (million)**	**Power (MW)**	**% of Total Power**	**Number (million)**	**Power (MW)**	**% of Total Power**	**Number (million)**	**Power (MW)**	**% of Total Power**	**Number (million)**	**Power (MW)**	**% of Total Power**
Animal power												
Human	131.10	9780	23.3	126.80	8385	18.7	146.80	10951	12.4	173.00	12906	10.1
Draught animal	80.43	30000	71.6	81.57	30426	60.5	84.60	31556	35.8	80.00	2984	23.3
Sub-Total		39780	94.9		39811	79.2		4257	48.2			
Mechanical power												
Tractor	0.03	671	1.6	0.11	2462	4.9	0.53	11861	13.4	1.14	25513	19.9
Power tillers	0	0	0	0.001	6	0	0.02	119	0.1	0.10	597	0.5
Diesel engines	0.23	856	2.1	1.54	5744	11.4	4.44	16561	18.8	5.10	19023	14.9
Electric motors	0.16	597	1.4	0.61	2275	4.5	4.60	17158	19.5	10.75	40093	31.3
Sub-total		2124	5.1		10487	20.8		45699	51.8		85226	65.5
Grand total		41904	100		50298	100		88206	100		129210	100
Net cropped area, (m ha)			133.20	140.30		140.00	141.00					
Available farm power, (kw/ha)			0.31	0.36		0.63	0.92					

Source: Mittal and Srivastava (1993).

If the green revolution has made the country self-reliant in grain production, the technologies developed in the areas of animal genetics, breeding, nutrition, health care and marketing have made White Revolution possible as an alternative model of rural change with little public investment and incentives. In contrast to Green Revolution, White Revolution narrowed the gap between the rich and the poor; for example, 80% of milk produced in the country comes from small and marginal farmers and landless labourers including women.

India has large livestock resources in the world with great diversity of germ-plasm, *viz.* camel, mithun, yak and Pashmina goat, to cite a few. Historically this country has contributed significantly to world animal production. The Red Indian Jungle fowl is the progenitor of the modern commercial layer and broiler breeds and the Indian Runner duck is a source of all the best breeds of ducks in the world. Jamunapari goat of India was the male parent of Anglo-Nubian goats. Sindhi, Sahiwal, Ongole, Kankrej and Gir cattle breeds of India have significantly contributed to the now famous tropical dairy and meat breeds of the world.

India with its formidable 14.23% (2011) ruminant population, the largest goat population and fifth largest sheep population in the world should have been a colossus in animal production and one of the most prosperous countries. In fact smaller countries such as Denmark, Holland and New Zealand have pulsating economies based largely on animal production. A small country, Israel, today is producing the best broiler stock and the best dairy cows in the world. Unlike India, Brazil and other countries that have imported our zebu stock and have achieved high level of prosperity through the establishment of meat and milk industries.

Animal production in India has been the victim of religious sentiments, prejudices, taboos and chronic dearth of funds for development. Surplus unproductive livestock number, indiscriminate breeding, gross deficiency of feed and fodder, poverty and illiteracy of the farmer have further compounded the problem. In India 70% of all livestock are owned by 67% of small and marginal farmers who produce 80% of the milk in the country. It has been estimated that with the growth rate up to 5% in dairy sector, 20% in poultry eggs and broiler, and overall employment growth of 10%–15%, this sector is poised for fast track development. Livestock sector can create the required employment and wealth by removing constraints and adopting new strategies, some of which are indicated below. India can generate sufficient employment potential and achieve higher rates of prosperity.

4.1 STRATEGIES

The annual growth rate of employment in animal husbandry sector has been approximately 5.4% compared to 2%–3% in crop agriculture sector. These figures do not account for large employment generated by the livestock sector of small and marginal farmers and landless labourers, particularly women who get part-time employment throughout the year in this sector. No appropriate methodology to collect this information has yet been evolved. Maintenance requirement of these animals has been estimated to be 31.6 million man days/year. Such estimates of employment opportunities through family farming in the household servicing sector of Dairy Industry, involving consumption and distribution of 85% of the milk produced in India, are yet to be made.

With 1950–51 as the base year, India's milk production in 2015 has increased by nine-fold, egg production has increased 39 times and wool production has doubled. Meat production from goats and other animals has almost tripled mostly during the last four decades. All these achievements were by and large through horizontal increase in the population putting a heavy load on feed burden, which is already large under shrinking land resources for fodder cultivation, pastgures and grazing.

Future strategies have therefore to be based on a mixed approach of reduction in population size of the livestock, elimination of unproductive stray cattle and other livestock to the tune of 200 million bovines within the next four years, and concentrate on vertical genetic improvement of animals and knowledge of intensive animal production through entrepreneurship, thus converting financially non-viable farms into financially viabling entrepreneurships.

Buffaloes and cattle fit in ideally in mixed farming system being practised by our farmers. This system has built-in safeguards against the vagaries of nature and farming situations and hence it is likely to continue in future. Milk being a perishable commodity creates special problems of disposal, particularly during glut season of October to March. During this period of surplus, market price gets depressed and this force the producers to sell milk at a price lesser than the production cost.

There is a need to develop more modern milk processing units to process the produce and pay a reasonable return on investment to the farmer. There is a lot of demand for the supply of animals and frozen semen, mainly of Murrah breed from abroad. It is necessary to explore the possibility of large-scale export of Murrah germ plasm (semen, ova, embryos) at a premium price along with buffalo milk products such as Mozzarella cheese, etc. While doing so care should be taken to retain the best animals. China is maintaining a small number of Murrah buffaloes for producing cross-breds (Murrah x swamp buffaloes). These cross-breds yield higher milk than local swamp buffaloes. They imported Murrah from India in the 50s and Nili from Pakistan in 60s. It has been reported that 3-breed crosses (Nili 50% x Murrah 25% x swamp 25%) are the most superior groups as it ideally combines all four traits (milk yield, draught, meat and docile temperament).

India has not paid adequate attention towards the research and development of buffaloes. The size of herds being kept by individual farmers is very small and there is vast variability in feeding and management practices. The infrastructure needed for the evaluation of sires based on performance does not exist as per need. The coverage and success of AI is very low. Thus, it has not been possible to realise the attainable target of doubling the milk yield per animal in India. Theoretically it is possible to enhance the genetic potential for lactation yield in Murrah breed at 1%–2% per annum through adoption of robust selection and strict culling/selective breeding based on progeny testing. The genetic potential for milk production of low yielding nondescript buffaloes can also be enhanced by 50% in first grade through grading up with improved breeds such as Murrah, Nili and Mehsana.

If we do not change our present situation and allow the population growth to continue at the present level of 3.19% in buffaloes (2012 Livestock Census) except reduction in indigenous cattle population (−4.1%), sheep (−9.07%), goats (−3.82%) and pig (−7.54%) non-removal of unproductive bovines resulted into heavy feed burden with consequences of land degradation, loss of grass cover and environmental pollution.

The present achievements will not sustain the kind of growth rate achieved during the 80s and 90s, and the new genotypes of cattle, sheep, poultry and other species developed in last decades after independence will become *desi/ mangaroles* leading to significant drop in animal products, particularly in milk, much to the distress of the large vegetarian population in India.

A significant change is needed in the policy frame and mode of its implementation to have optimum milk, egg and meat production with minimum livestock population. The carrying capacity of the Indian agriculture for large and small ruminants and other livestock cannot sustain the population of livestock if intensive culling is not adopted. The livestock perforce of circumstances has to be raised on crop residues and agricultural by-products, therefore, it is the technology which need to be brought into play.

Vertical genetic improvement of livestock and reduction of feed burden are two major strategic components needing imminent attention. In genetic improvement in livestock, the results of research so far have indicated that it is not possible, with the present literacy rates to put a performance recording system of livestock on the ground at the village level. Therefore, development of methodologies and projects based on performance recording of livestock should be given a low priority. In contrast, more funds should be invested in areas of molecular genetics and biotechnology which holds promise for selecting bulls, bull mothers based on associated markers (QTLs), for linked loci and structural genes responsible for economic traits. These need to be located by genome mapping of Indian breeds and species. This is critical to our system. As a corollary, large investment will have to be made in embryo transfer technologies and development of transgenics. ETT should be used as an adjunct to progeny testing programmes based on farm testing of sires using multiple ovulation and embryo transfer (MOET) in livestock (Bhat, NDRI).

For faster gains, the policy framework needs radical modification. At present, conducive conditions of bull production, particularly for cross-bred populations are non-existent. Resources available in Military Dairy Farms of approximately 30,000 heads of cross-bred female cattle have not been mobilised as had been recommended for progeny testing and development of a new strain of cattle called Frieswal. This programme needs intensification and to be backed up with molecular genetics support.

Genetic resource conservation of indigenous germplasm has not made any headway. The Government has yet to have a policy framework on conservation of livestock resources and funds must be clearly earmarked for this activity. This neglected area, if attended to, will have impact on future developments in Animal Husbandry in this country and in the world. India being a major centre of origin for livestock species, their conservation is most important. Methods based on DNA fingerprinting, location of structural genes and micro/mini satellites would be vital as most variability for improvement has to come from this large biodiversity in germ plasm as the genes for important traits are available in them. For instance, Baroola gene for **fecundity** in sheep is available in the Indian Garole sheep, but has been patented by Australians and is available to us at a cost. Such situations can only be avoided if our investment on genetic research on native livestock is enhanced. Moreover, improvement is possible only with people's participation which needs direction. Without direction to the livestock production system in the country, new technologies which have been put in place during the last 40 years will become self-defeating.

Vertical genetic improvement must be linked with reduction in animal numbers. Any selection programme can go on its own only till the frequency of genes for high production is increased in the population. This is dependent on reduction of genes for low productivity which calls for regular culling scheme in the selection process. Unfortunately this is not practised for any species of livestock for various reasons. If the reduction in cattle number is not made mandatory by a policy frame, all the feed resources will get exhausted by the turn of this century as it is already affecting existing feed availability in the country. The Indian farmers had over centuries developed various systems of reducing population size without hurting anyone's sentiments. This kept in check the cattle numbers up to the 70s. However, due to intensive media exposure, generating intolerance, this route is slowly becoming ineffective.

4.2. FEED BURDEN

Reduction in the livestock population size will obviously lead to a reduction in feed burden. Development and use of sexed semen in milch breeds of cattle (Sahiwal, Gir, Tharparkar, Rathi) and buffalo (Murrah, Nili-Ravi) to produce more females is attractive proposition and require adequate attention. Arguments for increasing area under cultivated forage crop rotation or introduction of one forage crop in the farming system cannot be sustained especially in view of growing food grain requirements of human population. It must be clearly realised that animal production must be sustained on feeding of crop residues and industrial by-products, particularly oilcake and oil-meals (from cultivated crops and forest produce). Research in plant biotechnology and breeding will definitely bring improvement in the quality and quantity of residues but that will make only a marginal impact of 20% by 2050 AD.

The present research on animal nutrition and physiology promoting traditional feeding system will no longer deliver high productivity levels as the basic raw materials have a low digestibility (2%–3%). However, a major reserve of energy is found in the untapped plant cell-wall in terms of ligno-cellulose complex and, if tapped, will yield energy enough to feed all the 400 million livestock or more. However, this will require large investments in research on feed biotechnology, and manipulating and constructing microbes which can digest the lignin. Current nutrition research has to focus on this area to generate and release carbohydrates from plant cell-wall.

The source of protein in oil-meals, oil-cakes etc. unfortunately is bound by various chemical moieties such as tannin and other antigrowth factors, and is not readily available to animals. Suitable technologies must be developed to remove these undesirable chemicals from cakes and meals so that the abundant protein available in these meals is available to livestock.

4.3 ANIMAL HEALTH

The animal health protection system under the state patronage is fairly organised and is the backbone of the total system but needs to be integrated at village level. Keeping pace with global research on animal health care system, it should lead to satisfactory levels of protection provided the knowledge intensive approach is integrated in the development process.

All these technologies when packaged need a delivery system and the recipient. Though, both are in existence but remain elusive. The package will get expensive as the knowledge is converted to wealth. The small and marginal farmers, and landless labour, user of this package, need to have the financial viability to pay for the package. This brings in question the need for improvement in capacity building of marginal and small farmers, gender issue and access to market at his level, so that he moves from sustenance farming to entrepreneurship. These and others related issues need to be addressed before the package is delivered.

If these problems as identified above are seriously addressed in the next 25 years, we can become a leading nation in livestock production. Let us close on that assumption and happy note.

■■■■

Chapter 5

Breeding Policy: A Critique

What has been discussed in Chapter 4 shows in detail that the British India Government did not take any special interest in livestock development as a development strategy except for horses which were part of army and law and order strategies until 1925. When Royal Commission was appointed in the year 1925 under the chairmanship of J.A. John Hope, who had great interest in Animal Husbandry as the Chairman with the following mandate: To examine and report on the conditions of agricultural and rural economy in British India and to make recommendations for improvement of agriculture and to promote welfare and prosperity of rural operations. The other members of the Commission were Sir Henry Staveley Lawrence, ICS: Sir Thomas Middleton: Sir Ganga Ram, Engineer, Agriculturist and Philanthropist from the Punjab: Sir James Mac Kenna, ICS, authors of 'The Wealth and Welfare of the Punjab' (1892): Raja Shri Krishna Chandra Ganapathi Narayan Deo Garu; Nagrendra Nath Gangulee, Professor of Agriculture and Rural Economy, Calcutta University; Lodhi Karim Hyder, Professor of Economics, Aligarh University and Balkrishna Sitaram Kamat.

The Commission recognised that 'India is still pre-eminently the land of the small holders'. The typical agriculturist is still the man who possesses a pair of bullocks and cultivates a few acres of land with the assistance of his family and occasionally hired labour, he requires all the help which science can afford and which organisation, education and training can bring within his reach. The report emphasised the importance of providing for a minimum standard of life in villages and controlling the rate of growth of population in promoting agricultural development. The modernisation of Indian agriculture was to be brought about though research, extension, greater research coordination of various departments dealing with agriculture and the development of co-operative institutions within its framework.

The commission made detailed recommendations in the field of agricultural research. For the improvement of Livestock, a policy of selective breeding within species, a general reduction in unwanted stock, the improvement of draught and milch cattle, the expansion of veterinary departments, the upgrading of veterinary education and greater attention to fodder cultivation were recommended. The encouragement to group marketing, the establishment of regular markets, the appointments of marketing staff at the Centre and in the provinces and the setting up of Central Bureau of Information on Irrigation to deal with matters arising from Hydro-electric development infrastructure.

The commission felt concerned that due to constitutional changes of 1919, *viz.* the Montague-Chelmsford Reforms, the Government of India divested themselves, except to a very limited extent, of all powers of superintendence, direction and control over the administration of 'transferred' subjects to the provincial governments, which include agriculture and veterinary departments. The administration of the Central agencies and institutions for research and for professional and technical training was retained as 'Central' subject, but no specific provision was made for coordinating the work of these with that of similar institutions in the provinces. Thus, the provincial departments had, in the all-important matter of research, been left without the stimulus of a Central organisation which could guide and coordinate their policy; however, no specific provision had been made in the Constitution of 1919 for coordinating research, either as between the central and provincial spheres or as between province and province.

The Commission felt that there was nothing inherent in that Constitution, which prevented appropriate machinery being devised for that purpose.

The problem of Breeding Policy of livestock and strategy remained the same as suggested earlier and its general principle of genetics and animal breeding has been discussed under Chapter 4, where a detailed analysis has been made of the results obtained so far and the reasons for the failure of the policy as laid out but for the non-availability of implementation platform so that gains, both genetic and economic, could not be obtained. Broadly the failure was due to the lack to identify the (i) stakeholder (ii) and the financial requirements needed for the deliverable. In the absence of these basic requirements, no breeding policy could be implemented.

In addition, the problem of surplus numbers was not put in the programme. Hence, in the name of breeding strategy the numbers of cattle increased without realizing impact of tractorisation of agricultural operations. As of 2015, we have 200 million cattle and buffalo which are surplus to the requirement for agricultural power.

In our view, if the breeding policy strategy is to the recommended first these two problems must be resolved: (i) What to do with the surplus cattle numbers? (ii) What to do with feed non-availability in the short and long run? and (iii) Infrastructure and other needs.

We suggest two specific lines of approaches: term one-long term; the first and foremost is to separate breeding of cattle and buffalo. There is no problem in resolving the issue of surplus buffalo males as we have provided a plan of the same in Chapter 4 for buffalo slaughter and once the investors/GOI puts in place

PT programme for 400 buffalo bulls, the entire cost of the project will be recovered from the export of buffalo meat. A project with 400 bulls in each unit can be started on buffalo for producing progeny tested (PT) bulls which should be mated with dams having average of milk yield of 2000 litres locked for five years. This first project will give you 40 PT bulls in a period 6 years and recovery of all the investment made will be made.

Regarding cattle if an effective selection is to be done the plan must eliminate unproductive cattle, the most humane method of elimination is through slaughter. We have to concentrate on an open breeding system as recommended in chapter 3 and 4 and eliminate the rest of the cattle from the system, but it has problems with Hindu religious sentiments. Once that sorted out, India will become the largest gainer in the field of milk and milk products in the world, and we have every faith if half of agricultural and industrial resources are handled by the present technologies, we will solve the problem of feed burden if we put single minded attention on it.

■■■■

RESOURCE PERSONS

Acharya R.M.	Former DDG (AS) ICAR, 784, Sector 9, Faridabad (Haryana)
Ahmad A.H.	Head, Dept. of Pharmacology & Toxicology, College of Veterinary and Animal Sciences, G.B. Pant University of Agriculture & Technology, Pantnagar – 263 145, (Uttarakhand)
Asokan S.A.	Dean incharge, Madras Veterinary College, Chennai – 600 007, Tamil Nadu, India
Bhat P.N.	Chairman, World Buffalo Trust, Flt No. 205, F-64, C/9, Sector-40, Noida
Bhat P.P.	Former ADG (AP&B), ICAR, 600A Hamilton Court, DLF Phase –IV, Gurugram
Bhatia B.B.	College of Vet. Science, G B P U A T Pantnagar (U.P)
Chander Suresh	Lala Lajpat Rai University of Veterinary and Animal Sciences, Hisar-125004
Chatterjee R.N.	Director, Directorate of Poultry Research, Rajendranagar, Hyderabad 500030
Deb S.M.	Director, National Research Centre on Yak, West Kameng District, Dirang- 790101, Arunachal Pradesh
Dutt N.S.	Ganganiwash 18, Aundh Road, Pune (MS)
Gaya Prasad	Vice Chancellor, SVBP Univ. of Agriculture and Technology, Meerut
Harikrishnan T.J.	Vice – Chancellor Incharge, Tamil Nadu Veterinary and Animal Sciences University, Chennai
Johri D.C.	Former Director, Central Avian Research Institute, IVRI, Izatnagar
Kalra A	Executive Director – Ayurvet Limited, 6th Floor, Sagar Plaza, District Centre, Vikas Marg, Laxmi Nagar, Delhi 110092
Kamra D.N.	National Professor, Division of Animal Nutrition, Indian Veterinary Research Institute, Izatnagar (U.P.)

Kataria J.M.	Director, Central Avian Research Institute, Izatnagar, Bareilly (U.P.)-243122
Khera S.S.	Former Head, Epidimology at IVRI, 29-B, Udham Singh Nagar, Civil Lines, Ludhiana (Punjab)
Kondaiah N	Former Director, National Research Center on Meat, Hyderabad
Krishna Lal	Former ADG(AH),ICAR, C-302, Exotica Elegane, Ahinsa Khand –II, Indrapuram, Ghaziabad (UP) 201010
Kulkarni V.V.	Director in charge, National Research Centre on Meat, Chengicherla, P.O. Box No. 19, Boduppal, Hyderabad 500092
Kumar Puneet	Incharge, Placement Cell and ZTN & BPD, Indian Veterinary Research Institute, Izatnagar-243 122, Bareilly
Mahapatra S.C.	Former Director, Central Avian Research Institute, IVRI, Izatnagar
Maru Ajit	55, Milan Park, Near Sandesh Press, Vastrapur, Ahmedabad-380054, Gujarat
Mishra B.P.	Jt. Director (Research) Indian Veterinary Research Institute, Izatnagar
Mukherjee S	In charge, PME Cell, National Research Centre on Mithun, Jharnapani, Medziphema, Nagaland – 797106
Narain Prem	B-3/27 A, Lawrence Road, Keshav Puram, Delhi 110035
Narayan K.G.	Head, Department of Vet., Public Health and Epidemiology, College of Vety and Animal Husbandry, Ranchi, Bihar
Naveena B.M.	Senior Scientist, National Research Centre on Meat, Chengicherla, Bod Uppal Post, Hyderabad, 500092, Andhra Pradesh
Nimbkar Chanda	Director, Animal Husbandry Division, Nimbkar Agricultural Research Institute, Phaltan, District Satara, Maharashtra
Pattnaik B	Project Director, Project Directorate on Foot-and-Mouth Disease, IVRI Campus, Mukteswar – 263 138, Distt: Nainital (Uttarakhand)
Pradhan K	Former VC, HAU, Hisar, C-24/HIG, Housing Board Colony, Baramunda, Bhubaneswar 751003, Orissa
Prakash B.S.	ADG, Animal Science Division, ICAR, New Delhi
Prakash Birham	Director, Central Institute for Research on Cattle, Grass Farm Road, Meerut Cantt
Prasad C.S.	Former Director, NIANP, 144/1, 8 Main, 14th Cross, Malleswaram, Bangalore 560003
Raghavendra Bhatta	Director, National Institute of Animal Nutrition and Physiology, Bangalore

Rajkhowa Chandan	Former Director, NRC on Mithun, Jharnapani, Medziphema, Nagaland-797106
Rakib A	Central Institute for Research on Goats, Makhdoom, Farah (UP)
Ranjan S.K.	Former FAO Expert, H-334, FF, New Rajinder Nagar, New Delhi
Ravikant K	Ayurvet Limited, 6th Floor, Sagar Plaza, District Centre, Vikas Marg, Laxmi Nagar, Delhi 110092
Reddanna P	National Institute of Animal Biotechnology, Miyapur, Hyderabad-500049
Roy D.S.	Director, Central Island Agricultural Research Institute, P.B. No. 181, Port Blair-744 101, Andaman and Nicobar Islands
Roy K.S.	Punjab Agricultural University, Ludhiana (Punjab)
Saxena M.J.	Managing Director, Ayurvet Ltd., 6th Floor, Sagar Plaza, Distt. Centre, Vikas Marg, Laxmi Nagar, Delhi.
Sharma Arjava	Director, National Bureau of Animal Genetic Resources, Karnal
Sharma Kusumakar	Former ADG (Edn), ICAR, Flat No- 05073, ATS Greens Paradiso, Tower 05, 7th Floor, Sector Chi-4, Greater Noida-201308, Distt. G.B. Nagar (UP)
Sharma M.C.	Former Director & Vice Chancellor, IVRI, Izatnagar, D-76, 2nd Floor, Sector -51, Noida – 201301 (U.P)
Sharma Nagendra	Former VC, SKUAT, House No.10, Shivnagar Colony, IVRI Road, Izatnagar, Bareilly-243122
Shinde A.K.	Principal Scientist, Central Sheep and Wool Research Institute, Avikanagar via Jaipur Rajasthan 304501
Singh R.K.	Director, Indian Veterinary Research Institute, Izatnagar-243 122, India
Singh S.N.	Managing Director, Biovet Pvt. Ltd., # 308, 3rd Phase, KIADB Industrial Area, Malur Taluk, Kolar District, Karnataka
Srivastava A.K.	Director & Vice-Chancellor, National Dairy Research Institute, Karnal-132 001, Haryana
Taneja V.K.	Former Vice Chancellor, Guru Angad Dev Veterinary & Animal Sciences University, Ferozpur Road, Ludhiana-141004
Thirumalesh T	Head, Project Planning and Monitoring Cell, KVAFSU, Bidar- 585 401. Karnataka
Tiwari S.P.	Dean, Faculty of Veterinary and Animal Science, Anjora, Durg-491001, C.G.
Tripathi B.N.	Director, National Research Centre on Equines, Hisar

Uppal P.K.	Former Director, NRC Equines, Hisar, T-56, Rajouri Garden, New Delhi
Verma Rishendra	Former Joint Director, Centre for Animal Disease Research and Diagnosis (CADRAD), Indian Veterinary Research Institute, Izatnagar-243 122 (UP)
Yadav M.P.	Former Vice Chancellor, SVBP Univ. of Agriculture and Technology, Meerut, H.No. 365, Sector-45, Gurgaon

Bibliography

Note: References: (a) as given in the Text, (b) references to materials edited by the authors and (c) extended reading material on the subject.

Abhi HL, Nagpal MP, TL Sharma and Grewal AS 1973. A study on the breeding behaviour of Murrah buffaloes and their breeding efficiency through A.I. under farm conditions. *Indian J. Dairy. Sci.* **26** 107

Abraham RJJ, Govindrarajulu M, Radhakrishnan KT, Ramamurthi R 1990 Short term preservation of cara beef using liquid nitrogen. *Cheriron* **19** 261-64

Acharya RM 1966 Genetic Analysis of a closed herd of Indian Cattle. Ph.D. Dissertation. IOWA State College, Library, Ames. IOWA, USA

Acharya RM 1974 Evaluation of native breeds of sheep for wool and mutton and scope for introduction of exotic inheritance. *Indian J. Genet.* 34 A: 945. (Paper presented in 2nd General Congress, SABRAO, New Delhi)

Acharya RM 1975 A model for cooperative sheep farming. *Indian Fmg* **25** (2):19

Acharya RM 1978 Breeding strategy for sheep in India—A discussion paper. Mimeo, CSWRI, Avikanagar

Acharya RM 1981 Monograph on Breeds of Sheep and Goats in India — FAO— Rome

Acharya RM 1982 Sheep and goat breeds of India. Animal Production and Health paper-30, Food and Agriculture Organization of the United Nations, Rome

Acharya RM 1982 Sheep and Goat Breeds of India. FAO Animal Production and Health paper 30, Food and Agriculture Organisation of the United Nations, Rome, Italy

Acharya RM 1982 Sheep and Goat Breeds of India. FAO Animal Production and Heath Paper No 30. FAO, Rome

Acharya RM 1985 Small Ruminants Production in Arid and Semi-arid Asia —A paper in small ruminants production in developing countries—FAO Animal Production and Health Paper 58—Food and Agriculture Organisation of the United Nations Rome, Italy

Acharya RM and Bhat PN 1983 Genetical Researches in Livestock—*XV International Congress of Genetics,* New Delhi India, December 12-21

Acharya RM and Bhat PN 1984 Livestock and Poultry Genetic Resources in India, Indian Veterinary Research Institute, Izatnagar, India

Acharya RM and Lush JL 1966 Genetic Progress through selection in a closed herd of Hariana Cattle *J. Dairy Sci.* **51** 1059-1064

Acharya RM and Mani Mohan 1978 Genetic considerations in crossbreeding for evolving new breeds of sheep—Results of a cross breeding experiment. *Indian J. Anim. Genet.* Breed. **1** 37

Acharya RM and Patnayak BC 1974 Role of Sheep in desert ecosystem and drought proofing through improved sheep production with special reference to Rajasthan. Monograph CSWRI

Agarwal KP, Mongha I and Bhattacharya NK 1982 Collection and transfer of embryos in goats by surgical method *Indian Vet. J.* **59** 298-303

Agnihotri M, Nageswara Rao SB and Pal UK 1995. Development of crisp snack from goat stomach. In "*Proceedings of Symposium on production and marketing of goat meat"* held at CIRG, Makhdoom, on Jan. 10-11. p.33

Agnihotri MK and Prasad VSS 1992 Chevonette: a dehydrated minced meat product for emergency rations. In: Recent advances in Goat Production. Proc. *V International Conference on goats*, held in N. Delhi, 2-8 March, 1989-92

Agrawal MC and Shah HL 1984 Stephanofilarial dermatitis in India. *Vet. Res. Commun.***8**(2) 93-102

Agricultural Research Council 1965. Nutrient Requirements of Farm Livestock No. 2. Ruminants London: Agricultural Research Council U.K.

AHD West Bengal 1965. Final Report of the scheme for development of mutton breed of goats in West Bengal. Department of Animal Husbandry, Calcutta, West Bengal

AHD, Kerala 1964 Final Report of the Scheme for developing a strain of goat for milk production. Department of Animal Husbandry. Trivandrum, Kerala

Ahmed H, Wanzala W, Zessin KH, Kyule NM, Baumann MPO, Mathias E 2005 Ethnoveterinary Medicine: A Critical Review of its Evolution, Perception, Understanding and the Way Forward. (CIPAV, 2005)

Ahmed NM, Irfan RA, Chaudhry and Ahmed W 1983 Reproductive efficiency of Nili-Ravi buffaloes in Pakistan. *Indian J. Anim. Sci.* **53** 1066-1068

Ahuja AK and Arora SP 1980 Sulphur requirements of growing and lactating crossbreds (Broen Swiss X Sahiwal) Z. Tierphys Tierermahr, *Futtermittelk* **44** 190-97

Ahuja AK and Arora SP 1981 Utilization of nitrogen and sulphur by rumen microbes and their excretion in calves. *Indian J. Anim. Sci.* **51** 1128-32

AICRP—1980. Goat, Mannuthy. Annual Report 1980 All India Coordinated Research Project on Goats for milk. Kerala Agricultural University, Mannuthy Kerala

AICRP—1985. Goat, Avikanagar. Annual Report 1985 All India Coordinated Research Project on Goats for meat, Central Sheep & Wool Research Institute, Avikanagar (Rajasthan)

Alim KA and Ahmed IA 1954 Month of calving, age at first calving and calving intervals of the buffaloes in a dairy herd in Egypt. *Emp. J. Exp. Agric.* **22**: 37-41 (fide Anim. Breed. Abstr. **22** 1438

Alur MD, Chawla SP and Nair PM 1992 Microbiological method to identify irradiated meat. *J. Food Sci.* **57** 593-595

Amble V N and Malhotra J C 1968 Statistical studies on Rambouillet x Rampur Bushair cross of sheep at Pipalkoti, U.P. *Indian J. Vet. Sci.* **38** 101

Amble V N, Khandekar NC and Garg JN 1967 Statistical studies on breeding data of Deccani and crossbred sheep. *Indian J. Vet. Sci.* **37** 305

Amble VN and Jain JP 1966 Review of cross-breeding work in cattle in India. Seminar on Animal Breeding. Haringhatta/Calcutta, pp. 16-63, Indian Council of Agricultural Research, New Delhi

Amble VN and Malhotra J 1966 Review of cross breeding of sheep for wool production in India. Proceedings of the Seminar on Animal Breeding, Calcutta, Haringhata ICAR, 219

Amble VN, Krishnan KS and Srivastava JS 1956. Statistical studies on breeding data on Indian herds of dairy cattle. *Indian J. Anim. Sci.* **26**. 33:62

Anandjiwala S, Kalola J, Rajani MJ 2006 AOAC Int. **89** 1467-1474

Andreu P, Paim GL, Castilho BA, Veho RF 2008 *Pharmacol Res.* **57** 332-338

Anjanayulu ASR, Kondaiah N and Murthy TRK 1995. Quality of buffalo meat suggests and rolls., *Fleischwritsh.* **75** 893-95

Anjaneyulu ASR 1988 Effect of salt and phosphate on quality of ground buffalo meat. Ph.D. thesis, IVRI, Izatnagar, India

Anjaneyulu ASR and Kondaiah N 1990 Quality of buffalo meat suggests and rolls containing edible byproducts. *J. Meat Sci. & Technol.* **3** 95-99

Anjaneyulu ASR and Sharma N 1989 Influence of phosphate on functional properties of buffalo meat and palatability of patties containing reduced salt. *Indian J. Meat Sci. & Technol.* **2** 72-75

Anjaneyulu ASR, Kondaiah N and Murthy TRK 1993 Effect of pork rind on the quality of suggests. In *"3rd International Food Convention"* held at Mysore, Sept. 7-12, p.23

Anjaneyulu ASR, Rao KH, Singh RRB and Yadav PL 1996 Effect of different milk proteins on the quality of smoked chicken sausages. XX World's Poultry Congress Proceedings, Vol. III, 199-204

Anjaneyulu ASR, Sharma N and Rondaiah N 1989 Evaluation of salt and polyphosphates and their blends at different levels on physicochemical properties of buffalo meat and patties. *Meat Sci.* **25**:293-306

Annual Report 2012 Project Directorate on Animal Disease Monitoring & Surveillance (PD_ADMAS), Bangalore

Annual Report 2013-14 National Institute of Veterinary Epidemiology and Disease Informatics, Bengaluru

Annual Reports (1947-48, 1948-49, 1949-50, 1959-60, 1965-66, 1966-67, 1967-68, 1970-71 to 1977-78, 1979-80 to 1980-81, 1984-85, 1999-2000). Indian Veterinary Research Institute, Mukteswar/Izatnagar

Annual Reports 1923-24, 1927-28 Imperial Veterinary Research Institute, Mukteswar/ Izatnagar

Anon 2012 Livestock technologies - ways to diversified agriculture, Animal Science Division, ICAR, New Delhi, pp1-180

Anon, 1988-1990 Annual Progress Reports of the All India Coordinated Research Project on Intracellular Blood Protista with particular reference to immunoprophylaxis and control

Anon, 2003 Tamil Nadu Veterinary and Animal Sciences University, Commemorative Centenary Volume

Anon, 2006-2007 Annual Report, Department of Animal Husbandry, Dairying and Fisheries, Govt. of India, Delhi

Anon, 2009-2010 Annual Report, Department of Animal Husbandry, Dairying and Fisheries, Govt. of India, Delhi

Anon, 2010-2011 Annual Report, Department of Animal Husbandry, Dairying and Fisheries, Govt. of India, Delhi

Anon, 2011 Annual Report, Project Directorate on Animal Disease Monitoring & Surveillance (PD_ADMAS), Bangalore

Anon, 2011-2012 Annual Report, Department of Animal Husbandry, Dairying and Fisheries, Govt. of India, Delhi

Anon, 2012 Annual Report Indian Veterinary Research Institute, Mukteswar/Izatnagar. Anon. 2012. Livestock technologies - ways to diversified agriculture, Animal Science Division, ICAR, New Delhi, pp1-180

Anon, 2012 Livestock technologies - ways to diversified agriculture, Animal Science Division, ICAR, New Delhi, pp1-18.

Anon, 2012-2013 Annual Report, Department of Animal Husbandry, Dairying and Fisheries, Govt. of India, Delhi.

Anon, Annual Reports. 2009-2010, 2010-2011, 2012-2013 of NDRI, Karnal

Anonymous 1994-95 Report of the Technical committee of Direction for improvement of Animal Husbandry and Dairying statistics. Dept. of A.H.&D. Ministry of Agricultural, Govt. of India.

Anonymous 1996 Report of the working group on Animal Husbandry and Dairying for Formulation of the Ninth Five Year plan. Ministry of Agriculture, Govt. of India. New Delhi.

Anthony-Das S and Ramanathan LA 1990 Dehydrated spiced rabbit meat mince. *Indian Food Packer*. **44**(4) 33-36

Arangasamy A 2010 Effect of stallionseminal plasma proteins on in vitro capacitation of equine spermatozoa. *Indian Journal of Animal Sciences* **80** 107-109

Arangasamy A, Bhure SK and Dwivedi S K 2008 Isolation and Characterization of Heparin and Gelatin Binding Stallion seminal plasma Proteins. *Indian Journal of Animal Sciences*, **78**(11) 1231-1234

Arora AL and Prince LLL (eds.) 2011 Network Project on Sheep Improvement. Project coordinator's Annual Report 2010-11. Central Sheep and Wool Research Institute. Avikanagar, Rajasthan. p.33

Arora CL 1983 Research highlights AICRP sheep breeding 191-1981. CSWRI, Avikanagar

Arora CL and Bhat PN 1998 *WCGAAP Conference* Australia

Arora R and Bhatia S 2006 Genetic Diversity of Magra Sheep from India Using Microsatellite Analysis. *Asian-Aust. J. Anim. Sci.* **19**(7): 938 -942

Arora SP 1982 Annual progress report of NDRI, Karnal

Arora SP, Bajpai LD and Muley AR 1962 Influence of dry period on milk yield and effect of season on lactation yield and lactation days of Murrah buffaloes. *J. Vet. Anim. Husb. Res.* **5**: 51-53 (fide Anim. Breed. Abstr. **32** 1858

Arora SP, Chabra A, Aterja PP and Sharma KNS 1978 Nutrients requirements by ad lib feeding system of buffalo calves. *Indian J. Dairy Sci.* **31** 9-13

Arora SP, Kaur H and Dass RS 1979 Dry matter and total digestible nutrient requirements of growing Karan Swiss calves. *Indian J. Dairy Sci.* **32** 205-207

Arora SP, Ludri RS, Singhal SK and Atreja PP 1974 Ammonium Sulphate as a nitrogen and sulphur sources for utilization by rumen microbes. *Indian J. Dairy Sci.* **97** 300

Arora SP, Parvinder Kaur, Khirwar SS, Chopra RC and Ludri RS 1975 Selenium levels in fodder and its relationship with degnala disease. *Indian J. Dairy Sci,* **28** 249-53

Arun Rasheed, Sravya Reddy B, Roja CA 2012 Review on standardization of herbal formulation. *International. J. of Phytotherapy*. Vol.**2**, Issue **2** 74-88

Ashfaq M and Mason IL 1954 Environmental and genetical effects on milk yield in Pakistani buffaloes. *Exp. J. Exp. Agric.* **22** 161-175

Attrey DP, Singh J and Sharma TR 1982 Effect of certain processing and storage variables on sensory quality of raw and pre-cooked freeze dried mutton. *Indian Food Packer*. **36**(4)71-78

Babu YH, James RM, Sreenivas CS and Khan MA 1984 The use of IPT to detect RP viral antigen in the infected tissue smears. *Ind. Vet. J.* **61** 271

BAHFS 2016 Basic Animal Husbandry and Fishery Statistics – 2016. Department of Animal Husbandry, Dairying and Fisheries, Ministry of Agriculture, Government of India, *AHS Series 17* New Delhi pp 166

Bala AK and Nagarcenkar R 1979 Evaluation of different cattle breed groups in hot humid tropics. Annual Report, pp. 162-63, NDRI, Karnal

Balaine DS, Singh J, Acharya RM and Kanaujia AS 1971 Post-weaning feedlot performance of Nali and Lohi sheep and their crosses with Nellore and Mandya. *Indian Vet J.* **48**: 52

Balakrishnan V, Philip Robinson J, Manickasamy A and Ravindran KC 2009 Ethno veterinary Studies Among Farmers in Dindigul District Tamil Nadu, India. *Global J.Pharmacol.* **3** (1)15-23

Balamurugan V, Sen A, Saravanan P, Bhanuprakash V, Patra RC, Swamp D, Kumar Mahesh, Yadav MP, Singh RK 2008. Antiviral effect of aqueous extract of Acacia arabica on peste des petits ruminants (PPR) virus replication. *J. Pharma. Biology* **46**(3)171-179

Balamurugan V, Sen A, Saravanan P, Singh RP, Singh RK, Rasool TJ and Bandyopadhyay SK 2006 One-step Multiplex RT-PCR Assay for the Detection of Peste des petits ruminants Virus in Clinical samples. *Veterinary Research Communications*, **30** 655–666

Bandyopadhyay S, Sasmal D, Dutta TK, Ghosh MK, Sarkar M, Sasmal NK, Bhattacharya M 2009 Seroprevalence of brucellosis in yaks (*Poephagus grunniens*) in India and evaluation of protective immunity to S19 vaccine. *Trop Anim Health Prod* **41** 587-592

Bandyopadhyay SK, Ganjendragad MR, Dhal NK, Gupta AR and Yadav MP 1984 Some observations on a goat pox virus isolated from an outbreak among black Bengal goats in Orissa. *Indian J. Anim. Sci.* **54** 961-964

Bansal RP and Rattan B 2003 Rinderpest. Publ. DIPA, ICAR, New Delhi. p 188

Bansal and Mondokhot UV 1988 Studies on differentiation of cattle meat from meat of other species of animals. II Comparative efficacy of different serological methods. *J. Food Sci. & Technol.* **25**(4) 232-'35

Bansal RP, Joshi RC and Sahi RS 1983 Counter immunoelectrophoresis test for detection of antibodies to rinderpest virus. *Ind. J. Anim. Sci.* **52** 139

Bansal RP, Uppal PK, Oberoi MS and Joshi RC 1981 The quick diagnosis of rinderpest by detection of antigen by CIE. *Ind. J. Anim. Sci.* **51** 766

Bansal RS, Pal Y, Pareek PK, Purohit GN, Legha RA and Gupta AK 2006 Progesterone profile during oestrous cycle, pregnancy and post foaling in donkey mares. *Veterinary Practitioner*, **7** 5-6

Barhat NK, Chowdhary NS 1978 Factors affecting some production traits in Rajasthani goats. *Indian Journal of Dairy Science* **31**:185-188

Barman C, Baruah A, Sarmah BK, Baruah KK, Sarmah BC, Deka BC, Dutta A, Chakravarty P, Khan MH and Bharti PK 2013 Superovulatory treatment response and embryo recovery in mithun cows. *Indian J Anim Sci.* **83**:272 -275

Barnes J 2003 Quality, efficacy and safety of complementary medicines: fashions, facts and the future. Part I. Regulation and Quality. *British Journal of Clinical Pharmacology*, **55** (3) 226-233

Baruah KK, Dhali A, Mech A, Bora B, Das J, Bora R, Mondal M, Sarmah BC, Deka BC, Rajkhowa C (2013) Effect of concentration and addition method of glycerol on the quality of cryopreserved mithun (*Bos frontalis*) spermatozoa. *J Anim Physiol Anim Nutr.* **97**:1051-1058

Basavaiah P 1978 Genetic studies of lactation curve and its use in sire evaluation in Murrah buffaloes. Ph.D. thesis, Punjab University, Chandigarh, India

Basu S 1962 Seasonal variation of fertility in Murrah buffaloes. *Indian Vet. J.* **39**: 433

Basu SB, Bhatnagar DS and Taneja VK 1982 Estimates of genetic parameters for first lactation performances in Tharparkar cattle. *Indian J. Anim. Sci.* **52**: 279-83

Basu SB, Nagarcenkar R and Sarma PA 1974 Selective breeding in Murrah buffaloes for economic characters. Annual Report, NDRI, Karnal, pp. 133- 135

Basu SB, Sarma PA, Sharma DD and Nagarcenkar R 1972 Evaluation of sires and dams for economic merit in Murrah buffaloes. Annual Report, pp. 101. NDRI, Karnal, India

Bauer S, Stomer E, Kaiser R, Tremblay PB, Brockmoller J, Roots I 1983 *Biomed Chromatography*.**16** 187-190

Bawa AS, Orr HL and Usbrone WR 1985 Effect of antemortem papain injection on tenderness of spent white leghorn hens. *J. Food Sci. Technol.* **2**:954-57

Baxby D and Hill BJ 1969 Buffalo pox virus. *Vet Rec.* **85**:315-6

Baxby D and Hill BJ 1971 Characteristics of a new poxvirus isolated from Indian buffaloes. Arch Gesamte Virusforsch. 1971.35:70-9., D. (1969). *Journal of Hygiene* (London), **67** 637-647

Belorkar PM, Khire DW, Kadu MS and Kaikini AS 1977 Studies on the effect of service period on lactation yield, lactation period and dry period in Nagpuri (Berari) buffaloes. *Indian Vet. J.* **54**: 384-388

Bera BC, Shanmugasandaram K, Sanjay B, Venkatesan G, Virmani N, Riyesh T, Gulati BR, Bhanuprakash V, Vaid RK, Kakker NK, MalikP, Bansal M, Gadvi S, Singh RV, Yadav V, Sardarilal, Nagarajan G, Alamurugan V, Hosamani M, Pathak KML and Singh RK (2011) Zoonotic cases of camel pox infection in India. *Vet Microbiol* **152**:29–38

Bhalaru SS and Dhillon JS 1978 Estimates of phenotypic and genetic parameters of some measures of efficiency of milk production. Journal of Agriculture Science, *Camb.*, **91** (1): 181-184

Bhalla JS 1988 History of the Remount and Veterinary Corps, 1794-1987. Additional Director, RVC, Quarter Master General Branch, Army Headquarters, New Delhi

Bhasin NR 1968 A study on economic characters of Nagauri cattle. *Indian Vet. J.* **45** (12): 1022-1026

Bhat PN 1967 Constraints in genetic improvement for livestock production in tropics. *Proc. IV International Congress of SABRAO*, pp. 51-65

Bhat PN 1974 Crossbreeding of cattle in India. Aims and Objectives of All India Coordinated Research Project on Dairy Cattle Breeding. Invitational paper, Round Table D "Heredity and Environment" *Proceedings of 1st World Congress on genetics applied to livestock production Madrid.* Vol. **II** pp 237

Bhat PN 1974 Evaluation of productivity of native cattle including buffaloes and scope for improvement by selection. *Proc. Second General Congress, SABRAO, New Delhi. Indian J. Genetics* vol. **34-A** pp 955-960

Bhat PN 1974 Prospects of cattle production research in Eastern Region, Andaman and Nicobar Islands, Goa and Lacadive Islands. Invitational paper presented at Regional Seminar, *Eastern Regional Conference, Shillong*

Bhat PN 1977 Infusion of exotic germ plasm in cows for enhanced and stable productivity under India's varying ecology. Invitational subject matter paper presented at Indian Dairy Conference. *Indian Dairyman XXIX* : 401-411

Bhat PN 1977 New and indigenous animal variability. Special invitational paper. *Proc. 3rd International Congress of SABRAO*, Canberra, Australia. **1(c)** 1-3

Bhat PN 1983 Constraints in genetic improvement for livestock production in tropics. *Proc. IVth International Congress of SABRAO*. pp 51-65

Bhat PN 1983 Performance of exotic cattle in India. *Proc. IVth International Congress of SABRAO*. pp 67-81

Bhat PN 1984 Introduction of exotic dairy breeds in India and the performance of their crosses with native breeds, stabilisation of exotic inheritance and further improvement of crossbreds through selection. Agenda Paper II. IVRI, Izatnagar, India

Bhat PN, Khanna RS 1970 Crossbreeding for higher milk production. *All India Workshop on Feed Production and Management of Large Livestock Farms*, Jabalpur, India

Bhat PN, Taneja VK and Garg RC 1978a Effect of crossbreeding on reproduction and production traits. *Indian J. Anim. Sci.* **48**: 71-78

Bhat PN, Taneja VK and Garg RC 1978b. Body weight changes in Holstein x Sahiwal crossbred in relation to level of Holstein inheritance. *Indian J. Anim. Sci.* **48**: 485-491

Bhatia D R, Mohan M, Patnayak BC and Sheikh QD 1973 Nutritive value of the available pasture in semi-arid areas of Rajasthan. *Indian J. Anim. Sci.* **43**: 838

Bhatnagar CS, Pathak KML, Kapoor M and Yadav MP 1997 Evaluation of various diagnostic techniques in donkeys experimentally infected with *Trypanosoma evansi. Indian J. Anim. Sci.* **67**(10): 836-838

Bhatnagar DS, Nagarcenkar R, Gurnani M and Sharma RC 1980 National Dairy Res. Inst. Karnal Ann. Rep. p. 134-142

Bhatnagar DS, Sharma RC and Sundaresan D 1975 Studies on comparative performance of Sahiwal and various Brown Swiss X Sahiwal crossbred group of dairy cattle at NDRI, Karnal, India. *Indian J. Dairy. Sci.* **28**: 77-84

Bhatnagar DS, Sharma RC and Sundaresan D 1976 Crossbreeding of cattle for increased milk production. NDRI Extn. Press Bulletin, Publication No. 143

Bhatnagar DS, Taneja VK, Basu SB and Murthy KMK 1983 Genetic parameters for some economic traits in Sahiwal cattle. *Indian J. Dairy Sci.* **36** 402-406

Bhatt N, Singh M and Ali A 2009 Effect of feeding herbal preparations on milk yield and rumen parameters in lactating crossbred cows. *International Journal of Agricultural Biology*, **11** 721–726

Bhattacharya P and Bhat PN 1982 Researches in Animal Production. Indian Council of Agricultural Research, Krishi Bhawan, New Delhi

Bhattacharya TK and Chatterjee RN 2013 Polymorphism of the myostatin gene and its association with growth traits in chicken. *Poultry Science*.**92** 910-915

Bhattacharya TK and Chatterjee RN and Priyanka M 2012 Polymorphism of Pit-1 gene and its association with growth traits in chicken. *Poultry Science* **91** 1057-1064

Bhattacharya TK, Chatterjee RN, Sharma RP, Niranjan M and Rajkumar U 2011 Associations between novel polymorphisms at the 5'-UTR region of the prolactin gene and egg production and quality in chickens. *Theriogenology* **75** 655-661

Bhattacharya TK, Chatterjee RN, Sharma RP, Niranjan M, Rajkumar U and Reddy BLN 2011 Association of polymorphism in prolactin promoter and egg quality traits in layer chicken. British Poultry Science. 52: 551-557

Bhattacharya TK, Chatterjee RN, Sharma RP, Niranjan M, Rajkumar U and Reddy BLN 2011 Identification of haplotypes in promoter of prolactin gene and their effect on egg production and quality traits in layer chicken. *Animal Biotechnology* **22** 71-86

Bhusan S, Sharma D, Bujarbaruah KM, Singh RV 2000 Annual Report, National Research Centre on Mithun, Jharnapani, Nagaland, pp 4-10

Binai-babu N, Kothandaraman P, Ramamurti R, and Sundradasu V 1979 Studies on tenderization of tough chicken meat with papain. *Indian J. of Poultry Sci.* **14** 13-19

Bodeker C, Bodeker G, Ong CK, Grundy CK, Burford G and Shein K 2005 WHO Global Atlas of Traditional, Complementary and Alternative Medicine. World Health Organization, Geneva.

Borimnejad V 2008 Niche Markets in the Agricultural Sector, Case Study: Iran, American-Eurasian *J. Agric. & Environ. Sci.,* **3** 893-899

Branton C, McDowell RE and Brown MA 1966 Zebu-European crossbreeding as a basis of dairy cattle improvement in the U.S.A. in Southern Coop. Ser. Bull. 114. Louisiana State Univ., Baton Rouge

Burgos RA, Hacke JL, Bertoglio JC, Aguirre V, Arriagada S, Calvo M and Caceres DD 2009 Efficacy of an *Andrographis paniculata* composition for the relief of rheumatoid arthritis symptoms: A prospective randomized placebo-controlled trial. *Clinical Rheumatology* **28**(8):931-946

Buvanendran V and Mahadevan P 1979 Crossbreeding for milk production in Sri Lanka. *Wld. Rev. Anim. Prod.* **15** 48-54

Cady RA, Shah SK, Scheymerhorn EC and McDowell RE 1983 Factors affecting performance of Nili-Ravi buffaloes in Pakistan. *J. Dairy Sci.* **66** 578-586

Chacko CT and George PK 1984 Evolution of a breed — Sunandini. Anim. Breed. *Abstr.* 53 (12): 7367

Chandra Rajesh, Garg SK and Yadav MP 1985 Characterization of virus isolates from cases of pox in buffaloes. *Indian J. Vet. Med.* **5** 101-106

Charoenteeraboon J, Ngamkitidechakul C, Soonthornchareonnon N, Jaijoy K, Sireeratawong S and Songklanakarin J 2010 *Sci. Technol.* **32**(6), 599-604

Chatteriee BP, Vaith P, Chattejee S, Karduck D and Uhlenbruck G 1979 *Int. J. Biochem*.**10** 321-327

Chatterjee AK 1969 The problems of manufacturing cured meat in India with special reference to ham, bacon and sausages. *Indian Food Packer* **23**(5) 17-28

Chattopadhyay RR, Bandopadhyay M 2005 *Indian J. Pharmacol.* **37** 184-185

Chattoraj DK, Bose AN, Sen M and Chatterjee P 1979 Physio-chemical studies of model meat emulsions in relation to the preparation of stable sheep and goat meat sausages. *J. Food Sci.* **44** 1695-99. 1702

Chaudhary RA and Ahmed W 1979 Buffalo breeds of Pakistan and programmes for their improvement. *Proceeding of Seminar on Buffalo Reproduction and Artificial Insemination*, FAO, Rome, pp. 173-181

Chaudhry TA and Shaw AO 1965 Production traits of Pakistani buffaloes. *Pakist. J. Sci* **17** 252-258 (fide Anim. Breed. Abstr. **35**: 3381)

Chaudhuri Pallab, Kumar S, Vinoth, Prasad Rajeev, Srivastava SK and Yadav MP 2005 Cloning and sequencing of 28 kDa outer membrane protein gene of *Brucella melitensis* Rev. 1. *Indian Journal of Experimental Biology*, Vol. **43** 838-840

Chauhan HV, Verma KC, Sah RL and Sharma BD 1984 Prediction of the occurrence of Marek's disease in chickens on the basis of the condition of cell mediated immunity. *Res. Vet. Sci.* **36** 133-135

Chawla DS and Nath T 1979 A note on the studies on meat production on goats. Effect of castration on dressed meat production in Beetal and its exotic crosses. *Indian Journal of Animal Science*, **49** 406-408

Chawla DS and OS Tomar 1986 Effect of year and month on milk yield and its components in Murrah buffaloes. Terminal progress Report of Buffalo Project, pp. 12 NDRI, Karnal, India

Chopra SC and Acharya RN 1969 Genetic and phenotypic parametric of body weight of Bikaneri sheep. *Anim. Prod.* **13** 343

Chouhan VS, Krishnan G, Ramesha KP, Kataktalware MA, Chakravarty P and Kandeepan G 2010 The hypo-osmotic swelling test for fresh yak semen. *Indian J Anim Sci* **80** 869–870

CLFMA 1996 CLFMA In Animal Production. The Compound Livestock Feed Manufacturers Association of India, Mumbai, p19

Confessor MV, Mendonca LE, Mourao JS, Alves RR 2009 Animals to heal animals: ethno- veterinary practices in semiarid region, Northeastern Brazil, *J. Ethnobiology and Ethno-medicine* **5** 37

Cunningham EP 2010 A time of change. *Journal of Animal Breeding and Genetics.* **127** 419-420

D Souza C, Bhat PN and Mukundan G 1978 Studies on growth, reproduction and production traits in Gir and its halfbred with Friesian. *Indian J. Dairy Sci.* **31** 395-397

Dandapat S, Pradhan HK and Mohanty GC 1994 Anti-idiotype antibodies to Marek's Disease - Associated tumour surface antigen in protection against Marek's Disease. *Vet. Immunol. Immunopathol.* **40** 353-366

Dandapat S, Banerjee D, Rajan, Kataria JM, Ram GC and Yadav MP 2003 Induction of mucosal immunity in poultry after oral immunization with a single dose of microencapsulated EDS-76 virus. *Indian J. Comp. Microbiol. Immunol. Infect. Dis.*, Vol. **24**, No. **2** (July-Dec.), 181-185

Das RB, Dabadghao PM, Marwaha SP and Debroy R 1963 Grazing capacity studies In grasslands of western Rajasthan. Ann. Arid Zone 2: (4)

Dass MN and Rajagopalan VR 1956 Studies on the scope for improvement of Bellary sheep. *Indian J Vet. Sci* **26** 571

Daubney R 1947 Recent development in Rinderpest control. *Bull. Office Int. Epizoot. Paris*. **28** 36-45.

de Groot B, Narayan Prasad RA, Soni RL, Nett P and Kropf W 1992 Performance of Sirohi goats under village conditions in Rajasthan, India. In Pre-conference Proceedings – Abstracts of Contributory Papers – *V International Conference on Goats*, New Delhi. Vol. **1** 53

De Smet PAGM 1999 Overview of herbal quality control, *Drug Inform, J.* **33** 717-724

Dedar RK, Yash Pal, Kumar S, Ghurai SK, Legha RA, and Singh RK 2011 Therapeutic Evaluation of Ivermectin against endoparasites of donkeys, *Vet Practitioner* **12**(1) 86-87

Deoprakash 1968 Avian leucosis complex: Demographic and pathoanatomical studies. M.Sc. Thesis Agra University, Agra

Deori S, Chakravarty P, Hussain M, Saikia A, Hazarika D, Baruah KK and Goswami J 2013 Efficacy of ovsynch protocol on ovarian response and progesterone profile in yak. *Indian J Anim Sci.* **83**:276 – 278

Dev DS and Tiwana MS 1978 Selection programmes in buffaloes in the past and the present. Paper presented in Indo-USSR Seminar held at Anand, India, 16th-18th October.

Devendra C 1981 The utilization of forages from cassava pigeon peas, leucaena and groundnut by goats and sheep. In Proc. of the symp. on Int. Nutr. systems and alinentation P.Mor. and Fehr A. Bourbouze and M. desimiane do La cloure (Nutrition and system of goat feeding) Vol. 1. pp 33845, INRAITOVIC, Tour (fide LOUCA, 1982)

Devendra C 1984 Nitrogen utilization and the requirements for maintenance of Malaysia Swamp Buffaloes MARDI. *Res. Bull* **12**, 2: 248-58

Devendra H, Shah, Bakshi CS, Rishendra Verma, Singh RK and Meenakshi Mallik 2005 Rapid differentiation of *Mycobacterium bovis* and *Mycobacterium tuberculosis* by a single tube multiplex PCR based on 12.7 kb fragment. *Veterinary Microbiol.* **109**: 211-216

Dhama K, Rathore R, Chauhan RS and Tomar S 2005 Panchgavya (cowpathy): an overview. *International Journal of Cow Science*, **1**(1): 1-15

Dhanda JS and Panda PC 1989 Retention of thiamine and riboflavin in duck meat during frozen storage and cooking. *Indian J. Meat Sci & Technol.* **2** 99-106

Dhanda MR, Uppal DR and Sharma GL 1967 Susceptibility of exotic breeds of cattle to Rinderpest lapinsed and lapinised avianised vaccines. *Ind. Vet. J.* **44** 1-6

Dilshad SMR, Rehman NU, Ahmad N and Iqbal A 2009 Documentation of ethno-veterinary practices for mastitis in dairy animals in Pakistan. *Pakistan Veterinary Journal*; **30**(3): 167-71

DiMasi JA, Hansen RW, Grabowski HG 2003 The price of innovation: new estimates of drug development costs. *Journal of Health Economics*, **22** 151-185

Doyle TM 1927 A hitherto unrecorded disease of fowls due to a filter-passing virus. *Journal of Comparative Pathology and Therapeutics*, **40** 144-169

Duc NV and Taneja VK 1984 Performance of pure-breed and crossbred grades for some reproductive traits in India. *Indian J. Anim. Sci.* **54** 141-144

Dutt M 1968 A preliminary note on some of the economic characters of Barbari goats. *Indian Journal of Animal Health*, **7**: 15-18

Dutta S 1954 The National Rinderpest Eradication Programme. *Ind. J. Vet. Sci. Anim. Husb.* **24** 1-9

Dwivedi SK and Dey S 2002 Fluoride concentrations in Thoroughbred Horses in India. *Veterinary and Human toxicology* **44** 292-293

Dwivedi SK, Sharma MC, Mukherjee SC, Jawahar Lal and Pandey NN 1986 Comparative efficacy of Liv-52 and *Andrographis paniculata* Nees. In experimental liver damage in rabbits. *Indian drugs*, **25** 1-4

Edwards JT 1927 Rinderpest: some points in immunity. Trans. *7th Congress Br. India.* **3** 707-717

Edwards JT 1927 Trans. far east Assoc. Trop. Med.,**3** 699

Edwards JT 1927/28 Active immunization of cattle by means of the "serum simultaneous" or "serum virus" method (third edition) Imp. Int. vet. Res., Mukteswar

Edwards JT 1930 The problem of Rinderpest in India. *Bull. Imp. Inst. Agr. Res. Pusa.* **199** 1-16

El-Arian MN 1981 *Studies on milk production'.* M.Sc. Thesis, Mansoura Univ., Mansoura, Egypt (A.R.E.)

El-Arian MN 1986 *Genetic analysis of Murrah buffalo herds*. Ph.D. Thesis, Kurukshetra University, Kurukshetra

El-Itriby AA and Asker AA 1958 Some production characteristics of native cows, Friesian, Shorthorn and their crosses in Egypt. *Emp. J. Expt. Agric.* **26** 314-322

EL-Kimary IS 1966 *Studies in Herd Records*. M.Sc. Thesis, Fac. Agric. Univ. Alexandria, Egypt (A.R.E.).

Eskinazi D, Blumenthal M, Farnsworth N and Riggins CW 1999 Botanical Medicine: Efficacy, Quality, Assurance and Regulation. Mary Ann Liebert, New York.

FAO 1984 FAO Production Year Book 1984. Food and Agriculture Organization of the United Nations, Rome (C.C. Bhattacharya and Khan, 1987).

FAO 1984 Production Yearbook 1983, Vol. **27**, Rome.

FAO 1985 Production Year Book. Food and Agriculture Organisation of the United Nations, Rome, Italy.

FAO 1986 Monthly bulletin of Statistics 9: 19-28.

FAO 2006 Feeding the future.

FAO 2013 World Livestock 2013 Changing disease landscapes. pp 1-101.

Gaedcke FW and Steinholf KB 2003 Herbal Medicinal Products. Stuttgart, Germany: CRC Press, ISBN: 0-8493-1023-7. p.1-30.

Gajendragad MR, Sharma B, Prabhudas K, Joshi RC and Bansal RP 1983 Immuno-peroxidase staining of rinderpest viral antigen in formalin fixed paraffin embedded tissue section. *Indian. J. Vet. Path.* **7** 79

Galav P, Anita Jain and Katewa SS 2013 Traditional veterinary medicine used by livestock owner of Rajasthan, India. *Indian journal of traditional knowledge* **12** 47-55

Gangwar PC 1980 Climate and reproduction in buffaloes (A Review). *Ind. J. Dairy Sci.* **33** 419-425

Ghalsasi PM and Nimbkar C 1993 The "Garole"- microsheep of Bengal, India. Animal Genetic Resources Information. FAO. No. 12. pp. 73-79.

Ghei JC, Uppal PK and Yadav MP 1994 Prospects of artificial insemination in equines. *Centaur* Vol. XI No. 1:7-13.

Ghosh S, Azhahianambi and Yadav MP 2007 Upcoming and future strategies of tick control: a review. *J. Vect Borne Dis* **44** 79-89

Gill BS 1991 Trypanosomes and trypanosomiases of Indian Livestock. ICAR Publication, New Delhi, 137p.

Gill BS, Bhattacharyulu Y, Kaur D and Singh A 1978 Chemoprophylaxis with tetracycline drugs in the immunization of cattle against theileriosis infection. *Int. J.Parasitol* 467-469.

Gill RS, Gangwar PC and Kooner DS 1973 Studies on estrus behaviour in buffaloes. *Indian J. Anim. Sci.* **43** 472

González JA, García-Barriuso M and Amich F 2011 Ethnoveterinary medicine in the Arribes del Duero, western Spain. *Veterinary Research Communications* **35(5)** 283–310

Gopalan R and Marutiram B 1974 Performance of Polworth x Rampur Bushair crosses. *Indian J Anim Sci.* **8** 65

Goswami SB and Nair AP 1964 Effect of some climatological factors on reproduction of buffaloes. *Indian J. Vet. Sci.* **34** 127

Govindaiah MG, Siddappa L and Mukundakrishna A 1983 Lactation parameters of Surti buffaloes. Sixth Workshop on A.I.C.R.P. on buffaloes. *Report of Dharwad Centre*, pp. 27-30.

Govindarajan R, Koteeswaran A, Venugopalan AT, Shyam G, Shaouna S, Shaila MS and Ramachandran S 1997 Isolation of peste des petits ruminant virus from an outbreak in *Indian buffalo (Bubalus bubalis) Vet.Rec.* **141** 573-4

Gowda RNS 1995 Experimental transmission studies on hydropericardium. *Ind. Poultry Sci. Conf.* IPSACON 95

Green S 2001 Quality assurance in herbal raw material supplies. Available (http:// www.samedanltd.com/members/archives/PMPS/Spring 2002/StuartGreen. htm).

Gulati BR, Kumar R, Mohanty N, Kumar P, Somasundaram RK and Yadav PS 2014 Bone morphogenetic protein-12 induces tenogenic differentiation of mesenchymal stem cells derived from equine amniotic fluid. Cells Tissue Org, DOI 10.1159/000358231

Gulati BR, Singh BK and Kumar D 2006 Development of a monoclonal antibody-based sandwich ELISA for detection of equine rotavirus from diarrhoeic foals. *Indian Journal of Microbiology*, **46** 349-354

Gulati BR, Singha H, Singh BK, Virmani N, Kumar S and Singh RK 2012 Isolation and genetic characterization of Japanese encephalitis virus from equines in India. *Journal of Veterinary Sciences*, **13(2)** 111-118

Gupta AK and Virmani N 2007 Equine influenza in India. CAB Reviews: Perspectives in Agriculture, Veterinary Science, Nutrition and Natural Resources, (PAVSNNR) **2**, No. 016, 9 pp

Gupta AK, Ahlawat A and Bhatia V 2003 Detection of equine influenza viral genome by RT-PCR and RNA:DNA hybridization. *Indian Journal of Bitechnology* **2** 214-219

Gupta AK, Chauhan M and Bhardwaj A 2013 Genetic diversity and bottleneck studies in endangered Bhutia and Manipuri pony breeds. *Molecular Biology Reports*, **30** 12: 6935-6943.

Gupta AK, Chauhan M, Bhardwaj A and Tandon SN 2012 Microsatellite markers based genetic diversity and bottleneck studies in Zanskari pony. Gene, **499** 357-361

Gupta AK, Chauhan M, Bhardwaj A, Gupta N, Gupta SC, Yash Pal, Tandon SN and Vijh RK 2014 Comparative genetic diversity analysis among six Indian horse breeds and English Thoroughbred. *Livestock Science*, **163** 1-11

Gupta AK, Chauhan M, Tandon SN and Sonia 2005 Genetic characterization and bottleneck studies in Marwari horse breed. *Journal of Genetics* **84** 295-301

Gupta AK, Kaur D, Rattan B and Yadav MP 2005 Molecular variability in different Indian isolates of equine herpesvirus-1. *Vet Res Communication.* **29(8)**:721-34

Gupta AK, Rattan B, Malik P, Kaur D, Singh BK and Yadav MP 2000 Experimental infection of donkeys with EHV-1 of horse origin – A study. *J. Equine Sci.*, **11 (2)** 29-34

Gupta AK, Sharma V and Yadav MP 2001 Detection of Equine Influenza virus by haemagglutinin (HA) Gene based RT-PCR. *Indian Vet. J.* **78** 383-385

Gupta AK, Singh BK and Yadav MP 1996 Application of polymerase chain reaction (PCR) for diagnosis of equine herpes virus-1 (EHV-1). *Indian J. Exptl. Biol.* **34** 1077-1080

Gupta AK, Singh R, Mamta, Yash Pal, Singh MK and Yadav MP 2003 Influence of Season on Characteristics of Jack Semen. *Indian J. Anim. Sci.*, **73 (9)** 986-991

Gupta AK, Varshney JP and Uppal PK 1993 Effect of vaccinations on biochemical profile in donkeys. *Int. J. Ani. Sci.* **8** 271-273

Gupta AK, Varshney JP and Uppal PK 1994 Comparative studies on Biochemical indices in different breeds of equines. *Indian Vet. J.*, **71** 26-30

Gupta AK, Varshney JP, Lal D Sharma YP and Yadav MP 2000 Mineral contents in healthy Thoroughbred female yearlings. *Indian Vet. J.* **77** 758-61

Gupta AK, Yadav MP and Uppal PK 1993 Single radial immunodiffusion potency test for standardization of indigenous equine influenza vaccine. *Indian J. Exptl. Biol.* **31** 944-47

Gupta AK, Yadav MP, Uppal PK, Mumford JA and Binns MM 1993 Characterization of equine influenza isolates from the 1987 epizootic in India by nucleotide sequencing of the HA1 gene. *Equine Vet. J.* **25**(2) 99-102

Gupta Ashok K, Yash Pal, Mamta and Yadav MP 1999 Effect of Feed Deprivation on Biochemical Indices in Equids. *J. Equine Sci.* **10**(2)33-38

Gupta PP,Tandon HD and Ramalingaswami V 1969 Spontaneously occurring vascular lesions in pigs. *J. Pathol.* **99** 19-28

Gupta M, Kondaiah Nand Anjanewyulu ASR 1993 The incorporation of egg Liquid into mutton and mutton+chicken sausages. *Fleischwirtsch.* **73** 66-68

Gurnani M 1983 Constraints of field progeny testing in buffaloes. Paper presented at a seminar on “Constraints to buffalo breeding” held at N.D.R.I., Karnal on Feb. 9, 1983, India.

Hadi MA 1965 A preliminary study of certain productive and reproductive characters of Marathwada buffaloes at Maharashtra State. *Indian Vet. J.* **42** 692-699

Hallen JHB, McLeod K, Charles JG, Keer HC and Allijan MM 1871 The cattle plague commission report to Govt. of India. Calcutta publication, June 30, 1871

Harikrishnan K, Kosalaramanan VR, Ramamurthi R and Radhakrishnan KT 1983 Studies on preparation of canned corned cara beef. *Cheiron.* **12** 69-74

Harvey WR 1966 Least squares analysis of data with unequal subclass numbers. United States Dept. Agric. ARS-20-8

Haunshi S, Niranjan M, Padhi MK, Reddy MR, Sujitha R, Rajkumar U and Panda AK 2012 Characterization of two Indian native breeds for production, egg and semen quality and welfare traits. *Poultry Science.* **90** 314-320

Haunshi S, Panda AK, Rajkumar U, Pathi M, Niranjan M and Chatterjee RN 2012 Effect of different levels of energy and protein on performance of Aseel breed of chicken during juvenile phase. *Tropical Animal health and Production* **44**(7) 1653-1658

Hayman RH 1974 The development of the Australian Milking Zebu. Wild Anim. Rev. (FAO), 11:31–35.

Hayman RH 1977 The development of the Australian milking zebu. FAO Animal Production and Health Paper No.1. Food and Agriculture Organization of the United Nations, rome.

Hazarika M and Biro G 1993 Effect of incorporation of blood proteins into sausage, *J. Food Sci. & Technol.* **30** 380-81

Hazary RC and Wishart G J 1999 Assessing the effect of mating ratio in broiler breeder flocks by quantifying sperm - egg interaction. *British Poultry Science* **40** 46-48

Hazary RC and Wishart GJ 2000 Assay of sperm quality in domestic fowl by MTT reduction. *British Poultry Science* **42** 111-114.

Hazary RC, Chaudhuri D and Wishart GJ 2001 Application of a simple tetrazolium reduction assay for assessing sperm quality and fertilizing ability of the domestic fowl sperm. *British Poultry Science* **42** 115-117

Hazary RC, Stains HJ and Wishart GJ 2000 Assessing of the efficiency of mating in broiler breeder flocks by enumerating the spermatozoa, which penetrate the inner previtelline layer over the germinal disc. *British Poultry Science* **41** 395-400

Hedge BP and AV Rai 1972 Studies on reproduction in Murrah buffaloes in coastal Mysore. *Indian Vet. J.* **49** 1120- 1125

Hirch and Schindler 1957 The syrian and dutch frisian cattle and their crosses in Israel. *Recc Agric Res sta.***7** 1-40

Honmode J 1971 Effect of megastrol acetate and progesterone on oestrus and fertility in cycllc Malpura and Chokla ewes. *Indian J Anim Sci.* **1**: 972

Hossain ASKE and Ahmed SU 1971 Seasonal variations of estrus in buffaloes of Bangladesh. *Bangladesh Vet. J.* **5** 47 (fide Anim. Breed. Abstr. 44: 611).

ICAR-IARI, 2012 Crop residues management with conservation agriculture: Potential, constraints and policy needs, Indian Agricultural Research Institute, New Delhi, vii+ 32 p.

Innes JRJ 1969 Veterinary pathology Retrospect and Prospects. *Vet. Rec.* **85** 730-741

Intrieri F, De Franciscis G and Meo G 1968 On the degree of association among milk production, butter fat production and lactation length in buffaloes in Caserta. *Acta Med. Vet., Napoli*, **14** 7-14 (fide Anim. Breed. Abstr. **37** 1297).

Ipe S 1979 Evaluation of Murrah buffalo sires on the basis of part lactation records. Ph.D. thesis, Kurukshetra University, Kurukshetra, India.

Iquebal MA, Sarika, Dhanda SK, Arora V, Dixit SP, Raghava GPS, Rai A and Kumar D 2013 Development of a model webserver for breed identification using micro-satellite DNA marker. BMC Genetics.14:118-125 doi: 10.1186/1471-2156-14-118.

Islam A, Ghosh R, Banerjee D, Nath P, Mazumder UK and Ghosa S 2008 *E.J. Biotechnol.* **11** 3417-3458.

Iyer GS and Dobson N 1940 A successful method of immunization against Newcastle disease of fowls. *The Veterinary Record*, **52** 879-894

Iyer PKR 1936 An unsual case of epizootic lymphangitis in a mule. *Indian J. Vet. Sci.,***6** 251

Jain SK, Rajya BS, Mohanty GC, Paliwal OP, Mehrotra ML and Sah RL 1975 Pathology of Chlamydial abortions in ovine and caprine. *Current Sci.* **44** 209-210

Jain LS and Kothari MS 1983 Factors affecting production traits in medium sized buffaloes. *Indian J. Anim. Sci.* **53** 1159-1161

Jain NC, Prasad G, Gupta Y and Mahajan NK 1988 Isolation of bluetongue virus from Culicoides species in India. *Sci. Tech. Rev. Off. Int. Epiz.* **17** 275-278

Jaman Lal 1989 *Studies on indigenous processing technologies of buffalo casings.* M.V.Sc. Thesis. IVRI Izatnagar.

Johri CB and Talpatra SK 1971 Growth studies with Jamunapari goats. III. Goat as a meat producing animal. *Indian Veterinary Journal*, **48** 602-604

Joshi RC, Shukla DC and Bansal RP 1972 Rinderpest diagnosis by agar gel diffusion test. *Indian. Vet. J.* **49** 449

Joshi RC and Bansal RP 1984 Transmission of rinderpest by contact from experimentally infected cattle to sheep and goats. *Ind. J. Comp. Microbiol. Immunol. Infect. Dis.* 5 27

Kadkol SB, Madhwaraj MS, Nair PR, Nair KKS, Dhanraj S, Govindarajan VS and Baliga BR 1980 Development of concentrated meat gravies. *J. Food Sci. & Technol.* **17** 146-149

Kadu MS, Khire DW, Belorkar PM and Kaikini AS 1978 Studies on reproductive efficiency in Nagpuri (Berari) buffaloes. *Indian Vet. J.* **55** 94-98

Kalicharan 1985 *Pathology and Immunological Reactivity of lung in pulmonary chlimydiosis of calves*. PhD. thesis. Rohilkhand Univ.

Kalra et al 1978 Degnala disease - Preview of veterinary pathology research (1970-75) Department of Vet. Pathology, HAU, Hissar.

Kamat AS and Nair MP 1995 Gamma Irradication as a means to eliminate histeria monocytogens from frozen chicken meat *J. Food* **69** 415-422.

Kandasamy N 2009 Breeding for improved wool and mutton production in Indian sheep: A retrospective. In *Proceedings of the X Annual Conference of the Indian Society of Animal Breeding and Genetics*, Chennai: Madras Veterinary College, Chennai. pp. 22-30.

Kanginakudru S, Metta M, Jakat RD and Nagaraju J 2008 Genetic evidence from Indian red jungle fowl corroborates multiple domestication of modern day chicken. *BMC Evolutionary Biology*. **8** 174

Kannaki TR, Reddy MR and Murugesan S 2011 Toll-like receptor mRNA expression, iNOS gene polymorphism and serum nitric oxide level in indigenous chickens. *Veterinary Research Communications*. **35**(5) 321-327

Kannaki TR, Reddy MR, Verma PC and Shanmugam M 2012 Toll-like receptors and molecular characterization of avian specific TLR15. *Molecular Biology reports*. **39** 8359-8349

Kannaki TR, Verma P and Reddy MR 2012 Different gene expression of antimicrobial peptides β defensins in the gastrointestinal track of Salmonella serovar Pullorum infected broiler chicken. *Veterinary Research Communications*. **36**(1) 57-62

Kannan G, Radhakrishnan KT and Sharmugum AM 1990 The effect of electrical stimulation on tenderness at spent chicken carcasses. *Cheiron.*, **19** 209-212

Kapila N, Kishore A , Sodhi M, Sharma A, Mohanty AK, Kumar P and Mukesh M 2013 Temporal changes in mRNA expression of heat shock protein genes in mammary epithelial cells of riverine buffalo in response to heat stress in vitro. *Int. J. Anim. Biotech*. **3** 5-9

Kapila N, Kishore A and Sodhi M 2012 Identification of Appropriate Reference Genes for qRT-PCR Analysis of Heat-Stressed Mammary Epithelial Cells in Riverine Buffaloes (*Bubalus bubalis*). ISRN Biotechnology. Volume 2013, Article ID 735053, 1-9.

Katpatal BG 1978 *The Indian Dairyman*, **30** 189–208.

Kaul and Prasad 1988 Experimental atherosclerosis in goats. *Res. Vet. Sci.* **44** 383-385

Kaul and Prasad 1988 Influence of experimental hypertension in induced atherosclerosis in goats - *Indian Vet. J.* **65** 13-16

Kaul and Prasad 1990 Hypocholestrolemia and antiatherosclerotic affects of garlic (*Allum satium*) in goats - An experimental study. *Ind. Vet. J.* **67** 1112-1115

Kaura RL 1943 Some common breeds of Indian sheep. *Indian fmg.* **6**(12) 605-609

Kaura RL 1943 Some common breeds of goats in India. *Indian Farming* **4** 459–552

Kearl LC 1982 Nutrient Requirements of Ruminant in Developing Countries. Utah Agric. Exp. Sta. Utah State Univ. Logan, Utah, U.S

Kehar ND and Murty VN 1953 Annual Report, Animal nutrition division, IVRI, Izatnagar

Kesteven KVL 1955 Rinderpest Vaccines: their production and use in the field. Food and Agriculture Organization of the United Nations, Rome, Italy

Khan BU and Sahni KL 1979 A study of some carcass characteristics in Jamunapari kids. *Indian Veterinary Medical Journal*, **48** 602-604

Khot SS, Mahal GS and Lall HK 1962 Rsearch In Animal Husbandry—Sheep and Wool—A Review of work done during 1929-30. Indian Council of Agricultural Research, New Delhi

Khurana SK and Malik Praveen 2009 Status of Mycoplasma equigenitalium among Indigenous equines. *Indian Journal of Veterinary Research*, **18** (1) 17-19

Khurana SK, Malik P, Virmani N and Singh BR 2009 Prevalence of Rhodococcus equi infection among foals with respiratory problems. *Indian Journal of Veterinary Research*, **18** (1) 20-22

Koil-Raja A, Krishnan KR and Shanmugam AM 1995 Studies on preparation of snack type meat food from rumen musculature of goat. *Proceedings of National Symposium on production and marketing of goat meat,* held at CIRG, Makhdoorn, Jan. 10-11, PP 33-34

Kolte AY, Ravikanth K, Rekhe DS and Maini S 2009 Role of polyherbal formulation in modulating rumen biochemical and growth performance parameters in calves. *The Internet Journal of Veterinary Medicine*, Vol. **6**(2)

Kondaiah N, Anjaneyulu ASR, Rao VK, Sharma N and Joshi HB 1985 Effect of salt and phosphate on the quality of buffalo and goat meats. *Meat Sci.* **15** 183-192

Kondaiah N, Panda B and Anjaneyulu ASR 1987 Evaluation of type and level of fat in chicken sausages from spent hen meat. *Indian J. Poult Sci.* **22** 30-34

Kondaiah N, Panda B, Anjaneyulu ASR and Singh RP 1988 Utilization of whole meat components from spent hens for chicken sausages production. *Indian J. Poultry Sci.* **23** 135-141

Kotagiri Ravikanth, Anil Kanaujia, Deepak Thakur, Parveen Singh and Bhupesh Gautam 2013 Validated high-performance thin-layer chromatographic methods for the quantitative estimation of bioactive constituents in Methiorep premix, a feed additive polyherbal formulation. *Journal of Planar Chromatography* **26** 4 312-315

Kowale BN and Rao VK 1995 Recovery of edible proteins from slaughter house waste. Proceedings of National Symposium on production and marketing of goat meat, held at CIRG, Makhdoom, Jan 10-11 pp.181-190.

Krishna Murty D and Adlakha SC 1962 Preliminary studies on outbreaks of swine fever in India. *Indian Veterinary Journal.* **39** 406 -419

Krishna L, Cattopadhyay SK, Iyer PKR and Sherma RM 1973 A pathological study on hepatocellular carcinoma in sheep. *Ind. J. Animal. Sci.* **43** 35-37

Krishnamurthi A 1970 The Wealth of India; Livestock, pp. 27 Publication and Information Directorate, C.S.I.R., New Delhi, India.

Krishnan G, Ramesha KP, Sarkar M, Chakravarty P, Kataktalware MA, Saravanan BC 2009 Modified temperature humidity index for yaks (*Poephagus grunniens* L.), *Indian J Anim Sci* **79** 788-790

Krishnan KR and Sharma N 1993 Studies on chilled and frozen stored ready to eat buffalo beef sausages prepared by incorporating skeletal and offal meats with 20% pork fat. *J. Food Sci. & Technol* **30** 301-302

Krishnan Nair BR and Kelath KR 1977 Studies on the first lactation yield of Brown Swiss X Zebu crossbred (Fl) cows in Kerala. *Kerala J. Vet. Sci.* **1** 1-8

Krishnaswamy S, Keshavamurthy BS, and Sunderajan S 1981 The use of the direct immuno peroxidase test to detect the multiplication of rinderpest virus in bovine kidney cell culture. *Vet. Microbiolo.* **6** 23

Kropf W, Narayan Prasad RA, Sharma OP, de Groot B, and Nieuwhof G 1992 A comparison of milk yield and reproductive performance of Sirohi goats with Alpine X Sirohi and Toggenburg X Sirohi crosses. In Pre-conference Proceedings – Abstracts of Contributory Papers – *V International Conference on Goats*, New Delhi. **1** 59

Kulkarani VV, Kondaiah N and Anjaneyulu ASR 1987 Incorporation of whole egg liquid in mutton patties. *2nd International Food Convention and Exhibition.* Abst. PRD. **5** 88

Kulkarni MA and Arankalle VA 2008 The detection and characterization of hepatitis E virus in pig livers from retail markets of India. *J. Med Virol.*, **80** 1387-1390

Kulkarni S, Kulkarni DK, Deo AD, Pande AB and Bhagat RL 2014 Use of Ethno-Veterinary medicines (EVM) from Vidarbha Region (MS) India. *Bioscience Discovery*, **5**(2) 180-186

Kulkarni VV, Kondaiah N and Anjaneyulu ASR 1988 Incorporation of whole egg liquid in mutton patties. In : Abstract of papers, *2nd Int. Food Convention and Exhibition.* Feb., Mysore, pp. 88. Assoc. Food Scientists Technologists (India) and CFTRI, Mysore.

Kulshreshtha SB, Yadav MP, Prasad AK, Verma KC and Kumar S 1973 Occurrence of a respiratory disease in poultry due to mixed infections of *E. coli*, fowl pox, infectious bronchitis and mycoplasma. *Indian J. Poult. Sci.* **8** 27-29

Kumar A, Kumar S, Yadav MP 1981 Anti-rickettsial property of *Allium sativum* Linn.: *in vivo* and *in vitro* studies. *Indian J. Anim. Res.* **15** 93-97

Kumar A, Kumar S, Yadav MP and Kakkar S 1981 Human milk as a source of Q-fever infection for breast-fed babies: a preliminary report. *Indian J. Med. Res.* **73** 510-512

Kumar A, Yadav MP, Chandra R and Garg SK 1987 Clinico-etiological features of a pox outbreak among buffaloes. *Indian J. Anim. Health.* **26**(11) 41-46

Kumar B 1984 *Index method of sire evaluation in buffaloes.* Ph.D. Thesis, Kurukshetra Univ., Kurukshetra, India

Kumar D, Bedi PMS, Verma GM, Subramanian B and Sharma TR 1976 Dehydration of pre-cooked mutton chunks and food stuffs by consinign the hot air drying and freeze drying techniques *Research and Industry-India* **21**(4) 257-262

Kumar D, Sharrna IR and Nath N 1977 Biochemical. Physical and organoleptic studies on production of enzyme tenderized accelerated freeze dried (AFD) goat meat chunts. *J. Fd. Sci. Technol.* **14** 257-60

Kumar P, A R Freeman, Loftus RT, Gaillard C, Fuller DQ and Bradley DG 2003 Admixture analysis of South Asian cattle. *Heredity*, **91** 43-50

Kumar R, Kumar S, Virmani N and Yadav SC 2014 Transplacental transmission of Trypanosoma evansi from experimentally donkey mare to neonatal foal. *Journal of Equine Veterinary Science*, **35** (4) 337-341

Kumar S 1976 African Horse Sickness, ICAR Tech Bull 15, p 34

Kumar S and Kumar R 2007 Diagnosis of Babesia equi infection: an update on the methods available. CAB Reviews: Perspectives in Agriculture, Veterinary Science, Nutrition and Natural Resources **2** (35) 14

Kumar S, Gupta J, Kumar N, Dikshit K, Navani N, Jain P and Nagarajan M 2006 Genetic variation and relationships among eight Indian riverine buffalo breeds. *Molecular Ecology* **15**(3)593-600

Kumar S, Kim JY, Yokoyama N, Inoue N, Xuan X, Igarashi I and Sugimoto C 2004 Expression of Babesia equi EMA-1 and EMA-2 during merozoite developmental stages in erythrocyte and their interaction with erythrocytic membrane skeleton. *Molecular and Biochemical Paraitology* **133** 221-227.

Kumar S, Kumar R and Sugimoto C 2009 A perspective on Theileria equi infection in donkeys. *Japanese Journal of Veterinary Research* **56** 171-180

Kumar S, Kumar R, Gupta A, Yadav SC, Goyal S, Khurana SK and Singh RK 2013 Development of EMA-2 recombinant antigen based enzyme-linked immunosorbent assay for seroepidemiological studies of Theileria equi infection in Indian equine population. *Veterinary Parasitology* **8** 10-17

Kumar S, Kumar R, Gupta AK and Dwivedi SK 2008 Passive transfer of Theileria equi antibodies to neonate foals of immune tolerant mares. *Veterinary Parasitology* **151** (1) 80-85

Kumar S, Yokoyama N, Kim JY, Bork-Mimm S, Inoue N, Xuan X, Igarashi I and Sugimoto C 2012 Theileria equi merozoite antigen-2 interacts with actin molecule of equine erythrocyte during their asexual development. *Experimental Parasitology* **132**(4) 508-12.

Kumar Subodh, Malik Praveen, Verma S, Vijaipal, Gautam V, Mukhopadhyay C and Rai GP 2011 Use of a Recombinant Burkholderia Intracellular Motility A Protein for Immunodiagnosis of Glanders. *Clinical and Vaccine Immnology* **18** 1456-1461

Kunle OF, Egharevba HO and Ahmadu PO 2012 Standardization of herbal medicines – A review. *International Journal of Biodiversity and Conservation* **4**(3) 101-112

Kurar CK and Mudgal VD 1977 Effect of plane of nutrition on the utilization of feed nutrients for milk production in buffaloes. Annual report NDRI, Karnal pp 118-20

Kurar CK and Mudgal VD 1980 Protein requirement of Murrah buffaloes in the early stage of lactation. *Indian J. Dairy Sci.* **33** 443-50

Kurar CK and Mudgal VD 1981 Maintenance requirement for protein in buffaloes. *Indian J. Anim. Sci.* **51** 817-24

Kuriean V 1987 Cited in Diary India 1992.

Lakshmanan V and Mendiratta SK 1996 Indigenous processing of buffalo calf runners. In "*Annual Conference of IAAVR and National Symposium on prospects of livestock and Poultry development in 2lst Century*", held at CARI Izatnagar on Feb. 23-24 p.74.

Lal SM and Singh IP 1977 Buffalo pox a review, *Trop. Anim. Health Prod.* **9** 107-112

Lingard A 1897 Preliminary Note on Rinderpest, Imperial Bicteriological Laboratory, Mukteswar, by Prof. R. Koch, 4th to 15th June, 1987

Lingard A 1903 Annual Report, Imperial Bacteriological Laboratory, Mukteswar, 1903-1904

Lingard A 1905 Camel tuberculosis. Annual Report of Imperial Bacteriologist, India, s.n. 190 tou. 06

Lingard A 1906 Annual Report, Imperial Bacteriological Laboratory, Mukteswar, 1906-1907

Little Wood 1933 Cited by Bhat and Bhattacharya 1982

Livestock census 1987 Agricultural Statistics at a glance 1993. Directorate of economics & statistics. Department of Agriculture and Cooperation, Ministry of Agriculture. Govt. of India, New Delhi

Livestock Census 2012 19th Livestock Census. Government of India, Ministry of Agriculture, Department of Animal Husbandry and Dairying, Krishi Bhawan, New Delhi (http://www.dahd.nic.in)

Madan ML, Singla SK, Jailkhani S and Ambrose JD 1991 In vitro fertilization in buffalo and birth of first ever IVF buffalo calf. In: *Proc. 3rd World Buffalo Congress, Varna, Bulgaria*, 15–19 May. 7: 11–17.

Madhwaraj MS, Nair KKS, Kadkol SB and Baliga BR 1983 Corned buff from buffalo meat. *Indian Food Packer* **37**(4) 83-87

Madsen O 1975 Analysis of performance of pure-breed and crossbred dairy cattle and recommendation for breeding policy in the Bangalore area, Karnataka State, India. Report to DANIDA

Madsen O and Vinther K 1975 Performance of purebred and crossbred dairy cattle in Thailand. *Anim. Prod.* **21** 209–216

Mahajan P and Panda PC 1991 Effect of post exsanguination electrical stimulation on meat quality and attributes of mutton. *Indian Food Packer* **45**(3) 9-14

Mahendra-Prasad, Chandiramani NK and Mondokhot UV 1986 Thermonuclease as an indicator of *S. aureus* and production of enterotoxin, *J. Food Sci. & Technol* **23** 1-5

Malik P and Kalra SK 2009 Variability and protective efficacy of M protein of streptococci of equine origin. *Indian Journal of Animal Sciences* **79** 466-469

Malik P, Khurnan SK and Dwivedi SK 2010 Re-emergence of Glanders in India-Report of Maharashtra state. *Indian Journal of Microbiology* **50** 345-348

Malossini F 1966 Variations in the production of milk and its constituents during lactation and their interrelationships in buffalo cows. *Annali, Sper. Agr. N.S.* **18** 243-273 (fide Anim. Breed. Abstr. **34**: 922)

Mamta Gupta AK and Dhillon S 2004 Genetic Characterization of Indian Spiti Horses. *Journal of Genetics* **83** 291-295

Mann HS and Ahuja LD 1975 Range management research in a rid zone of India. Paper presented at the *Annual Meeting Society of Range management USA*, Mexico City, Feb. 1-13

Manoharan S, Ramesh S, Parthiban M, Koteeswaran A, Chandran NDJ and Reddy M Ranga 2004 Effect of a Poly Herbal Ingredient on Day Old Chick Quality by Feeding in Parent Flocks. *International Journal of Poultry Science* **3** 773-778

Manuja A, Kumar P, Kumar R, Kumar B, Singha H, Sharma RK and Yadav SC 2014 CpG-ODN class C-mediated immunostimulation and its potential against Trypanosoma evansi in equines. *International Immunopharmacol* **22**(2) 366-370

Manuja A, Kumar S, Dilbaghi N, Bhanjana G, Chopra M, Kaur H, Kumar R, Kumar B, Singh SK and Yadav SC 2014 Quinapyramine sulphate-loaded sodium alginate nanoparticles show enhanced trypanocidal activity. *Nanomedicine* **9**(11) 1625-1634

Manuja BK, Manuja A, Dahiya R, Singh S, Sharma RC and Gahlot SK 2014 Diversity of interferon inducible Mx gene in horses and association of variations with susceptibility vis-à-vis resistance against equine influenza infection. *Infection Genetics Evolution* **27** 142-148

Mazars G 1994 Traditional veterinary medicine in India. *Rev. sci. tech. Off. int. Epiz.* **13** (2) 443-451

Meckenzie 1956 Cited in Report NCA, 1976.

Mehrotra and Rajya 1980 *In vitro* observation on IBR-IPV virus employing FAT. *Ind. J. Comp. Vet. Microbiol. Immunol. Inf. Dis.* **1** 11-13

Mehrotra ML, Rajya BS and Kumar S 1976 Infectious bovine rhinotracheitis (IBR) kerato conjunctivitis in calves. *Indian J. Vet. Path* **1** 70-73

Mehta SC, Mishra BP and Sahani MS 2006 Genetic differentiation of Indian camel (*Camelus dromedarius*) breeds using random oligonucleotide primers. *Animal Genetic Resources Information* **39** 77-88

Mendiratta SK and Panda PC 1995 Chemical and histological changes caused by pressure and enzyme treatment for tenderization of spent hen meat. *J. Food Sci. & Technol* **32** 336-38

Mendiratta SK, Lakshmanan V, Kesava Rao V and Yadav PL 1995 Effect of injection of different concentrations of calcium chloride on tenderization of goat meat. *Indian J. Anim. Sci.* **65** 1361-64

Meyn K and Wilkins JV 1973 National cattle breeding programs in Kenya. In: *Third World Conference on Animal Production. Pre-conference* volume No.3, pp.5(c)1-5(e)10.Melbourne, Australia

Meyn K and Wilkins JV 1974 Breeding for milk in Kenya with particular reference to the sahiwal stud. *Wld Anim . Review (FAO)* **11** 24-30

Meyn K and Wilkins JV 1975 National cattle breeding programmes in Kenya. Proc. *3rd World Conference on Animal Production*, 595-602

Minakshi P, Singh R, Ranjan K, Kumar P, Joshi CG, YKM Reddy and Prasad G 2012 Complete genome sequence of bluetongue virus serotype 16 of goat origin from India. *J. Virol* **86** 8337-38

Mirmahmoudi R and Prakash BS 2012 The endocrine changes, timing of ovulation and the efficacy of the Double synch protocol in the Murrah buffalo (*Bubalus bubalis*). *General and Comparative Endocrinology* **177** 153-159

Mirmahmoudi R, Souri M and Prakash BS 2014 Endocrine changes, timing of ovulation, ovarian follicular growth and efficacy of a novel protocol (Estradoublesynch) for synchronization of ovulation and timed artificial insemination in Murrah buffaloes (*Bubalus bubalis*). *Theriogenology* **81** 237-242

Mishra L, Mishra P, Pandey A, Sangwan RS, Sangwan NS, Tuli R 2008 Phytochemistry 69 1000-1004.

Mishra M, Mishra PK and Nayak JB 1978 Relation of body measurement with some economic traits in dairy cows. *Indian J. Dairy Sci.* **31** 135-140

Mishra OP, Singh AK and Kumar S 2011 Indigenous Knowledge of Bihar Farmers. *Journal of Community Mobilization and Sustainable Development* **6**(1) 46-49

Misra B 1966 The Râjamârtanda of Sri Bhojarâja with the 'Vidyotinî' Hindî Commentary. Chowkhamba Vidyabhawan, Varanasi, pp.108-112

Misra RC, Kushwaha NS and Singh R 1983 Studies on the influence of environmental factors upon lactation milk yield in Murrah and Bhadawari buffaloes. *Indian Vet. J.* **7** 135-139 (fide Anim. Breed. *Abstr.* **53** 1206).

Misra RK 1983 *Genetic analysis of growth, survivability, efficiency of feed conversion and carcass yield and composition in Sirohi (goat) and its crosses with Beetal.* Ph.D. thesis. Kurukshetra University, Kurukshetra, Haryana

Misra RK 1985 All India Coordinated Research Project on Goats. Terminal Report 1985. Central Institute for Research on Goats, Makhdoom, Farah.

Misra RN 1988 Project Coordinator's report-1988. All India Co-ordinated research project on goats. Farah, Mathura (U P)

Mittal JP 1979 A note on the effect of certain growth attributes on milk production traits in Barbari goat. *Indian Journal of Animal Sciences*, **49** 757-759

Mittle AK, Kowle BN, Kesava Rao V and Murari Prasad 1990 Fractionation and Characterisation of neets foot oil from buffalo hooves. *Indian J. Meat. Sci & Technol.* **3** 107-115

Mohanty and Rajya 1977 Growth and morphological characteristics of CVT cells in vitro. *Vet. Pathol.* **14** 420

Mohanty N, Gulati BR, Kumar R, Gera S, Kumar P and Yadav PS 2014 Isolation, characterization and tri-lineage differentiation of equine mesenchymal stem cells derived from umbilical cord blood. *Cytotechnology*, DOI: 10.1007/s10616-014-9831-z.

Mohanty N, Gulati BR, Kumar R, Gera S, Kumar P, Somasundaram RK and Kumar S 2014 Immunophenotypic characterization and tenogenic differentiation of mesenchymal stromal cells isolated from equine umbilical cord blood. *In Vitro* Cellular & Development Biology – Animal, **50** (6) 538-548

Mohanty N, Gulati BR, Kumar R, Gera S, Kumar P, Somasundaram RK and Kumar S 2014 Immunophenotypic characterization and tenogenic differentiation of mesenchymal stromal cells isolated from equine umbilical cord blood. In Vitro Cell. Dev. Biol. Animal, DOI 10.1007/s11626- 013-9729-7.

Mondal M, Baruah KK, Rajkhowa C 2014 Mithun: An Animal of Indian Pride. Livestock Research for Rural Development. Volume 26, Article #6 (Retrieved on August 23, 2015, from http://www.lrrd.org/lrrd26/1/mond26006.html).

Mondal M, Karunakaran M, Rajkhowa C, Prakash BS 2008 Development and validation of a new method for visual detection of estrus in Mithun (*Bos frontalis*). Appl Anim Behav Sci. doi:10.1016/j. applanim.2008.02.009.

Mondal M, Rajkhowa C, Prakash B S 2006a Relationship of plasma estradiol-17ß, total estrogen and progesterone to estrous behavior in mithun (*Bos frontalis*) cows. *Hormones and Behaviour* **49** 626-633

Mondal M, Rajkhowa C, Prakash B S 2006b Development of a biotin-streptavidin amplified enzyme immunoassay for oxytocin and its application during milk ejection and reproductive cycle in the mithun (*Bos frontalis*). *Zoolo Sci* **23** 633-639

Mondal M, Rajkhowa C, Prakash BS 2007a Development and validation of a highly sensitive economic enzyme immune assay for prolactin determination in blood plasma of mithun (*Bos frontalis*) and its application during milk let down and cyclicity. *Anim Reprod Sci* **99** 182-195

Mondal M, Rajkhowa C, Prakash BS 2007b Highly sensitive second antibody format enzyme immunoassays for determination of estradiol-17ß concentration in blood plasma of mithun (*Bos frontalis*). *Zoolo Sci* **24** 408–413

Mostageer A, Morsy MA and Sadek RR 1981 The production characteristics of a herd of Egyptian buffaloes. *Zeitschrift fur Tierzuchtung and Zuchtungsbilogie*, **98** 220-236 (fide Anim. Breed. *Abstr.* **50** 710)

Mourad KA 1978 *Some productive and reproductive characters of Egyptian buffalo.* M.Sc. Thesis, Fac. Agric., Cairo, Univ, Cairo, Egypt. (A.R.E).

Mudgal VD and Pradhan K 1988 Animal feed resources and current pattern of utilization in India. Proc. of consultations on Non-conventionally Feed Resources and Fibrous Agricultural Residues held by IDRC-ICAR. Hisar p 139-145

Mudgal VD 1987 Feeding and management: of the livestock during the drought "Navabharat Times, 15th Sept. 1987, New Delhi.

Mudgal VD and Kurar CK 1978 Comparaive utilization of dietary energy in lactating crossbred cows (Brown Swiss X Sahiwal) and Buffaloes during early lactation. Paper presented at the *XX Int. International Dairy Congress*, Paris.

Mudgal VD, Singhal KK and Sharma DD 1995 *Advances in Dairy Animal Production.* International Book Publ. House Lucknow.

Mukesh M, Sodhi M, Bhatia S, Mishra BP 2004 Genetic diversity of Indian native cattle breeds as analysed with 20 microsatellite loci. *Journal of Animal Breeding and Genetics* **121**(6) 416 - 424

Mukherjee A, Mukherjee S, Longkumer I 2014b Cytogenetic architecture of Indian mithun (*Bos frontalis*), mithun cattle crossbred by conventional staining, G-, C and R-banding techniques. *Chromosome Res* **22** 427

Mukherjee A, Mukherjee S, Longkumer I, Adebambo AO 2014a Molecular cytogenetic Studies on Indian mithun, gaur, mithun cattle crossbred and cattle. *Chromosome Res* **22** 410

Mukherjee A, Mukherjee S, Longkumer S, Rajkhowa C 2013a Genetic parameters of growth and reproductive traits in mithun (*Bos frontalis*). *Indian Vet Journal* **90** 34-36

Mukherjee A, Mukherjee S, Sahoo NR, Longkumer S, Moonmoon M, Rajkhowa C 2013b Characterization of leptin gene and analysis of genetic polymorphism in mithun. *Indian J Anim Sci* **83** 851-853

Mukherjee Pulok K 2002 Quality Control of Herbal Drugs, An Approach to Evaluation of Botanicals; New Delhi: *Business Horizons.*

Mukherjee S, Mukherjee A, Longkumer I, Moonmoon M, Kobu, KP, Perumal Rajkhowa C 2014 Morphomertic Characterization of Indian mithun. *Indian J Anim Sci* **84** 699-702

Mukherjee. S.C. 1977 Personal communication

Mukhopadhyaya GN 1926 History of Indian Medicine, Vol. II. University of Calcutta, pp. 356-424,400-424,491-495.

Muralidhara Rao N. 1995 Slaughter house by-products of goat. *Proceedings of National symposium on production and marketing of goat meat, held at CIRG. Makhdoom* on Jan. 10-11. pp.162-171

Murlimanohar and Sunderraj 1955 A report on an outbreak of rift valley fever - like disease in sheep in Tamilnadu. *Indian Vet. J.* **72** 662-664

Murthy TRK 1995 Bacterial growth on goat meat at room temperature and the effect of some chemical dip treatments. Proceedings of National symposium on production and marketing of goat meat held at CIRG, Makhdoom, Jan. 10-11. pp.30-36.

Nagarajan K, Saikumar G, Arya RS, Gupta A, Somvanshi R and Pattnaik B 2010 Influenza A H1N1 virus in Indian pigs and its genetic relatedness with pandemic human influenza A 2009 H1N1. *Indian J. Med. Res.* **132** 160-167

Nagarcenkar R 1979 Breeding programme for buffalo in India: situation and prospects. Paper presented at FAO Expert consultation on "Dairy Cattle Breeding in the Humid Tropics" pp. 92-101

Nagarcenkar R and Rao MK 1982 Performance of Tharparkar x Exotic crosses for productive and reproductive traits. *Indian J. Anim. Sci.* 52 129-138

Nagarcenkar R, Bhatnagar DS, Gurnani M and Rao MK 1981 Crossbreeding of Tharparkar with exotic breeds. Annual Report, pp. 68-69, NDRI, Karnal.

Naidu KN and Desai RN 1965 Genetic studies on Holstein-Friesian x Sahiwal cattle for their survivability in Indian tropical conditions as dairy animals. *Indian J. Vet. Sci.* **35** 197-203

Nair PNR 1973 Evolutionary crossbreeding as a basis for cattle development in Kerala State, India. Unpublished thesis presented to University of Zurich for Doctor of Veterinary Medicine, Juris Druckt Veriag, Zurich, p. 115.

Nakula 1952 Asvasâstra. Tanjore Saraswathi Mahal Series, No. 56, Tanjore

Nanda PN and Singh G 1948 Improvement of wool quality by selective breeding In Bikaneri and Lohi sheep. *Indian J Vet Sci* **18** 195

Narahari D, Thangovel A, Babu M, Prabhakaran K and Sundarasasu V 1988 Studies on relative processing and cut up yields of chicken, duck and Japanese quail at eight weeks of age. *Cheiron* **17** 24-29

Naresh R, Mahadeviah M, Gowramma RV and Modhavraj MS 1986 Stability of the tin free steel cans for packing meat and poultry products. In Production and processing of meat and poultry products", held at CFTRI, Mysore. Jan. 17-18, p.91.

Nath K, Rajagopal S, and Garg AH 1983 Water washed neem seed kernel cake as cattle feed. *J.Agric, Sci, Camb* **101** 325-326

NCA 1976 National Commission of Agriculture, Vol. VII. Government of India, Ministry of Agriculture New Delhi.

NCA 1976 Report of the National Commission on Agriculture, Part VII Animal Husbandry, Ministry of Agriculture Irrigation, Govt. of India, New Delhi

NDRI 1980 Annual Report, Annual Report, National Dairy Research Institute, Karnal

Nem Singh 1997 Personal communication

Nimbaker C 2014 Personal Communication

OECD 2012 Inclusive green growth: for the future we want. pp 1-43

Ohri SP and Singh M 1970 A study of economic characters of Rathi cattle. *Indian Vet. J.* **47** (1 l) 952-958

Ohri SP and Singh M 1971 Factors affecting milk production in Rathi cows. *Indian Vet. J.* **48** (3) 252-257

OIE 2004 Terrestrial Animal Health Code. Chapter 3.8.2. Appendix. Surveillance system for Rinderpest. OIE, Paris. www.oie.int

Oloufa MM 1968 Some aspects of reproduction efficiency in Egyptian cattle and buffaloes. Egypt. *Vet. Med. J.* **15** 173-177 (fide Anim. Breed. **Abstr. 37** 3538)

Omnawar S, Rao JR, Singh RK and Butchaiah G 2001 DNA amplification in *Trypanosoma evansi* isolates defined by random amplified polymorphic DNA-PCR. *Vet. Rec.***148** 244-246

Operation Flood - Third phase Dairy India 1992

Osburn B, Scott C, Gibbs P 2009 One world--one medicine--one health: emerging veterinary challenges and opportunities. *Rev Sci Tech.* **28**(2) 481-486

Oyagbemi AA, Saba AB, Arowolo 2008 ROA, *Int. J. Poultry Sci.* 574-578

Padda GS and Sharma N 1982 Processing and preservation of meat Pickles (Meat Achar) *Indian Food Packer* **36** 51-54

Padda GS, Keshri RC, Rao VK, Sharma BD and Sharma N 1986 Organoleptic and physico-chemical properties of cara beef (buffalo beef) patties with different fat combinations. *Indian J. Anim. Sci.* **56** 478-81

Padda GS, Keshri RC, Sharma N, Sharma BD and Murthy TR 1988 Physico-Chemical and Organoleptic Properties of Patties From Hot, Chilled and Frozen Goat Meat, *Meat Sci* **22** (4) 245-253

Padda GS, Keshri RC, Sharma N, Sharma BD and Murthy TRK 1988a Physico-chemical and organoleptic properties of patties from hot chilled and frozen goat meat. *Meat Sci.* **24** 245-55

Padda GS, Kesllri RC, Sharma BD and Sharma N 1988b The effect of some extender/ binders on organoleptic and chemical properties of pork kofta. *Indian J. Meat Sci. & Technol.* **1** 9-13

Padda GS, Sharma N and Bisht GS 1989 Effect of some vegetable extenders on organoleptic and physico-chemical properties of goat meat balls. *Indian J. Meat Sci. & Technol.* 2116-122

Pal UK and Agnihotri MK 1993 Storage stability of chevon pickle at room temp. In *3rd International Food Convention held at CFTRI*, N sore 7-12 Sept. p.31

Pal Y and Gupta AK 2005 Oestrous cyclicity, follicle development and progesterone profile in indigenous donkeys and ponies. *Indian Journal of Animal Sciences* **75** 1398-1400

Paliwal OP, Charan K and Lalkrishna 1972 A case of hepatoma in sheep. *Indian Vet. J.* **49** 975

Panda AK, Lavanya G, Rama Rao SV, Raju MVLN, Niranjan M and Reddy MR 2012 Effect of nutrient density on production performance, egg quality and humoral immune response of brown laying (Dahlem Red) hens in the tropics. *Tropical Animal health and Production* **44**(2) 1293-1299

Panda AK, Lavanya G, Reddy E, Pradeep Kumar, Rama Rao SV, Raju MV.N and Shyam Sunder G 2012 Utilization of quality protein maize in the diet of White Leghorn layers. *Animal Feed Science and Technology* **173** 210-216

Panda AK, Raju MVLN, Rama Rao SV, Lavanya G, Reddy E, Pradeep Kumar and Shyam Sunder G 2012 Nutritional evaluation and utilization of quality protein maize, Nityashree hybrid maize and normal maize in broiler chicken. *British Poultry Science* **52** (5) 632-638

Panda AK, Rama Rao SV, Raju MVLN and Bhanja SK 2005 Effect of phytase on production performance with White Leghorn layers fed on a diet low in non-phytate phosphorus. *British Poultry Science.* **46** 464-469

Panda AK, Rama Rao SV, Raju MVLN and Chatterjee RN 2008 Effect of vitamin E and C supplementation on production performance, immune response and antioxidant status of White Leghorn layers during summer stress. *British Poultry Science* **49** 592-599

Panda AK, Reddy MR, Rama Rao SV and Praharaj NK 2003 Production performance, serum/yolk cholesterol and immune competence with White Leghorn layers as influenced by supplemented with probiotics. *Tropical Animal health and Productio* **35** 85-94

Panda PC 1985 Studies on utility of frozen debored turkey meat for production of meat block. *Indian J. Poult. Sci.* **20** 78-82

Pandey HN 1987 Personal Communication.

Panisup AS, Mohanty GC and Verma KC 1988 Ultra structural changes of Rou's Sarcoma exhibiting spontaneous occurrence of A-type particle in chickens *J. Vet. Med.* B **35** 760-764

Panisup AS, Mohanty GC and Verma KC 1994 Occurrence of avian leukosis-sarcoma virus infection in poultry in India *Indian J. Compo Microbiol. Immunol. Infect. Dis.* **15** 31-34

Parihar NS, Rajya BS and Gill BS 1975 Occurrence of malignant catarrhal fever in India. *Indian Vet. J.* **52** 857-859

Parmar OS, Dev DS and Dhar NC 1980 Interse mating among Jersey and Hariana crossbred cattle. *Indian J. Dairy Sci.* **33** 465-467

Pathak KML and Chhabra MB 2011 Trypanosomosis of livestock in India: A review of two decades. *Indian J. Anim. Sci.,* **81** 653-60

Pathak KML, Bhatnagar CS, Yadav MP 1998b Comparative evaluation of different diagnostic techniques for *Trypanosoma evansi* infection in donkeys. In: *Proc. 8th International Conference on Equine Infectious Diseases.* Dubai March 22-26, 1998.

Pathak KML, Singh Y, Meiryenne NV and Kapoor M 1997 Evaluation of various diagnostic techniques for *Trypanosoma evansi* infections in naturally infected camels. *Vet Parasitol.* **69** 49-54

Patir H and Upadhyay RC 2010 Purification, characterization and expression kinetics of heat shock protein 70 from *Bubalus bubalis. Research in Veterinary Science,* **88** 258–262

Patle BR and Mudgal VD 1975 Maintenance requirements for energy in crossbred cattle. *Br. J. Nutr.* **33** 127-39

Patle BR and Mudgal VD 1976 Protein requirements of crossbred lactating cows. *Indian. J. Dairy Sci.* **29** 247-52

Patle BR and Mudgal VD 1977 Utilization of dietary energy for maintenance, milk production and lipogenesis by lactating crossbred cows during their mid stage of lactation. *Br. J. Nutr.* **37** 23-33

Pattabiraman D 1955. Milch goats of Malabar and their development. *Indian Veterinary Journal* **32** 398-402

Pattnaik B, Subramanian S, Sanyal A, Mohapatra JK, Dash BB, Ranjan R and Rout M 2012 Foot and mouth disease: Global status and future road map for control and prevention in India. *Agri. Res.*,**1** 132-147

Pawar HN, Brah GS, Agrwal RK and Verma R 2012 Molecular and Immunological Characterization of Heat Shock Proteins 70 (HSP70) Gene from Buffalo. *Proc. Natl. Acad. Sci. India. Sect. B. Biol. Sci.* DOI 10.1007/s40011-012-0113-2.

Pease HT 1905a Report on investigations in connection with the disease 'dourine'. Government Central Press, Shimla.

Pease HT 1903 Note on dourine or maladie du coit.

Peer F, Sharma MC and Prasad MC 1990 Efficacy of Liv-52 AND *Tinospora cordifolia* in experimental CCI, hepatopathy in goats. *Indian J. Anim. Sci.* **60** 526-531

Pica G, Pica-Ciamarra U and Otte J 2008 The Livestock Sector in the World Development Report 2008: Re-assessing the Policy Priorities. PPLPI Research Report

Pieroni A, Howard P, Volpato G, Santoro RF 2004 Natural remedies and nutraceutical used in ethnoveterinary practices in Inland Southern Italy. *Vet Res Commun* **28** 55-80

Plowright W and Ferris RD 1959 Studies with Rinderpest virus in tissue culture. II Pathogenicity for cattle of culture passaged virus. *J. Comp. Path.*, **69** 173

Plowright W and Ferris RD 1962 Studies with Rinderpest virus in tissue culture. The use of attenuated culture virus and its use as a vaccine for cattle. *Res. Vet. Sci.* **3** 172

Polihronov D 1965 The effect of service period on the milk production and lactation length of buffaloes in Bulgaria. Zhivot Nauk. **2**: 537-544 (fide Anim. Breed. *Abstr.*, **34** 1173)

Polihronov D 1969 Comparison of production characters in Murrah and Bulgarian buffaloes. Zhivot. Nauk. **6** 95-106 (fide Anim. Breed. *Abstr.* **37** 3390)

Porwal ML, Singh MP and Karandikar GW 1981 Studies on reproductive performance of Murrah buffaloes. *Indian Vet. J.* **58** 295-299

Prabhakar K and Ramamurthi R 1990 A study on preparation of Intermediate moisture meat. *J. Food Sci. Technol.* **27** 162-64

Prabhakara Reddy K and Narashari D 1990 Effect of polyphosphate treatment of quail carcasses on drip volume and composition. *Indian Veterinary J* **67** 537-541

Prabhudas K and Sambamurthy B 1976 A note on fluorescent antibody technique for rapid diagnosis of rinderpest. *Ind. J. Anim. Sci.* **46** 454

Pradhan HK, Mohanty GC and Mukit A 1985 Marek's disease in Japanese quails (*Cotumix cotumix japonica*): a study of natural cases *Avian Diseases* **29** 575-582

Pradhan HK, Mohanty GC, Lee WY, Kaul L, and Kataria JM 1988 Immune complex mediated glomerulopathy in Marek's Disease. *Vet. Immunol. Immunopathol.* **19** 165-171 (http://dx.doi.org/10.1016/0165-2427(88)90007-4)

Pradhan K, Bhatia SH and Sangwan DC 1991 *Relative Rumen Ecosystem and Nutrient Digestibility in Cattle and Buffalo, Fed High Fibrous Diets*. Tech. Bull, Dept. of Anim. Nutr. H.A.U. Hisar 125004 (Haryana)

Pradhan NR and Dey NK 1996 Induced hepatopathy in calves and therapeutic efficacy of a herbal liver tonic (AV/LTP/14). *Indian Journal of Animal Sciences*, **66** 1238-1241

Praharaj NK, Gross WB, Dunnington EA, Nir I and Siegel PB 1996 Immuno-responsiveness of fast growing chickens as influenced by feeding regimes. *British Poultry Science* **37** 979-987

Prasad 1996 *National Symp. on Impact of Environ. Pollution on health and production in livestock, poultry and wild life and 13th Annual convention of IAVP*, Dept. of Pathology College VEty Scoence, G.B. Pant Univ. of Agri. and Tech. Pantnagar pp. 32-35

Prasad and Iyer 1986 Enzootic bovine haematuria - clinical and pathomorphological studies Acta *Vet. BRNO* **55** 343-351

Prasad CS and Arora SP 1978 Influence of dietary zinc on the activity of alcohol dehydrogenase in different tissues. *Indian J. Anim. Sci.* **48** 582-84

Prasad CS and Arora SP 1979 Influence of dietary zinc on B-carotene conversion and on the level of retinol binding protein in the blood serum. *Indian J. Dairy Sci.* **32** 275-79

Prasad et al 1988 Final report on fetal pathology in hypothyroid goats. Division of Pathology, IVRI, Izatnagar.

Prasad et al 1970 Pathology of bovine theileriosis. *Indian J. Anim. Sci.* **40** 622-624

Prasad MC, Rajaya MC and Mohanty MC 1972 Caprine arterial diseases: I Spontaneous aortic lesions *Experimental and Molecular Pathology* **18** 14-28

Prasad MC, Rajaya MC and Mohanty MC 1973 Caprine arterial diseases: II Spontaneous aortic lesions *Experimental and Molecular Pathology* **19** 328-338

Prasad G, Hari Singh and Minakshi 2014a "*A novel process of detection of cattle tissue using cattle specific novel primer pair*" Patent Number 261685 dated 11-7-2014; CCS Haryana Agricultural University, Hissar-125004

Prasad G, Hari Singh and Minakshi 2014b "*A Novel Process of Detection of Buffalo (Bubalus bubalis) Tissue using Novel set of Primers*" Patent Number 261631 dated 3-7-2014; CCS Haryana Agricultural University, Hissar-125004

Prasad G, Minakshi, Malik Y and Maan S 2000 Trends in bluetongue virus diagnostics: a review. *Indian J. Anim* **70** 103-109

Prasad VSS 1989 Annual report, Central Institute for Research on Goat, Farah, Makhdoom, Mathura (U.P.)

Proceedings of National Symposium on production and marketing of goat meat held at CIRG, Makhdoom, Jan. 10-11. p.32.

Project Coordinator's Report 1986 All India Coordinated Research Project on Cattle, Indian Veterinary Research Institute, Izatnagar.

Puri DS 1976 *Inheritance of service period and dry period and their effect on subsequent lactation yield.* M.Sc. Thesis, Punjab Agric. Univ., Ludhiana, India.

Pushpa-Paul Venugopal V and Nair PM 1990 Shelf life enhancement of lamb meat under refrigeration by gamma irradiation *J. Food Sci* **55** 865-66

Puttarajappa P, Nair KKS and Kadkol SB 1990 Development of meat preserve. *J. Food Sci & Technol* **27** 102-103

Quality Standards of Indian Medicinal Plants 2000 published by ICMR, New Delhi, India, Vol.**8** 55

Radha Krishna K, Vijaya Rao D, Jayath kilakan, K. Souza Jd. and Sharma TR 1988 Studies on freeze drying of a rehydratable poultry product as large pieces. *Indian Food Packer* **42**(5) 41-52

Radhakrishnan KT and Ramamurthi R 1987a Effect of antemortem treatment with papain an organoleptic changes in mutton. *Cheiron* **16** 261-62

Radhakrishnan KT and Ramamurthi R 1987b Studies on the quality of hog casing treated with chemicals. *Cheiron* **16** 123-124

Rai GS 1966 Season of calving in buffalo heifers and cows maintained in village and government farms. *Indian Vet. J* **43** 228-231

Raina B, Desai RN and Bhat PN 1973a Studies In crossbred sheep. I. analysis of weight at birth of various genetic groups produced by crossing Rambouillet with Karna, Poonch and Kashmir Valley. *Indian J Anim Prod.* **4** 49

Raina BL, Desai RN and Bhat PN 1973 b Studies In crossbred Sheep. II. Mortality in various genetic groups produced by crossing Rambouillet with Karna, Poonch and Kashmir Valley. *Indian J Anim Prod.* 93

Raina BL, Desai RN and Bhat PN 1973c Studies in crossbred sheep. III. Genetic and phenotypic parameters of body weight and fleece weight in different crosses of Rambouillet x Karnah, Rambouillet x Poonch and Rambouillet x Kashmir Valley. *Indian J Anim Prod.* 99

Rajan 1977 Personal communication

Rajkhowa S, Rahman H, Rajkhowa C and Bujarbaruah KM 2005 Seroprevalence of brucellosis in mithuns (*Bos frontalis*) in India. *Preventive Veterinary Medicine* **69** 145-151

Rajkhowa S, Rajkhowa C and Bajrbauah KM 2003 Diseases of mithun (*Bos frontalis*). *Vet Bull* **73** IR-6R

Rajkhowa S, Rajkhowa C and Chamuah J 2008 Seroprevalence of *Toxoplasma gondii* antibodies in free-ranging mithuns (*Bos frontalis*) from India. Zoonoses Public Health, 55:32-322.

Rajkhowa S, Rajkhowa C and Hazarika GC 2007a Serological evidene of coronavirus infections in mithuns (*Bos frontalis*) from India. *Rev. Sci. Tech.* **26** (3) 747-753

Rajkhowa S, Rajkhowa C, Rahman H and Bujarbaruah KM 2004 Seroprevalence of infectious bovine rhinotracheitis in mithun (*Bos frontalis*) in India. *Rev Sci Tech Off Int Epiz* **23** 821-829

Rajkhowa S, Sarma DK, Hazarika GC and Rajkhowa C 2007b A preliminary study on *Mycobacterium paratuberculosis* infection in mithun from India. *Indian Vet J* **84** 988-989

Rajkhowa TK and Saikumar G 2012 Occurrence of postweaning multisystemic wasting syndrome (PMWS) in cross-bred pigs of Uttar Pradesh and Uttarakhand. *Indian J. Anim. Sci.* **82** 1122-1127

Rajkhowa S, Rajkhowa C and Bujarbaruah KM 2008 *Diseases of Mithun (Bos frontalis),* National Research Centre on Mithun, ICAR

Rajkumar U, Sharma RP, Pathi MK, Rajaravindra TK, Reddy BLKN, Niranjan M, Bhattacharya TK, Haunshi S and Chatterjee RN 2011 Genetic analysis of juvenile growth and carecass traits in a full diallel mating in selected broiler lines. *Tropical Animal health and Production* **43** 1129-1136

Rajmane BV and Sonawane NS 1997 Biotechnologija-u-stocarstvu (Yugoslavia) **13** 215-218

Raju MVLN and Devegowda G 2000 Influence of esterified-glucomannan on performance and organ morphology, serum biochemistry and haematology in broilers exposed to individual and combined mycotoxicosis (aflatoxin, ochratoxin, T-2 toxin). *British Poultry Science* **41**(5) 640-650

Raju MVLN, Rama Rao SV, Chakrabarti PP, Rao BVSK, Panda A K, Prabhavathi Devi BLA, Sujitha V, Reddy JRC, Shyam Sunder G and Prasad RBN 2011 Rice bran lysolecithin as a source of energy in broiler chicken diet. *British Poultry Science* **52**(6) 769-774

Raju MVLN, Rama Rao SV, Radhika K and Panda AK 2005 Effect of amount and source of supplemental dietary vegetable oil on broiler chickens exposed to aflatoxicosis. *British Poultry Science* **45** 587-594

Rajya BS and Singh CM 1961 Pathology of JD in sheep. *Am. J. Vet. Res.* **22** 189-203

Rajya BS and Singh CM 1964 Occurrence of Jaagziekte and Maedi in sheep and goats in India *Am J. Vet. Res.* **25** 61-67

Rama Rao S, Prakash B, Raju MVLN, Panda AK and Moorthy OK 2014 Effect of supplementing non-starch polysaccharide hydrolyzing enzymes in guar meal based diets on performance, carecass variables and bone mineralization in Vanaraja chickens. *Animal Feed Science and Technology* **188** 85-91

Rama Rao SV, and Ramasubba Reddy V 2003 Relative bio-availability of phophatic fertilizers as source of phosphorus in broiler and layers. *British Poultry Science* **44** 96-103

Rama Rao SV, Praharaj NK, Reddy MR and Panda AK 2003 Interaction between genotype and dietary concentration methionine for immune function in commercial broilers. *British Poultry Science* **44** 104-112

Rama Rao SV, Praharaj NK, Reddy MR, and Sridevi B 1999 Immune competence, resistance to E. coli and growth in male broiler parent chicks fed different levels of crude protein. *Veterinary Research Communications* **23**(6) 323-336

Rama Rao SV, Prakash B, Kanya Kumari, Raju MVLN and Panda AK 2013 Effect of supplementing different concentrations of organic trace minerals on performance, antioxidant activity, and bone mineralization in Vanaraja chickens developed for free range farming. *Tropical Animal health and Production, DOI* 10.1007/s 11250-013-013-0384-5

Rama Rao SV, Raju MVLN, Panda AK and Reddy MR 2006 Sunflower seed meal as a substitute for soyabean meal in commercial broiler chickens diet. *British Poultry Science* **47** 592-598

Rama Rao SV, Raju MVLN, Panda AK, Poonam NR, Shyam Sunder G and Sharma RP 2008 Utilization of sesame (*Sesamum indicum*) seed meal in broiler chicken diets. *British Poultry Science* **49** 81-85

Rama Rao SV, Raju MVLN, Panda AK, Poonam NS and Shyam Sunder G 2011 Effect of dietary α-tocoperol concentration on performance and some immune responses in broiler chicken fed on diets containing oils from different sources. *British Poultry Science* **52** 97-105

Rama Rao SV, Raju MVLN, Panda AK, Shyam Sunder G and Sharma RP 2009 Performance and bone mineralization in broiler chicks fed on diets with different concentration of cholecalciferol at a constant ratio of calcium to non-phytate phosphorus. *British Poultry Science* **5** 528-535

Rama Rao SV, Ravindera Reddy V and Ramasubba Reddy V 1999 Enhancement of phytate phosphorus availability in the diets of commercial broilers and layers. *Animal Feed Science and Technology* **79** 211-222

Rama Rao SV, Ravindera Reddy V, Ramasubba RV 1999 Non-phytate phosphorus requirement of commercial broilers and white leghorn layers. *Animal Feed Science and Technology* **80** 1-10

Rama Rao SV, Ravindera Reddy, V and Ramasubba Reddy V 1999 Utilization of different phosphorus source in relation to their fluorine content for broilers and layers. *British Poultry Science* **42** 376-383

Rama Rao SV, Reddy MR, Praharaj NK, Shyam Sunder G 2000 Laying performance of broiler breeder chicken fed various millets or broken rice as a source of energy at a constant nutrient intake. *Tropical Animal Health and Production* **32** 329-338

Rama Rao SV, Shyam Sunder G, Reddy MR, Praharaj NK, Raju MVLN and Panda AK 2001 Effect of supplement of chloride on the performance of broiler breeders fed on different energy sources. *British Poultry Science* **42** 362-367

Rao PR, Rajya BS Parihar NS, Mehrotra ML and Mohanty GC 1975 Granular vulvovaginitis (GVV) in cattle - immunofluorescent studies *Vet. Rec.* **96** 361

Ramesh KP, Kumar KVP, Chandrashekar K, Das S, Saravanan BC, Pourouchottamane R and Kataktalware M 2001 Genetic variability of Indian yaks using random amplified polymorphic DNA markers. *African Journal of Biotechnology* **10** (43) 8558-8561

Ramesha KP, Das S, Kataktalware MA, Saravan BC, Pourouchottamane R, Rajkhowa J, Bandopadhyaya S, Krishnan G and Sarkar M 2010 Association of polymorphism of defensin genes with milk somatic cell count in yaks and related species. *Indian J Anim Sci* **80** 181-182

Ramesha KP, Kataktalware MA, Das S, Pourouchottamane R, Bandopadyaya S, Saravanan BC, Krishnan G, Sarkar M, Bhattacharya M (2008) Performance and body measurements of yaks and their hybrids under field condition, *Indian J Anim Sci* **78** 1034-1035

Ranjhan SK 1976 *Prospects for the commercial utilization of Agro-industrial by-products for ruminants in India.* Anim. Nutr. Div. IVRI (Memograph)

Ranjhan SK 1990 *Animal nutrition and feeding practices.* Fifth revised edition, Vikas Publication House, New Delhi

Ranjhan SK 1991 Animal *nutrition in the tropics.* Third revised edition, Vikas Publication House, New Delhi

Ranjhan SK and Khera RC 1976 *Feeding of Farm Animals during scarcity.* ICAR New Delhi

Ranjhan SK and Pathak NN 1979 *Management of Feeding of buffaloes.* Vikas Publishing house New Delhi

Ranjhan SK, Faylon PS, Momongan and Cruz LC 1987 Husbandry Swamp Buffaloes in Philippines, PACARD, Lon Banos, Philippines pp 59

Ranjhan SK, Krishna Mohan DVG and Singh R 1975 Energy and protein requirements of Holstein Friesian and Holstein-friesian × Haryana Calves for maintenance production. *Indian J. Anim. Sci* **45** 717-22

Ranjhan SK, Sawhney PC and Jayal MM 1974 *Application of life saving research in animal feeding*, Farm bulletin, ICAR, New Delhi.

Rao and Iyer 1976 Mammary introductory carcinoma in a pig. *Indian Vet. J.* **53** 92-93

Rao and Paliwal 1993 Hepatoma in Pig Immunofluoreascence studies. *Indian Vet. J.* **70** 461

Rao AR, Sastry AP, Reddy KK and Rajulu PV 1909 Studies on reproductive characters of Ongole cattle. I. Age at first calving, intercalving period and sex-ratio. *Indian Vet. J.* **46** 679-684, 1969

Rao BR, Patel UG and Tahman SS 1973 Seasonal trend in reproductive behaviour of Surti buffaloes, service period and post-partum oestrus interval. *Indian Vet. J.* **50** 413-417 (fide Anim. Breed. *Abstr.* **42** 150)

Rao CH, Rao CV and Naidu KN 1976 A note on the effects of off-season breeding in buffaloes. *Indian J. Dairy Sci.* **29** 237-240

Rao VK, Anjaneyulu ASR, Lakshmanan V, Sharma N and Kondaiah N 1985 Meat and fat characteristics of male buffalo calves. *Indian J. Anim. Sci.* **55** 696-698

Rao VP 1979 *The role of genotype and environment in sire evaluation.* Ph.D. Thesis, Kurukshetra University, Kurukshetra

Rao VP and Taneja VK 1982 Genetic evaluation of Holstein-Sahiwal crossbred grades: First lacation milk yield. *2nd World Congress on Genetics Applied to Livestock Production*, Madrid, 4-8 Oct. 1982

Rarotra JR, Yadav MP and Sethi MS 1978 Sero-epidemiology of Q-fever in Poultry. *Avian Diseases*, **22** 167-169

Rattan B and Yadav MP 1994 Occurrence of Rota viral diarrhoea among thoroughbred foals in India. *Ind. J. Anim. Sci.* (**9**) 353-355

Rattan B and Yadav MP 2001 Antibody Isotypes in Response to equine Influenza Vaccine. *Indian J. Com. Microbiol. Immunol. Infect. Dis.* Vol. **22**, No. **2** (July-Dec.) 100-102

Rattan B, Gupta AK, Yadav MP, Uppal PK and Malik Praveen 1998 Lower susceptibility of donkeys to equine herpes virus and equine infectious anemia virus in comparison to horses. In "*Proc.3rd Coloquio Int. Working Equines*". Mexico, p 112-116

Rattan B, Khurana SK, Singh BK and Yadav MP 1999 An outbreak of abortions in Thoroughbred mares due to equine herpes virus-1. *Int. J. Anim. Sci.*, **13** 233-235

Rattan B, Rao AK, Khurana SK and Yadav MP 1999 An outbreak of perinatal foal mortality due to equine herpes virus-1. *Indian J. Virol.* **15** 81-83

Rattan B, Uppal PK, Singh BK and Yadav MP 1994 Studies on growth kinetics of equine herpes virus-1 in donkey kidney cell culture. *Indian J. Virol.* **10** 104-108

Rattan B, Yadav MP and Uppal PK 1993 Comparison of enzyme linked immunosorbent assay and Coggins immunodiffusion test for diagnosis of equine infectious anaemia. *Indian J. Anim. Sci.* **63**(9) 910-913

Rattan B, Yadav MP and Uppal PK 1993 Immunodipstick and strip antigen capture ELISA's for detection of equine influenza virus (H3N8) infection *International J. Anim. Sci.* **8** 275

Rattan B, Yadav MP and Uppal PK 1997 Production of equine infectious anemia viral antigen in hybrid cells derived from infected horse's leukocytes and vero cells *Indian J. Comp. Microbiol. Immunol. Infect. Dis.* **18** 18-23

Rattan B, Yadav MP and Uppal PK 1998 Biotin-Avidin ELISA for diagnosis of equine infectious anaemia *Indian J. Comp. Microbiol. Immunol. & Infect. Dis.,* **19**(2) 79-81

Raut KC and Singh S 1971 Classification and productivity of milch animals *Ind. J. Anim. Sci* **41** 1-6

Ray SN and Mudgal VD 1968 Research on Nutrition of Cattle and Buffalo in India *Indian. J. Vet. Sci.* **38** 117-133

RBS College, Agra 1979 *Studies on the combining ability of desirable characters of important goat breeds for milk and meat separately and in combination* Technical Progress Report. 1974-79. Raja Balwant Singh College, Bichpuri, Agra.

Reddy AO 1985 *Studies on certain behavioural patterns in Murrah buffaloes under loose housing system* Ph.D. Thesis, Dairy Science College, NDRI, Karnal

Reddy KS and Mandokhot UV 1987 Studies on SO2 treatment of minced goat meat. Effect of storage life and organoleptic qualities *J. Food Sci. & Technol.* **24** 299-303

Reddy VK, Srinivas Reddy RA, Srujana G, Surekha M and Reddy SM 2011 *Int. J. Pharm. Bio. Sci.* **2** 492-498

Reddy YKM, Murali Manohar B, Pandey AB, Narsimha Redday Y, Prasad G and Chauhan RS 2010 Development and evaluation of inactivated pentavalent adjuvanted vaccine for bluetongue. *Indian Veterinary Journal* **87** 434-436

Report 1968 Final Report of the Scheme for Sheep and Wool Improvement on Regional Basis-Dry Northern Region-Joria North Gujarat, Patan, Distt Mehsana, Gujarat

Report 1972 Annual Report, Central Sheep and Wool Research Institute, Avikanagar

Report 1973 Annual Report, Central Sheep and Wool Research Institute, Avikanagar

Report 1974 Annual Report, Central Sheep and Wool Research Institute, Avikanagar

Report 1975 Annual Report, Central Sheep and Wool Research Institute, Avikanagar

Report 1976 Annual Report, Central Sheep and Wool Research Institute, Avikanagar

Report, ICRETT, Fellowship 1980 International Fellowship against cancese, Genewa

Rishendra Verma and Jaiswal TN 1997 Protection, humoral and cell-mediated immune responses in calves immunized with multiple emulsion haemorrhagic septicaemia vaccine *Vaccine* **15** 1254-1260

Rout PK, Joshi MB, Mondal A, Laloe D, Singh l and Thangaraj K 2008 Microsatellite based phylogeny of Indian goats. *BMC Genetics* **9** 11

Roy A 1957 Egg coagulating enzyme in the semen and cowpeas gland of the goats *Nature* 179 318-319

Roy A, Sengupta BP and Misra MS 1964 Effect of varying environment on semen quality, cardio-respiratory activity, milk production and female fertility of buffalo. *Arid Zone Res.* **24** 275

Roy AK, Yadav MP and Sengar OPS 2004 Effect of season on the biochemical characteristics of donkey (*Equus asinus*) semen *Indian Journal of Animal Sciences* **74**(4) 345-347

Roy AK,Yadav MP and Sengar OP 2003 Effect of season on the physical characteristics of donkey (*Equus asinus*) semen *Indian J. Anim. Sci.* **73** (6) 620-622

Roy N, Nageshan RK, Pallavi R, Chakravarthy H and Chandran S 2010 Proteomics of *Trypanosoma evansi* infection in rodents. PLoS ONE 5(3): e9796.doi:10.1371/ journal.pone.0009796.

Roy N, Nageshan RK, Rani Pallavi, Chakravarthy H, Chandran S, Rajender Kumar, Gupta AK, Singh RK, Yadav SC and Tatu U 2010 Proteomics of *Trypanosoma evansi* infection in rodents. PloS ONE, **5** 1-10.

Royal Commission on Agriculture 1928 *Report of the Royal Commission on Agriculture in India.* Presented to Parliament by command of His Majesty, June, 1928.

Roychoudhury PN and Succi G 1969 Effect of month of calving on the length of service period in Italian buffaloes. *Indian Vet. J* **46** 1059-1062

Roychoudhury PN, Mathur BS and Deshmukh SN 1971 Some environmental factors affecting calving interval and lactation length in Italian buffaloes *Indian Vet. J* **48** 711 -716

Sachdev AK, Ram Gopal Verma SS, Kapoor KN and Kulshreshtha SB 1994 Quality of chicken pickle during processing and storage *J. Food Sci. & Technol.* **31** 32-35

Sahin K, Onderci M, Sahin N, Gursu MF and Kucuk O 2003 Dietary vitamin C and folic acid supplementation ameliorates the detrimental effects of heat stress in Japanese quail *J. Nutr.* **133** 1882-1886

Sahni KL and Roy A 1972 A note on seasonal variation In semen production of Corriedale and Corriedale x Bikaneri (halfbred) rams under tropical conditions *Indian J Anim Sci.* **42** 99.

Sakhare PS, Harne SD, Kalorey DR, Warke SR, Bhandarkar AG, and Kurkure NV 2007 Effect of Toxiroak polyherbal feed supplement during induced aflatoxicosis, ochratoxicosis and combined mycotoxicoses in broilers *Vet. Arhiv* **77** 129-146

Salahuddin M, Sharma N and Bisht GS 1993 *Quality of fermented mutton sausages at various stages of processing.* In 3rd International Food Convention held at Mysore, Sept. 7-12. p.23.

Salahuddin WI 1988 Studies on effect of phosphate and some technologies on the quality of kabab from different Meats. M.V.Sc. thesis. IVRI, Izatnagar.

Salerno A 1961 On the length of calving interval in the buffalo. 1st Zootec. Geng. Univ. Stud., Naples. Reprinted from Ann. Fac. Agric. Protici. Ser. III 26: 18 (fide Anim. Breed. *Abstr.* **29** 1418).

Samoon AH and Sharma N 1988 Processing and preservation of Nate-yakhni, An indigenous meat products of Kashmir *Indian J. Meat Sci. & Technol* **1** 48-57

Samoon AH and Sharma N 1991 Studies on processing and preservation of Goshtaba *J. Food Sci. & Technol* **28** 212-215

Samoon AH and Sharma N 1994 Studies on processing and refrigerated storage of Rista' *Meat Sci.* **37** 347-368

Saravanan P, Balamurugan V, Sen A, Sarkar J, Sahay B, Rajak KK, Hosamani M, Yadav MP and Singh RK 2007 Mixed infection of peste des petits ruminants and orf on a goat farm in Shahjahanpur *India. Vet Rec.* **160**(12) 410-12

Sasaki M 1992 A review on Rinderpest situation (Part I). Asian Livestock Vol. XVII 5 49-54

Sasaki M 1992 A review on Rinderpest situation (Part II). Asian Livestock Vol. XVII 6 61

Sastry NSR 1980 *Problems of seasonal peaks and troughs in buffalo milk production.* Summer Institute on Buffalo Management Systems held at HAU, Hissar. pp. 121-128

Satyavati GV, Gupta AK 1987 eds. *Medicinal plants of India*, vol. II New Delhi; Indian Council of Medicinal Plants Research 1–22

Sawhney PC and Kehar ND 1948 Annual report, Animal nutrition division, IVRI, Izatnagar

Sawhney PC, Bedi SPS and Bisaraya SC 1972 *2nd J. Animal Sciences* **42** 656

Schneeberger CP, Wellington KE and McDowell RE 1962 Performance of Jamaica Hope cattle in commercial dairy herd in Jamaica *J. Dairy Sc.* **65** 1364-1371

Schneider BH 1994 Breeding for milk production in India. *Allahabdad Fmr* **16** 2-26

Schulz V, Hansel R, Tyler VE 2001 *Medicinal Plants, phytomedicines and phytotherapy.* In: Rational Phytotherapy: A physician's guide to herbal medicine. 4ed. New York, Berlin: Springer, 2001. Cap.1, p.1-39

Sekhon KS and Bawa AS 1991 Effect of frozen storage and quality of meat tikkas from culled hens and broiler breeder male *J. Food Sci & Technol* **28** 296-300

Selvaraj R, Ramaswamy AM, Asumugum ALP and Ramamurthy R 1989 Influence of storage period on certain organoleptic qualities of carried and canned mutton and chevon *Cheiron* **18** 165-68

Selvaraju A, Ayyanar M, Rathinakumar SS and Sekar T 2011 Plants used in ethno-veterinary medicine by malayali tribals in Salem district of Tamil Nadu, India *Medicinal Plants* **3**(3) 1-7

Sen AR and Sharma N 1996 Studies of selected mitochondrial enzymes for differentiating fresh and frozen thawed chicken muscles. *Proceedings of XX World's Poultry Congress*, Vol. IV p.428

Sen KC 1953 *Animal nutrition research in India.* Macmillan and Co., Calcutta

Sen KC and Ray SN 1964 *The Nutritive Value of Indian Cattle Feeds and Feeding of Animals* 5th edn. Indian Council Agric. Res. New Delhi

Sen KC, Ray SN and Ranjhan SK 1978 *The Nutritive Value of Indian cattle feeds and feeding of Animals* 7th edn Bull ICAR, New Delhi

Sengupta PP and Yadav MP 1997 Occurrence of parasitic infection in ponies of Tarai region, Uttar Pradesh *Indian. J. Anim. Sci.* **67** 460-62

Sengupta PP and Yadav MP 1998 Incidence of gastro-intestinal parasites in organized and unorganized equid farms of Haryana *Indian J. Anim. Sci* **68**(12) 1218-1220

Sengupta PP, Virmani N and Yadav MP 1999 Pathological changes in the donkeys experimentally infected with Babesia equi *Indian J. Anim. Sci* **69**(3) 169-170

Sethi MS and Singh VB 1980 *Reservoirs of zoonotic diseases of Uttar Pradesh.* Final Technical Report, P.L. 480 Project No. IN-ARS-75.

Sethi MS, Bhupender Singh and Yadav MP 1978 Experimental infection of *Coxiella burnetii* in chicken: Clinical Symptoms, serologic response and transmission through egg *Avian Diseases* **22** 391-395

Sethi MS, Rarotra JR and Yadav MP 1978 A note on the role of dairy animals in the epidemiology of Q-fever *Indian J. Anim. Sci* **48** 477-478

Shah KV 1964 'Investigation of African horse sickness in India. II. Reaction in non-immune horses after vaccination with the polyvalent African horse-sickness vaccine' *Indian J. Vet. Sci.* **34** 75

Shahin M.A, El-Itribv A and Barradu MS 1966 Thirty two years of European cattle breeding in UAR. A mortality rates and culling *Agri. Res. Rev.* **4** 18-40

Shalu Verma, Tiwari SC and Yadav MP 1999 Use of polyherbal immunomodulator in conjunction with inactivated vaccine against equine herpes virus-1 *Indian J. Virology* **15** 63-67

Shankar H and Yadav MP 1986 Occurrence of paralytic syndrome in equines due to equine herpes virus-1 *Rev. Sci. Tech. Off. Int. Epiz.* **5(1)** 15-21

Shankar H and Yadav MP 1987 Seroprevalence of EHV-1 and group B arboviruses in nervous disorders in equines, *Indian J. Anim. Sci* **58** 1-5

Shankar H and Yadav MP 1987 Experimental infection of sheep with BHV-1 (IBR/IPV virus) and its possible role in epizootiology *Indian Vet. Med. J* **11** 71-76

Shanmugam AM and Ranganathan M 1972 Differentiation of mutton and beef by precipitation test *Indian Vet. J.* **49** 1024-28

Shanmugam M and Rama Rao SV 2013 Effect of dietary ellagic acid supplementation on semen quality parameters in chickens *Animal Production Science* DOI: 10.1071/ AN13110

Shanmugam M, Rajkumar U, Reddy MR and Rama Rao SV 2012 Effect of age on semen quality in Naked neck and Dwarf chicken under tropical climatic conditions *Animal Production Science* **52** 964-968

Shanmugam M, Vinoth A, Rajaravindra KS and Rajkumar U 2014 Evaluation of semen quality in roosters of different age during hot climatic condition *Animal Reproduction Science* **145** 81-85

Sharma AK, Karthikeyan S, Kaur P, Singh TP and Yadav MP 1996 Purification, crystallization and preliminary crystallographic analysis of mare lactoferrin *Acta Cryst. D* **52** 1196-1198

Sharma AK, Rajashankar KR, Yadav MPand Singh TP 1999 Structure of mare apolactoferrin: The N and C lobes are in the closed form *Acta Crystallogr D Biol Crystallogr* **55**(6)1152-1157

Sharma Ashwani K, Paramasivam M, Srinivasan A, Yadav MP and Singh Tej P 1999 Three-dimensional Structure of Mare Diferric Lactoferrin at 2.6 A Resolution *J. Mol. Biol* 1-15

Sharma B, Lahiri DK, Joshi RC and Bansal RP 1986 A comparison of immunofluorescence and immunoperoxidase tests for detection of rinderpest antigen in acetone fixed smears *Ind. J. Anim. Sci.* **56** 487

Sharma B, Negi BS, Pandey AS, Bandyopadhyay SK, Shankar H and Yadav MP 1988 Detection of goat pox antigen and antibody by counter immunoelectrophoresis (CIE) test using inactivated antigen *Trop. Anim. Hlth. Prod.* **20** 109

Sharma BD, Padda GS, Keshri RC and Sharma N 1988 Acceptability studies on different stuff binders in chevon samosa *Indian J. Meat Sci. & Technol* **1** 32-36

Sharma DD, Sehgal JP, Singhal KK and Gupta MK 1995 *Project report on growing male buffalo calves for quality meat production* NDRI Alkabeer consultancy project. NDRI Karnal

Sharma DN 1972 PA complex in sheep and goats Agra Univ. *J. Res.* **21** 85-86

Sharma N and Mukhopadhyaya R 1995 Processing of fermented sausages *Fleischwirtchaft.* **75** 452-54

Sharma N, Gandemer G, Goutefongea R and Kowle BD 1986 Fatty acid composition in buffalo meat *Meat Sci.* **16** 237-243

Sharma N, Keshri RC, Padda GS and Sharma BD 1988 Influence of hot boning on meat yield and physical characteristics of goat carcass *Indian J. Anim. Sci* **58** 1122-24

Sharma N, Mahadevan TD and Panda B 1973 Studies on preservation of poultry meat by curing and smoking *Indian Food Packer* **25**(5)1220

Sharma N, Padda GS and Kondaiah N 1982 Preparation of buffalo sausages *Indian Food Packer* **36** 50-52

Sharma N, Sharma BD, Keshri RC and Padda GS 1988 Effect of particle size on processing of goat meat patties *J. Food Sci. & Technol* **25** 249-50

Sharma NK, Srivastava VA, Gill JPS and Joshi DV 1994 Differentiation of meat from food animals by enzyme assay *Food Control* **5** 613-630

Sharma PV 1983 *Caraka-Samhitâ* (text with English translation), Vol. II. Chaukhambha Orientaba, Varanasi & Delhi; pp. 665-666

Sharma R and Saikumar G 2008 Porcine circovirus 2 associated reproductive failure in Indian pigs. *Indian J. Anim. Sci* **78** 1238-1240

Sharma RC, Mehta SC, Bansal RS and Pathak KML 2009 PCR-RFLP profile of MHC-DRB3 class II genes in Marwari horses *Indian Journal of Animal Sciences* **79** 1036-39

Sharma V, Gupta AP, Bhandari P, Gupta RC and Singh B 2007 *Chromatographia* **66** 801-804

Shas N Keshn RC, Sharma BD, Padda GS and Kondaiah N 1989. Processing and palatability properties of goat meat tikka *Indian J. Anim. Sci* **59** 202-296

Shedduck JA 1985 Changes and challenges in Veterinary Pathology *Cornell. Vet.* **75** 77

Sherikar AT, Karkare UTD, Khot JB, Jayarao BM and Bhilegaonkar KT 1993 Studies on thermostable antigens, production species-specific antigens sera and comparison of immunological techniques in meat speciation *Meat Sci.* **33** 121-36

Sherikar AT, Khot JB, Jayarao BM and Pillai SR 1988 Differentiation of organs of meat animals and identification of their flesh in chicken using antigens BE sera *Indian J. Anim. Sci* **58** 565-573

Shilston AW 1917 The vitality of rinderpest virus outside the animal body under natural conditions. Agri. Res. Instt. Pusa, vet. Ser. **3** 32

Shinde VM, Dhalwal K, Potdar M , Mahadik KR 2009 Application of quality control principles to herbal drugs. *International Journal of Phytomedicine* **1** 4-8

Shrestha SL 1985 Some reproductive traits in Murrah buffaloes *Indian J. Anim. Sci* **55** 177 (fide Anim. Breed. *Abstr.* **53** 6359)

Siddiquee GM, Tajane KR and Pande MB 1984 Effect of year and parity on some of the reproductive traits in Mehsana buffaloes *Livestock Adviser* **9** 23-25

Singh B 1982 *Divergence among genetic grades of crossbred animals*. Ph.D. Thesis, Rohilkhand University, Bareilly

Singh and Iyer 1973 Mammary intraductal carcinoma in goats *Path. Vet.* **9** 441-446

Singh B 1939 Goat virus vaccination in central provinces and Bihar *Indian J. Vet. Sci.* **9** 279

Singh BB and Singh B 1974 Performance of Jamunapari goats *Indian Veterinary Journal* **51** 326-33

Singh BK, Gulati BR, Tewari SC and Yadav MP 2002 Antigenic differentiation of equine herpes virus-1 (EHV-1) isolates of Indian origin using monoclonal antibodies *Indian J. Biotech* Vol. **1** 170-174

Singh BK, Sinclair R, Gupta AK, Uppal PK, Yadav MP and Mumford JA 1993 Production of monoclonal antibodies against equine influenza A/Equi-2 (H3N8) isolate *Indian J. Exp. Biology* **31** 774-75

Singh BK, Tandon SN and Virmani N 2006 Immune response to inactivated oil adjuvanted equine herpes virus-1 using different emulsifiers in horses *Indian Journal of Biotechnology* **5** 42-46

Singh BK, Vrimani N and Gulati BR 2009 Assessment of protective immune response of inactivated equine herpesvirus-1 vaccine in pregnant BALF/c mice *Indian Journal of Animal Sciences* **79**(4) 345-349

Singh BK, Yadav MP and Tewari SC 2001 Neutralizing and complement-fixing monoclonal antibodies as an aid to the diagnosis of equine herpesvirus-1 infection *Veterinary Research Communications* **25**(8) 675-686

Singh BK, Yadav MP and Uppal PK 1995 Comparison of enzyme linked immunosorbent assay, virus neutralization and complement fixation tests for measurement of antibodies against equine herpes virus-1 infection in equine sera *Indian J. Anim. Sci.* **65** (1) 1-5

Singh BK, Yadav MP, and Uppal PK 1992 Sero-epidemiology of equine herpes virus-1 infection in Indian horses *Int. J. Anim. Sci* **7** 43-46

Singh CSP and Singh B 1984 Incidence of estrus in buffaloes *Indian J. Animal Sci* **54** 259

Singh DK, Singh LB and Deb AR 1989 Annual Progress Report. All India Co-ordinated Research Project on Goat for Meat Production. Birsa Agricultural University, Ranchi, Bihar. p. 93

Singh IP and Singh SB 1967 Isolation and characterization of the aetiological agent of buffalo pox. *Journal of Research (Ludhiana),* **4** 440-448

Singh JN, Ranjan A and Archna Kumari 1995 Bleeching and chromicization of wet casings for preparation of plain and chromic catgut.

Singh M and Acharya R M 1977 A note on the mode of heat dissipation in different type of sheep. *Indian J Anim Sci* **47** 367

Singh MK, Gupta AK and Yadav MP 2007 Performance evaluation of donkeys in arid zone in India. *Indian J. Anim. Scie* **77**(10) 1017-1020

Singh MK, Yadav MP and Mehta NT 2002 Breed characteristics of Marwari and Kathiawari horses *Indian J. Anim. Sci* Vol. **72** (4): 319-323.

Singh MK, Yadav MP and Mehta NT 2004 Factors influencing body weight and physical traits in Kathiawari horses *Indian Journal of Animal Sciences* **74** 673-75

Singh N and Gangwar PC 1976 Effect of oxytocin on conception rate in buffaloes *Indian J. Anim. Sci.* **46** 8

Singh N, Rajya BS and Mohanty GC 1974 Granular vulvovaginitis (GVV) in goats associated with Mycoplasma agalactiae *Cornell Vet.* **64** 435–442

Singh NP, More T and Sahni KL 1976 Effect of water deprivation on feed intake, nutrient digestibility and nitrogen retention in sheep *J agric Sci, Camb* **86** 431

Singh RK, Hosamani M, Balamurugan V, Bhanuprakash V, Rasool TJ and Yadav MP 2007 Buffalo pox: an emerging and re-emerging zoonoses *Anim. Hlth. Res. Rev* **8**(1) 105-114

Singh RN and Acharya RM 1983 Beetal goat Genetics Selection for maximizing life time milk production. *International Goat and Sheep Research* **1** (3) 226-233

Singh RN, Acharya RM and Biswas DK 1970 Selection Index for goats: expected direct and correlated response *Acta Agric. Scandinavia* **20** 61-64

Singh RN, Acharya RM and Dutta OP 1982 Differences in carcass characteristics of Beetal and its crosses with Alpine and Anglo-nubian *Indian Journal of Animal Sciences* **52** 708-711

Singh RP and Verma SS 1995 Effect of hot and cold deboning on the yield and quality of Quail meat *J. Food Sci. & Technol* **32** 233-235

Singh RP, Sreenivasa BP, Dhar P, Roy RN and Bandyopadhyay SK 2000 Development of a monoclonal antibody based Competitive ELISA for the detection of Rinderpest antibodies. *Rev. Sci. Tech. Off. Int. Epiz.* **19** (3) 754-763

Singh RV 1987 *Genetic studies on short lactations in buffaloes*. M.Sc. Thesis, Kurukshetra University, Kurukshetra

Singh SB and Desai RN 1962 Production characters of Bhadawari buffalo cows *Indian Vet. J* **32** 332-343

Singh SV, Singh AV, Singh R, Sandhu KS, Singh PK, Sohal JS, Gupta VK and Vihan VS 2007 Evaluation of highly sensitive indigenous milk ELISA kit with faecal culture, milk culture and faecal-PCR for the diagnosis of bovine Johne's disease (BJD) in India *Comp. Immunol. Microbiol. Infect. Dis.* **30** 175–186

Singh UB, Verma DN, Verma A and Ranjhan SK 1974 *J. Agric, Sci.* **83** 13

Singh VK 1986 Revised Terminal Report. All India Coordinated Research Project on Sheep Breeding. Bikaner: CSWRI. p.356

Singha H, Goyal SK, Malik P, Khurana SK and Singh RK 2013 Development, evaluation, and laboratory validation of immunoassays for the diagnosis of equine infectious anemia (EIA) using recombinant protein produced from a synthetic p26 gene of EIA virus *Ind. J .Virol* **24** (3) 349-356

Singha H, Goyal SK, Malik P, Singh RK 2014 Use of heavy water (D2O) in developing thermostable recombinant p26 protein based enzyme linked immunosorbent assay (ELISA) for serodiagnosis of equine infectious anemia virus infection *Scientific World Journal* 620-906

Singha H, Gulati BR, Kumar P, Singh BK, Virmani N and Singh RK 2013 Complete genome sequence analysis of Japanese encephalitis virus isolated from a horse in India *Archives of Virology* **158** 113-122

Singhal RA and Kaushik SN 1980 Comparative performance of some Indian breeds of cattle *Dairy Guide* **2** 25-28

Sinth JN, Saigal RP, Bhatia BS and Sekhan KS 1985 Effect of sodium chloride. Trisodium polyphosphate and papain on histological structure of spent hen meat *J. Food Sci. & Technol* **22** 110-114

Sivaiah K and Mudgal VD 1978 Effects of feeding different levels of protein and energy on feed utilization for growth and milk production in buffaloes (*Bos bubals*) Annual Report, NDRI, Karnal pp. 145-46.

Sivaiah K and Mudgal VD 1983 Protein requirements for maintenance and milk production in buffaloes (*Bubalus bubalis*) in first lactation *Indian J. Anim. Sci* **53** 236-42

Somvanshi R 1993 *Prachin Bharat main Pashu Palan evum Pashu Chikitsha Vigyan* (In Hindi). Part I. Rajbhasha Anubhag, Indian Veterinary Research Institute, Izatnagar, Uttar Pradesh, India

Somvanshi R, Biswas JC, Yadav MP and Koul GL 1987 Epidemiological observations on outbreaks of contagious ecthyma in pashmina goats *Indian J. Anim. Health* **26** 167-170

Somvanshi R, Sharma B, Koul GL, Biswas JC and Yadav MP 1988 The delayed cutaneous hypersensitivity and humoral immune response in spontaneous bovine papillomas *Indian Vet. Med. J.* **12** 7-14

Sreenivasa BP, Dhar P, Singh RP and Bandyopadhyay SK 2000 Evaluation of an indigenously developed homologous live attenuated cell culture vaccine against peste des petits ruminant infection of small ruminants. *Proc. XX Annual Conference of Indian Association of Veterinary microbiologists, Immunologists and Specialists in Infectious Diseases and National Symposium on Trends in Vaccinology for Animal Diseases*, College of Veterinary Science, G.B. Pant University of Agriculture and Technology, Pantnagar, Uttaranchal, India. 84 pp

Srinivasan M, Gabriel-Raj A and Kothandaraman P 1983 Effect of smoking time in polyphosphate on the quality attributes and keeping quality of ready to cook broilers *Indian J. Poultry Sci* **18** 211-214

Srivastava VK and Roy A 1969 The Influence of season of birth and age on the carcass quality of Magra Lambs *Indian J Anim Sci* **39** 294

Srivastava VK, Raizada BC and Kulkarni V 1968 Carcass quality of Barbari and Jamunapari kids *Indian Veterinary Journal* **45** 219-225

Stephen M, Mukundan G, Iype S and Chacko CT 1985b Comparison of milk production of Jersey x local and Brown Swiss x local crosses in heavy rainfall area *Indian J. Anim. Sci* **55** (6) 485-487

Stephen M, Mukundan G and Iype S 1985a Age at first calving of Brown Swiss × local and Jersey × local crosses in Kerala *J. Vet. Sci* **16** 1-5

Stonaker HH, Agarwala OP and Sundaresan D 1953 Production characteristics of crossbred backcross and pure-breed Red Sindhi in Gangetic plain region *J. Dairy Sci* **36** 678-687

Sudharshana KJ, Suresh KB and Rajasekhar M 1998 *Rinderpest seromonitoring and sero-surveillance. Annual Report, 1998-1999*, Project Directorate-Animal Disease Monitoring & Surveillance, Bengaluru

Sundaresan D 1980 *Buffalo and dairy development.* Summer Institute on buffalo management Systems, pp. 254-262, 2-30 June

Sundaresan D 1985 Buffalo: India's milk machine. *Dairy India.*

Surve AN, Shericar AT, Bhilegaonkar KN and Karkare UD 1991 Preservative effect of combinations of acetic acid with lactic or propionic acid on buffalo meat stored at refrigeration temperature *Meat Sci* **29** 309-322

Swarup D and Singh SK (eds) 2011 *All India Coordinated Research Project on Goat Improvement* Project Coordinator's Report. 2009-10 and 2010-11. Central Institute for Research on Goats, Makhdoom, Mathura (U.P.). p. 60

Tajane KR and Vyas AP 1984 Production performance of Kankrej cattle *Indian J. Anim. Sci* **54** 363-364

Taneja GC 1965 Effect of varying the frequency of watering during summer on the temperature, respiration, body weight and packed cell volume of blood of sheep *Indian J. Exptl. Biol* **3** 259

Taneja VK 1973 *Genetic Analysis of Holstein x Zebu crosses*, Ph.D. Thesis, Agra University, Agra

Taneja VK 1987 A note on salient results and trends of AICR Project on cattle and programmes envisaged in the seventh plan under Project Directorate on cattle IVRI, Izatnagar.

Taneja VK and Bhat PN 1974 Estimation of additive and non-additive effects in Sahiwal × Friesian crossbreds. *1st World Congress on Genetics Applied to Livestock Production*, Madrid, Oct. 7 to 11, 1974, III, 611–615

Taneja VK and Bhat PN 1978 Additive and non-additive genetic effects for various economic traits among Sahiwal and Sahiwal x Holstein crossbred grades *Indian J. Anim. Sci* **48** (12) 843-850

Taneja VK and Bhat PN 1986 Milk and beef production in tropical environment. *Proc. 3rd Wld. Congr. Genet. Appl. Livest. Prod., Nebrasaka, Lincoln.* USA, **9** 73-91

Taneja VK and Bhat PN 1987 Perspective 1990 — technological breakthroughs in the field of breeding and milk production *Indian Dairy* **39** 49-58

Taneja VK and Chawla DS 1978 Comparative study of economic traits of Brown Swiss × Zebu crossbreds *Indian J. Dairy Sci* **31** 188-190

Taneja VK and Bhat PN 1972 Genetic and non-genetic factors affecting body weights in Sahiwal × Friesian crosses *Anim. Prod* **14** 295-298

Taneja VK, Basu SB, Bhatnagar DS and Patro RR 1981 A note on performance of non-descript and Holstein × non-descript crossbreds *Indian J. Anim. Sci.* **51** 1075-1078

Taneja VK, Bhat PN and Garg RC 1979 Genetic Divergence in Various Sahiwal, *Theor. Appl. Genet.* **54** 69-74

Taneja VK, Bhat PN and Garg RC 1979 Genetic divergence in various Sahiwal x Holstein crossbred grades. *Theor. and Appl. Genet.* **54** 69-74

Taneja VK, Nautiyal LP, Saxena MM, Nivsarkar AE and Dwivedi VK 1987 Genetic and economic evaluation of two and three-breed crosses. Personal communication

Tantia MSR, Vijh K, Bhasin V, Poonam S, Vij, Kataria PK, Mishra RS, Yadav BP, Pandey SP, Sethi AK, Joshi BK, Gupta SC and Pathak KML 2011 Whole genome sequence assembly of the water buffalo (*Bubalus bubalis*) *Indian J. Anim. Sci* **81** (5) 38-46

Taylor WP, Bhat PN and Nanda YP 1995 The principles and practice of rinderpest eradication *Vet. Microbiol* **44** 359

Taylor WP, Roeder PL and Rweyemamu MM 2006 *Use of Rinderpest vaccine in international programme for the control and eradication of Rinderpest* Chapter 13: 261-283. Cited from Book Rinderpest and Peste des Petits Ruminants Virus Plagues of Large and small Ruminants by Thomas Barrett, Paul-Pierre Pastoret and William P. Taylor. Publisher Academic Press an imprint of Elsevier.

Thathoo, 1986 Ph.D. thesis IVRI, Izatnagar

The Ayurvedic Pharmacopoeia of India. GOI. Part-I Volume-I. Pp 1-171

The New Encyclopaedia Britannica, 2007 *Inc. Encyclopaedia Britannica.*, ISBN-10: 0852299613. Publisher: William Benton and Helen Hemingway Benton; 19533.

Tierney ML 1985 Proceedings, 1985 ASAP/NZSAP Dairy Production Conference, Albury-Wodonga. p. 466

Tiwana MS 1985 Research highlights of PAU, Ludhiana centre of AICRP on buffalo. Presented at the *seventh workshop of AICRP On buffalo held at NDRI, Karnal*, from January 14-16.

Tiwana MS, Arora BS, Bhullar MS, Singh N, Sidhu SS and Singh J 1986 Buffalo breeding at PAU, Ludhiana (1911-85). *SARAS J. Livestock and Poultry Production* **1** 48-54

Tomar NS 1966 Effect of season of insemination on conception rates in Haryana cows and Murrah buffaloes *Ind. J. Dairy Sci.* **19** 14

Torri MC 2013 Traditional Veterinary in Rural Tamil Nadu, Linking Medicinal Plants with Local Livelihoods and Human Health Care *Journal of Developing Societies* **29** (1) 23-46

Tripathi VN 1986 *Managemental practices for reducing seasonal variation in milk production.* Invited paper presented at IDA (North Zone) Seminar held at Banaras Hindu University, Varanasi on August 23-24

Tripathi VN and Reddy KM 1986 *Terminal Progress Report of Buffalo Project*, NDRI, Karnal, India

Tule G and Windock 1979 Mycoplasma; Human and animal Mycoplasma **2** 104

Uppal DR and Seetharaman C 1957 Some observations on the large-scale production of freeze-dried goat adapted Rinderpest Virus Vaccine *Ind. Vet. J* **34** (3) 157-165

Uppal PK 2011 National Testimonies under the Global Rinderpest Eradication Programme. Project Report submitted to FAO.

Uppal PK and Yadav MP 1989 Occurrence of Equine Infectious Anaemia in India *Vet. Record* **124** 514-15

Uppal PK and Kataria JM 1987 Epidemic of Salmonellaabortus equi abortion in horses *J. Repord. Fert. Suppl* **35** 669-70

Uppal PK and Yadav MP 1987 Outbreak of equine influenza in India *Vet. Rec.* **121** 569-70

Uppal PK and Yadav MP 1992 Application of competitive ELISA in the diagnosis of equine infectious anaemia *Current Science* **63** 332-33

Uppal PK and Yadav MP 1993 Electron microscopic observations of the Indian isolate of equine infectious anaemia virus grown in equine leucocytes *Current Science* **64** 115-117

Uppal PK, Kumar S, Das SK and Bhat PN 1975 Foot and mouth disease vaccination in new born calves *Indian Vet. J* **52** 282-288

Uppal PK, Oberoi MS and Bansal RP 1978 *Annual Report* (1978-79) Division of Bacteriology and Mycology, Indian Veterinary Research Institute, Izatnagar

Uppal PK, Oberoi MS and Bansal RP 1983 A note on use of CIE technique with *Mycoplasama arginini*, *Aspergillus fumigatus* and rinderpest virus antigens-antibody system *Indian Vet. J* **60** 767

Uppal PK, Plateau E and Yadav MP 1993 Restriction enzyme digestion patterns of DNA from an Indian isolate of equine herpes virus-1 *Current Science* **64** 682-684

Uppal PK, Singh BK and Yadav MP 1991 Observations on an abortion outbreak in mares caused by herpes virus-1 *International J. Anim. Sci* **6** 60-63

Uppal PK, Singh IP and Kumar S 1972 Role of sheep in the epidemiology of foot and mouth disease *Indian J. Anim. Prod.***3** 123-125

Uppal PK, Yadav MP and Gupta AK 1994 Characterization of structural proteins of A/Equi-2/Ludhiana/87 *Curr. Sci* **67** 79

Uppal PK, Yadav MP and Manchanda VP 1990 Observations on strangles and purpura haemorrhagica as sequelae to equine influenza infection *Indian J. Anim. Sci* **60** (10) 1149-1153

Uppal PK, Yadav MP and Oberoi MS 1989 Isolation of A/Equi-2 virus during 1987 equine influenza epidemic in India *Equine veterinary Journal* **21** 364-366

Uppal PK, Yadav MP and Sharma SN 1987 Occurrence of equine influenza outbreak in India *Indian J. Comp. Microbiol. Immunol. Infect. Dis.* **8** 91-94

Uppal PK, Yadav MP, Singh BK and Prasad S 1989 Equine Coital Exanthema (EHV-3 virus) infection in India *J. Vet. Med. B* **36** 786-788

Uppal PK, Yadav MP, Tripathi BN and Kharole MU 1997 Studies on transmission of equine infectious anaemia in donkeys: Clinico-pathological changes *Indian J. Virology* **13** 47-52

USP 2013 Monographs in the Herbal Medicines Compendium Guidelines Document Version 1.0: May 20, 2013.

Varshney JP and Uppal PK 1991 Effect of dexamethsone on haematological indices of non-descript donkeys *Indian Vet. Med. J.* **15** 217-218

Varshney JP and Uppal PK 1992 Efficacy of ivermectin against psoroptic mange in donkeys *J. Vet. Parasitol* **6** (1) 47-48

Varshney JP and Uppal PK 1992 Incidence periodicity and haemato-biochemical changes in equines naturally infected with microfilariae of Setaria equine *J. Vet. Parasitol* **6**(2)

Varshney JP, Gupta AK and Yadav MP 1995 Occurrence of ataxia cystitis syndrome in horses fed of Sorghum vulgare in India *Indian Vet. J* **73**(9) 985-986

Varshney JP, Yadav MP and Uppal PK 1994 Clinico-heamatological findings in equine infectious anaemia *Indian J. Vet. Med.* **14** 9-12

Vasundhara TS, Kumudavally KV, Souza JD, Phanindra Kumar HS and Sharma TR 1988 Chemical Indicators for determining the presterilization quality of canned curried meat; factor affecting the quality of the chemical indices *Food chemistry* **99** 133-142

Vegad and Katiyar 1995 The acute inflammatory response in the chicken *Vet. Bull.* **65** 399-409

Vegad JN 1995 *Textbook of general Veterinary Pathology* Vikas Publishing House, New Delhi

Venkateshwarlu M, Singh BP, Tomar SP and Kapri BD 1973 Genetic studies on Ongole cattle. II. First lactation milk yield, peak yield and days taken to reach Peak yield *Indian Vet. J* **50** 325-331

Venkayya D and Anantakrishnan CP 1957 Effect of season of calving, weight at freshening and length of service period on milk yield *Indian J. Dairy Sci* **10** 123-130

Venkayya D and Anantakrishnan CP 1964 Influence of the length of preceding dry period on the subsequent lactation performance of Murrah buffaloes *Indian J. Dairy Sci.* **17** 87-88

Verma AK 1974 Ph.D. Thesis of Agra University, Agra.

Verma S and Singh SP 2008 Current and future status of herbal medicines *Veterinary World* **1**(11) 347-350

Vij PK 1984 *Genetic analysis of a closed herd of buffaloes*. Ph.D. Thesis, Punjab Agric. Univ., Ludhiana, India

Vijay Kumar M, Ravikanth K, Shivi Maini amd Adarsj 2014 Alternate methionine supplementation option in broilers: a field study. *International Journal of Current Research*, **6**(09) 2014.

Vijaya Kumar, James, Govinda-Rajan CN and Ramamurthi R 1990 Studies on tenderizing effect of electrical stimulation in mutton carcasses *Indian Veterinary J.* **67** 343-346

Vijh RK, Tantia MS, Mishra B, Bharani KS 2008 Genetic relationship and diversity analysis of Indian water buffalo (*Bubalus bubalis*) *J Anim Sci.* **86** 1495-1502

Virmani M, Gupta AK, Virmani Nitin and Garg SK 2008 Haemagglutination inhibition assay for Equine Chorionic Gonadotropin (ECG) detection in pregnant mare serum *Journal of Immunology and Immunopathology* **10** (1) 55-59

Virmani N, Bera BC, Gulati BR, Shanmugasundaram K, Singh BK, Vaid RK, Kumar S, Kumar R, Malik P, Khurana SK, Singh J, Manuja A, Dedar R, Gupta AK, Yadav

SC, Chug PK, Narwal PS, Thakur VLN, Kaul R, Knani A, Rautmare SS and Singh RK 2010 Descriptive epidemiology of equine influenza in India (2008-2009): temporal and spatial trends. *Veterinaria Italiana*, 46: 449-458

Virmani N, Bera BC, Shanmugasundaram K, Singh BK, Gulati BR and Singh RK 2011 Genetic analysis of the matrix and non-structural genes of equine influenza virus (H3N8) from epizootic of 2008-09 in India *Veterinary Microbiology*, doi: 10.1016 ld. vetmix. 2011.04.011

Virmani N, Bera BC, Singh BK, Shanmugasundaram K, Gulati BR, Barua S, Vaid RK, Gupta AK and Singh RK 2009 Equine influenza outbreak in India (2008–09): Virus isolation, sero-epidemiology and phylogenetic analysis of HA gene *Veterinary Microbiology* (doi:10.1016/j.vetmic.2009.12.007).

Virmani N, Singh BK, Gulati BR and Kumar S 2008 Equine Influenza outbreak in India *Veterinary Record* **163**(20) 607-608

Virmani N, Singh BK, Verma PC, Batra M and Panisup AS 2005 Immune responses to primary infection with indigenous strain of equine herpes virus-1 in pregnant BALB/c mice *Indian Journal of Animal Sciences* **75** 1385-1387

Virmani Nitin, Singh BK, Batra Munish, Verma PC and Panisup AS 2008 Comparative pathology and tissue tropism of two indigenous strains (H-90 and R-98) of equine herpes virus 1 in pregnant BALB/c mice towards occurrence of respiratory affections and abortions subsequent to intranasal inoculation *Journal of Immunology and Immunopathology* **10** (1): 14-19

Wani MS 2007 Herbal medicine and its standardization *Pharma. Info* **1** 6

Wankar AK, Shirbhate, Bahiram KB, Dhenge SA and Jasutkar RA 2009 Vet. World 2: 396-397

Wanzala W, Zessin KH, Kyule NM, Baumann MPO, Mathias E, Hassanali A 2005 Ethnoveterinary medicine: a critical review of its evolution, perception, understanding and the way forward. Livestock Research for Rural Development; 17(11)

Ware F 1940 *Progress of Veterinary Research in India during the past 75 years.* (Miscellaneous Bulletin No.35). The Imperial Council of Agricultural Research. Manager of Publication, Government of India, Delhi

Wellington KE, Mahadevan P and Roache KL 1970 Production characteristics of the Jamaica Hope breed of dairy cattle *J. Agric, Sci* **74** 463–468

WHA Resolution: WHA41.19: Traditional medicine and medicinal plants

WHA Resolution: WHA56.31: Traditional medicine

WHA Resolution: WHA61.21: Global strategy and plan of action on public health, innovation and intellectual property

WHA Resolution: WHA62.13: Traditional medicine.

WHO 2002 WHO Traditional Medicine Strategy 2002-2005, World Health Organization, Geneva

WHO Guidelines for the appropriate use of herbal medicines. 1998, pp 1-88

WHO World Health Organization. Quality Control Methods for Medicinal Plant Materials. Geneva: Organizacao Mundial de Saude. 1992

Wieser M, Schneider F and Wälty S 2000 Capitalisation of Experiences in Livestock Production and Dairying (LPD) in India (CAPEX). Bern: Intercooperation

Wright NC 1937 Report on the development of cattle and dairy industries in India. Manager of Publication, Delhi

Yadav CL, Varshney JP, Uppal RP and Uppal PK 1993 Efficacy of ivermectin, levamisole and thiophenate against strongyle infection in equines *Indian Vet. J* **70** 554-556

Yadav MP 2011 "Laboratory Contributions for Rinderpest Eradication in India'. Report submitted to FAO in December, 2011 under the Global Rinderpest Eradication Programme (GREP) (GCP/GLO/302/EC). PP. 58.

Yadav MP and Kumar S 1975 Development of live avian infectious laryngotracheitis (ILT) vaccine *Indian J. Anim. Sci* **45 I**: 563-71

Yadav MP and Malik BS 1971 Antibody response of Escherichia coli in hens and control of colisepticaemia in young chicks by natural passive immunity *Indian J. Anim. Sci* **41** 508

Yadav MP, Bhambani BD and Kumar S 1975 Virus isolations from respiratory and intestinal tracts of poultry in India *Indian J. Anim. Sci* **45**(11) 899-902

Yadav MP, Dutta SK and Kumar S 1974 Biochemical and biophysical characterization of avian adenoviruses isolated in India *Avian Diseases* **18**(2) 220-225

Yadav MP, Pandey AB, Negi BS, Sharma B, Shankar H 1986 Studies on inactivated goat pox vaccine *Indian J Virol* **2** 207-221

Yadav MP, Parihar NS and Kumar S 1975 Pathogenicity of avian adeno viruses in chickens *Indian J. Anim. Sci* **45** (1) 28-35

Yadav MP, Sharma VD and Sethi MS 1977 An outbreak of fowl cholera due to *Pasteurella gallinarum* in Uttar Pradesh (India), *Avian Diseases* **21**(2) 323-325

Yadav MP, Singh BK and Uppal PK 1992 Sero-prevalence of burcellosis in equines in some of the states in India *Indian J. Anim. Sci* **62** 41-42

Yadav MP, Uppal PK and Ahmed SN 1990 Humoral response in equine influenza after vaccination and natural infection *Indian Vet. J* **67** 293-298

Yadav MP, Uppal PK and Mumford JA 1993 Physico-Chemical and biological characterization of A/Equi-2 virus isolated from 1987 equine influenza epidemic in India *Int. J. Anim. Sci* **8** 93-98

Yadav MP, Uppal PK, Rao JR 2016 Animal Sciences. In: Singh, R.B. (ed.) *100 Years of Agricultural Sciences in India*. NAAS, New Delhi, India. Pp 158-258

Yadav MP, Rattan B, Uppal PK and Malik P 1998 Occurrence and control of equine infectious anaemia in Maharashtra *Indian J. Comp. Microbiol. Immunol. & Infect. Dis.,* **19**(1) 50

Yadav SC, Kumar R, Kumar V, Jaideep, Gupta AK, Bera BC and Tatu U 2013 Identification of immuno-dominant antigens of *Trypanosoma evansi* for detection of chronic trypanosomosis using experimentally infected equines *Research in Veterinary Science* **95** 522-528

Yash Pal, Dedar RK, Ravi SK, Legha RA, Gupta AK and Singh RK 2013 Effect of dietary antioxidant supplementation on fresh semen quality in indigenous jacks *Indian Veterinary Journal* **90** (3) 135-136

Yashpal, Ghei JC, Gupta AK and Yadav MP 1998 Comparative studies on heat tolerance in ponies and donkeys *Indian Vet. J* **75**(1) 78-79

Yongyu Z, Shujun S, Jianye D, Wenyu W, Huijuan C, Wu Jianbing and Xiaojun G 2011 Quality Control Method for Herbal Medicine - Chemical Fingerprint Analysis, Quality Control of Herbal Medicines and Related Areas, Prof. Yukihiro Shoyama (Ed.), 2011, ISBN: 978-953-307-682-9

Ziauddin M, Phansalkar N, Patki P, Diwanay S, Patwardhan B 1996 Studies on the immunomodulatory effect of Asgandh *J Ethnopharmacol* **50** (2) 69-76.

■■■■